Lecture Notes in Bioinformatics   8623

Subseries of Lecture Notes in Computer Science

More information about this series at http://www.springer.com/series/5381

Clelia di Serio · Pietro Liò
Alessandro Nonis · Roberto Tagliaferri (Eds.)

Computational Intelligence Methods for Bioinformatics and Biostatistics

11th International Meeting, CIBB 2014
Cambridge, UK, June 26–28, 2014
Revised Selected Papers

 Springer

Editors
Clelia di Serio
CUSSB
Universitá Vita-Salute San Raffaele
Milano
Italy

Pietro Liò
The Computer Laboratory
University of Cambridge
Cambridge
UK

Alessandro Nonis
CUSSB
Universitá Vita-Salute San Raffaele
Milano
Italy

Roberto Tagliaferri
Dipartimento di Informatica
Universitá degli Studi di Salerno
Fisciano, Salerno
Italy

ISSN 0302-9743
Lecture Notes in Bioinformatics
ISBN 978-3-319-24461-7
DOI 10.1007/978-3-319-24462-4

ISSN 1611-3349 (electronic)

ISBN 978-3-319-24462-4 (eBook)

Library of Congress Control Number: 2015950439

LNCS Sublibrary: SL8 – Bioinformatics

Springer International Publishing AG Switzerland is part of Springer Science+Business Media
(www.springer.com)

Preface

This volume contains a revised version of the proceedings of the International Meeting on Computational Intelligence Methods for Bioinformatics and Biostatistics (CIBB 2014), which was in its 11th edition this year.

At a time of rapid information exchange through the Internet, conferences represent an important opportunity of direct interactions between scientists and fresh, informal communication (forum) of scientific findings. CIBB is a lively conference that started 11 years ago in Italy. It maintains a large Italian participation in terms of authors and conference venues but has progressively become more international and more important in the current landscape of bioinformatics and biostatistics conferences.

The topics of the conferences have kept pace with the appearance of new types of challenges in biomedical computer science, particularly with respect to a variety of molecular data and the need to integrate different sources of information. This year the conference saw an impressive array of invited speakers covering the two main sections: biostatistics and bioinformatics (Biostatistics Free Topics: Monica Chiogna, University of Padova, Italy; Chris Holmes, University of Oxford, UK; Jean Michel Marin, University of Montpellier II; Causality in Genomics: Carlo Berzuini, University of Manchester, UK; Stephen Burgess, University of Cambridge, UK; Vanessa Didelez, University of Bristol, UK; Florian Markowetz, Cancer Research UK - CRI, Cambridge, UK; Lorenz Wernisch, MRC - BSU, Cambridge, UK. Bioinformatics Free Topics: Ezio Bartocci, Vienna University of Technology, Austria; Francesco Falciani, University of Liverpool, UK; Jasmin Fisher, University of Cambridge, UK; David Gilbert, Brunel University, Uxbridge, UK; Syed Haider, University of Oxford, UK; Jeanine J. Houwing-Duistermaat, Leiden University Medical Centre, The Netherlands; Marta Kwiatkowska, University of Oxford, UK; Pedro Mendes, Manchester Institute of Biotechnology, UK; Marie-France Sagot, Université Claude Bernard, France; Marco Viceconti, University of Sheffield, UK. Special Session: Spatial Problems in the Nucleus: Julien Mozziconacci, Pierre and Marie Curie University, Paris, France).

This year 42 papers were selected for presentation at the conference, and each paper received two reviews or more. A further reviewing process took place for the 25 papers that were selected to appear in this volume. The authors are spread over more than 22 countries (all the continents were represented). The editors would like to thank all the Program Committee members and the external reviewers both of the conference and post-conference versions of the papers for their valuable work. We are also indebted to the chairs of the very interesting and successful special sessions (Computational Intelligence Methods for Drug Design; Spatial Problems in the Nucleus, Large-Scale; HPC Data Analysis in Bioinformatics: Intelligent Methods for Computational, Systems and

Synthetic Biology; Computational Biostatistics for Data Integration in Systems Biomedicine), which attracted even more contributions and attention.

A big thanks also to the sponsors, Bioinformatics Italian Society, Source Code for Biology and Medicine, and in particular to the University of Salerno, Vita-Salute San Raffaele University, and the University of Cambridge Computer Laboratory, which made this event possible. Finally, the editors would also like to thank all the authors for the high quality of the papers they contributed.

June 2015   Clelia di Serio
                                                        Pietro Liò
                                                   Alessandro Nonis
                                                  Roberto Tagliaferri

Organization

CIBB 2014 was jointly organized by:
The Computer Laboratory, University of Cambridge, UK; CUSSB Centro Universitario di Statistica per le Scienze Biomediche, Vita-Salute San Raffaele University, Milano, Italy; Dipartimento di Informatica, University of Salerno, Italy; INNS International Neural Network Society, Bioinformaticcs and Intelligence SIG; and the IEEE-CIS-BBCT Task forces on Neural Networks and Evolutionary Computation.

Organizing Committee

General Chairs

Clelia Di Serio	Vita-Salute San Raffaele University, Milan, Italy
Pietro Liò	University of Cambridge, UK
Sylvia Richardson	University of Cambridge, UK
Roberto Tagliaferri	University of Salerno, Italy

Biostatistics Technical Chair

Ernst Wit	University of Groningen, The Netherlands

Bioinformatics Technical Chair

Claudia Angelini	IAC-CNR, Italy

Publicity Chair

Francesco Masulli	University of Genoa, Italy and Temple University, PA, USA

Local Organizing Committee Chairs

Mohammad Moni	University of Cambridge, UK
Alessandro Nonis	Vita-Salute San Raffaele University, Milan, Italy

Special Session and Tutorial Chairs

Pedro Ballester	EBI, Cambridge, UK
Yoli Shavit	University of Cambridge, UK

Publication Chair

Riccardo Rizzo	ICAR-CNR, Italy

Finance Chair

Elia Biganzoli   University of Milan, Italy

Steering Committee

Pierre Baldi   University of California, Irvine, CA, USA
Elia Biganzoli   University of Milan, Italy
Alexandru Floares   Oncological Institute Cluj-Napoca, Romania
Jon Garibaldi   University of Nottingham, UK
Nikola Kasabov   Auckland University of Technology,
                                        New Zealand
Francesco Masulli   University of Genoa, Italy and Temple
                                        University, PA, USA
Leif Peterson   TMHRI, Houston, Texas, USA
Roberto Tagliaferri   University of Salerno, Italy

Program Committee

Fentaw Abegaz   University of Groningen, The Netherlands
Qurrat Ain   University of Cambridge, UK
Marco Aldinucci   University of Turin, Italy
Federico Ambrogi   University of Milan, Italy
Claudia Angelini   National Research Council, Naples, Italy
Claudio Angione   University of Cambridge, UK
Sansanee Auephanwiriyakul   Chiang Mai University, Thailand
Krzysztof Bartoszek   Uppsala Universitet, Sweden
Giulio Caravagna   Università degli Studi di Milano-Bicocca, Italy
Gennaro Costaiola   University of Salerno, Italy
Michele Fratello   Second University of Naples, Italy
Christoph Friedrich   University of Dortmund, Germany
Javier González   University of Sheffield, UK
Marco Grzegorczyk   University of Groningen, The Netherlands
Sean Holden   University of Cambridge, UK
Elena Marchiori   Radboud University, The Netherlands
David Marcus   National Research Council, Naples, Italy
Ivan Merelli   National Research Council, Milan, Italy
Danilo Pellin   Vita-Salute San Raffaele University, Milan,
                                        Italy
Naruemon Pratanwanich   University of Cambridge, UK
Vilda Purutçuoğlu   Middle East Technical University, Turkey
Stefano Rovetta   University of Genoa, Italy
Marco Scutari   University College London, UK
Diego Sona   Istituto Italiano di Tecnologia - IIT, Genoa,
                                        Italy
Massimo Torquati   University of Pisa, Italy
Giorgio Valentini   University of Milan, Italy

Sponsors

Bioinformatics ITalian Society www.bioinformatics.it

Centro universitario di Statistica per le Scienze biomediche (CUSSB) www.cussb.unisr.it

Dipartimento di Informatica, Università degli Studi di Salerno www.di.unisa.it

Source Code for Biology and Medicine www.scfbm.org

Contents

Regular Sessions

Special Session: Computational Biostatistics for Data Integration in Systems Biomedicine

Special Session: Computational Intelligence Methods for Drug Design

Special Session: Large-Scale and HPC Data Analysis in Bioinformatics: Intelligent Methods for Computational, Systems and Synthetic Biology

Regular Sessions

GO-WAR: A Tool for Mining Weighted Association Rules from Gene Ontology Annotations

Giuseppe Agapito, Mario Cannataro, Pietro H. Guzzi, and Marianna Milano

Department of Medical and Surgical Sciences,
Magna Graecia University, Catanzaro, Italy
{agapito,cannataro,hguzzi,m.milano}@unicz.it

Abstract. The Gene Ontology (GO) is a controlled vocabulary of concepts (called GO Terms) structured on three main ontologies. Each GO Term contains a description of a biological concept that is associated to one or more gene products through a process also known as annotation. Each annotation may be derived using different methods and an Evidence Code (EC) takes into account of this process. The importance and the specificity of both GO terms and annotations are often measured by their Information Content (IC). Mining annotations and annotated data may extract meaningful knowledge from a biological stand point. For instance, the analysis of these annotated data using association rules provides evidence for the co-occurrence of annotations. Nevertheless classical association rules algorithms do not take into account the source of annotation nor the importance yielding to the generation of candidate rules with low IC. This paper presents a methodology for extracting Weighted Association Rules from GO implemented in a tool named GO-WAR (Gene Ontology-based Weighted Association Rules). It is able to extract association rules with a high level of IC without loss of Support and Confidence from a dataset of annotated data. A case study on using of GO WAR on publicly available GO annotation dataset is used to demonstrate that our method outperforms current state of the art approaches.

Keywords: Gene Ontology, Weighted Association Rules.

1 Introduction

The production of experimental data in molecular biology has been accompanied by the accumulation of functional information about biological entities. Terms describing such knowledge are usually structured by using formal instruments such as controlled vocabularies and ontologies [1]. The Gene Ontology (GO) project [2] has developed a conceptual framework based on ontologies for organizing terms (namely GO Terms) describing biological concepts. It is structured into three ontologies: Molecular Function (MF), Biological Process (BP), and Cellular Component (CC) describing different aspects of biological molecules.

© Springer International Publishing Switzerland 2015
C. di Serio et al. (Eds.): CIBB 2014, LNCS 8623, pp. 3–18, 2015.
DOI: 10.1007/978-3-319-24462-4_1

Each GO Term may be associated to many biological concepts (e.g. proteins or genes) by a process also known as annotation. The whole corpus of annotations is stored in publicly available databases, such as the Gene Ontology Annotation (GOA) database [3].

In such a way records representing the associations of biological concepts, e.g. proteins, and GO terms may be easily represented as $P_j, T_1, \ldots, T_n$, e.g. {P06727, GO:0002227, GO:0006810, GO:0006869} or {ApolipoproteinA-IV, innate immune response in mucosa, transport, lipid transport}.

The whole set of annotated data represents a valuable resource for analysis. Currently, there are different methods for the analysis of annotated data. From those, the use of association rules (AR) [4,5,6] is less popular with respect to other techniques, such as statistical methods or semantic similarities [7]. Existing approaches span from the use of AR to improve the annotation consistency, as presented in [8], to the use of AR to analyze microarray data [9,10,11,12,13,14,15] (see [16] for a detailed review).

As we pointed out in a previous work [5], the use of AR presents two main issues due to the Number and the Nature of Annotations [17]. Regarding the **Number of Annotations** it should be noted that, due to the different methods and the different availability of experimental data, the number of annotations for each protein or gene is highly variable within the same GO taxonomy and over different species as we depict in Figure 1.

Regarding the **Nature of Annotations**, it should be evidenced that the association among a biological concept and its related GO Term can be performed with 14 different methods. These methods are in general grouped onto two main categories: experimentally verified (or manuals) and Inferred from Electronic Annotation (IEA). IEA are usually derived using computational methods that analyze literature. Each annotation is labeled with an evidence code (EC) to keep track of the method used to annotate a protein with GO Terms. Manual

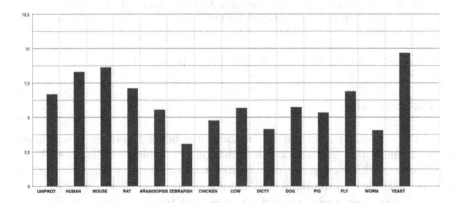

Fig. 1. Average Number of Annotations per Protein in Different Species. Each bar represents a different species. The height of the bar represent the average number of annotations.

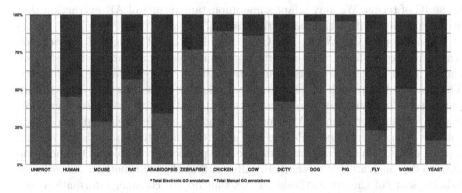

Fig. 2. Ratio of Electronic Inferred Annotations with respect to Manual ones. This picture depicts the differences on the average number of annotations. For each bar, green color represents the fraction of non-IEA annotations, while blue color represents the fraction of IEA annotations.

annotations are in general more precise and specific than IEA ones (see [1]). Unfortunately their number is generally lower and the ratio among IEA versus non-IEA is variable.

A considerable number of genes and proteins is annotated with generic GO terms (this is particularly evident when considering novel or not well studied genes) and the problem is also referred to **Shallow Annotation Problem**. The role of these general annotations is to suggest an area in which the proteins or genes operate. This phenomenon affects particularly IEA annotations derived by using computational methods.

Considering the problems discussed so far, the application of classical AR methods to the analysis of annotated data may yield to the extraction of rules with low specificity, with generic terms or with inconsistent annotations, following the true path rule (i.e. rules in which both a term t and its ancestors are included) [8,5]. For these reasons, Faria et al. [8] proposed the manual filtering of ancestors and of low specific terms proposing a definition of specificity in terms of descendant. Nevertheless, the measurement of specificity of a term following only topological information may yield to incorrect results as noted by Alterovitz et al. [18].

Consequently, we here propose to select a more stringent definition of specificity by considering the information content (IC) of a term [19]. There are different ways of calculating IC that are subdivided into **intrinsic** ones, e.g. approaches that estimate the IC of concepts by only considering structural information extracted from the ontology, and **extrinsic** ones, e.g. approaches that measure the IC starting from annotated corpora. Independently from the calculation, the main results of the use of IC is that we may associate to each GO term a measure of IC, i.e. a weight of its specificity yielding to the IC-weighted annotation as represented in the following: {P06727, GO:0002227 (16.77), GO:0006810 (5.18), GO:0006869 (10.73)} where the number into brackets is the IC of terms. We may adapt some important results of AR extraction that are able to deal with weighted attributes [20].

We developed GO-WAR, i.e. Gene Ontology-based Weighted Association Rules Mining, a novel data-mining approach able to extract weighted association rules starting from an annotated dataset of genes or gene products. The proposed approach is based on the following steps starting from the input dataset of annotated biological terms: (i) initially we calculate the information content for each GO term, (ii) then, we extract weighted association rules by using a modified FP-Tree like algorithm able to deal with the dimension of classical biological datasets. We use publicly available GO annotation data to demonstrate our method. Results confirm that our method outperforms existing state of the art methods for AR that do not consider information content. We also provide a web site containing supplementary materials and results `https://sites.google.com/site/weightedrules/`.

The rest of the paper is structured as follows: Section 2 discusses main related work, Section 3 presents main concepts related to the Gene Ontology and its annotations, Section 4 discusses GO-WAR methodology and implementation, Section 5 presents results of the application of GO-WAR on a biological dataset, finally Section 6 concludes the paper and outlines future work.

2   Related Work

Here we first compare GO-WAR with respect to three state of the art approaches that use Association Rule Mining on GO Annotations, (Faria et al [8], Benites et al [12], and Manda et al [13]); after we compare the GO-WAR tool with respect to some popular off-the-shelf data mining platforms.

Table 1 summarizes the comparison of the approaches. We considered on the comparison the following parameters: (i) aim of the approach, i.e. the application context for which the algorithm has been designed, (ii) the ontologies on which it has been tested, (iii) the mining strategy, i.e. which learning approach has been used, (iv) the preprocessing strategy, i.e. the removal of annotations prior to the learning of rules, (v) the used definition of interestingness of rules, (vi) the kind of rules, and finally (vii) the presence of a supporting tool.

Considering the aim, we should report that other algorithms are in general designed for a specific application (e.g. evaluation of Annotation consistency, mining of rare association, or analysis of multi-ontology association), while the GO-WAR approach is generalizable. Considering the ontologies, we note that Faria et al. and Benites et al. report the analysis of only MF, while Manda et al. focus on a multi-ontology approach. GO-WAR is more flexible since we used MF for comparison with Faria and in general it may use all the ontologies. Considering mining strategy, we report that GO-WAR uses FP-GROWTH tree for building rules [21] that guarantees better results in terms of time and memory usage with respect to Apriori algorithm [22].

First of all we should note that the GO-WAR approach is currently the only one that has a supporting tool enabling the user to easily apply the methodology

Table 1. Comparison of GO-WAR approach with respect to state of the art approaches.

Algorithm	Aim	Ontologies	Mining Strategy	Preprocessing	Interestingness	Rules	Software
Faria [8]	Annotation Consistency	MF	Apriori	Manual Filtering	Standard	$t_1 \leftarrow t_2$	No
Benites et al [12]	Rare Associations	MF	Apriori	No	AD HOC	$t_1 \leftarrow t_2$	No
Manda et al [15]	Generating Annotation Candidates Deriving new relationships among terms in GO -	MF e Anche Multi Ontology	Apriori	Insertion of all the ancestors	AD HOC	$t_1 \leftarrow t_2$	No
GO-WAR	Generalizable	Multi Ontology	FP-Growth	Based on IC	Weighted Support	$t_1,...t_n \leftarrow t_{n+1}$	Yes

on a different case-study. The approach of Faria et al, for instance, is based on the manual extraction of rules by using the GO database and SQL language, while Benites et al. and Manda et al. do not provide any supporting tool.

3   Gene Ontology and Its Annotations

This section presents the main concepts related to GO and its annotations, after which a deep discussion on the calculation of information content of annotations is presented.

3.1   Gene Ontology and Its Annotations

Gene Ontology [2] (GO) is one of the main resources of biological information since it provides a specific definition about protein functions. GO is a structured and controlled vocabulary of terms, called GO terms. GO is subdivided in three non-overlapping ontologies: Molecular Function (MF), Biological Process (BP) and Cellular Component (CC), thus, each ontology describes a particular aspect of a gene or protein functionality. GO has a specific structure, Directed Acyclic Graph (DAG), where the terms are the nodes and the relations among terms are the edges. This allows for more flexibility than a hierarchy, since each term can have multiple relationships to broader parent terms and more specific child terms [23]. Genes or proteins are connected with GO terms through annotations by using a procedure also known as annotation process. Each annotation in the GO has a source and a database entry attributed to it. The source can be a literature reference, a database reference or computational evidence. Each biological molecule is associated with the most specific set of terms that describe its functionality. Then, if a biological molecule is associated with a term, it will connect to all the parents of that term [23]. 18 different annotation processes exist that are identified by an evidence code, the main attribute of an annotation. The evidence codes available describe the basis for the annotation. A main distinction among evidence codes is represented by Inferred from Electronic Annotations (IEA) ones, i.e. annotations that are determined without user supervision, and non-IEA ones or manual annotations, i.e. annotations that are supervised by experts.

3.2   Measures of Specificity of GO Terms

There are two approaches for computing the IC of a given term belonging to an ontology: extrinsic (or annotation based) and intrinsic (or topology-based) methods.

Intrinsic approaches rely on the topology of the GO graph exploiting the positions of terms in a taxonomy that define information content for each term. Intrinsic IC calculus can be estimated using different topological characteristics such as ancestors, number of children, depth (see [19] for a complete review). For example, Sanchez et al. [24] computes the IC of terms exploiting only the number of leaves and the set of ancestors of a including itself, $subsumers(a)$ and introducing the root node as number of leaves max_leaves in IC assessment. Leaves are more informative than concepts with many leaves, roots, so the leaves are suited to describe and to distinguish any concept.

$$IC_{Sanchez\ et\ al.}(a) = -log\left(\frac{\frac{|leaves(a)|}{|subsumers(a)|} + 1)}{max\_leaves + 1}\right) \qquad (1)$$

We here use the formulation of Harispe et al. [19] that revises the IC assessment suggested by Sanchez et al. considering all the leaves of a concept when a is a root and evaluating max_leaves as the number of inclusive ancestors of a node. In this way, the specificity of leaves according to their number of ancestors is distinguished.

$$IC_{Harispe\ et\ al.}(a) = -log\left(\frac{\frac{|leaves(a)|}{|subsumers(a)|})}{max\_leaves}\right). \qquad (2)$$

As a general principle, since the structure of GO is periodically updated [23], the IC of terms reflects this evolution, thus the calculation is subject to the variation of GO structure.

4 The GO-WAR Framework

Here we present the GO-WAR Framework. Initially the paper discusses the weighted association rule algorithm, then the implementing architecture and main software modules are described.

4.1 GO-WAR Algorithm

The rationale of the GO-WAR algorithm is to take into account the relevance of items, i.e. GO Terms, balancing, thus, relevance and frequency as explained in [20]. Following the same approach we here formulate the problem of the extraction of weighted association rules introducing main concepts.

Definition 1 (Weighted Item). *A Weighted Item (wi) is a pair (x, w), where $x \in I$, i.e. a GO Term belonging to the set of the items, and $w \in R$, i.e. the associated real-valued item. For example, the GO term GO:00152 relatives to the transcription corepressor activity has an IC value of 11.876, conveyed as GO:00152, (11.876).*

Definition 2 (Weighted Transaction). *A weighted transaction WT is a set of weighted items. For instance, the line*
    P41226, GO:0005524 (10.07), GO:0005829, (10.07)
represents the protein P41226, its annotations GO:0005524, GO:0005829 and their weights. A set of WT is hereafter referred as Weighted Transaction Database WTB.

Definition 3 (Weighted Support). *The Weighted Support, (WS), is the product of the support of an item, calculated using the classical formulation, and its weight.*

Definition 4 (Weighted Minimum Support). *The weighted minimum support (wminSupp) of a weighted item is defined as:*

$$wminSupp = \sum_{i=1}^{n} \left(\frac{WS(x_i)}{n} \right) * p$$

where n is the number of transactions, and p is a user defined threshold.

The algorithm takes as input a transaction database T, after which it computes the weight for each item producing a weighted transaction database. Consequently, it generates candidates itemsets by applying a modified FP-GROWTH algorithm. These frequent itemsets and the minimum confidence constraint are used to form rules as represented in the following algorithm 1.

The generation of candidate itemsets follows a FP-Growth (Frequent Pattern) [21] like approach employing a two step strategy. In the first pass the algorithm counts the weighted occurrence of items (i.e. the product of the frequency of occurrences and their weight) in the datasets. All the results are stored in a table. Then it builds a FP-Tree structure by adding instances of items that have a weighted support greater than *wminSupp*. Once that the FP-Tree has been created, all itemsets with desired support coverage have been found, and association rule creation may start.

GO-WAR iteratively analyzes the $FP - Tree$ to mine significant rules [25] using a recursive methodology. We defined an *inverted DFS* (inverted Depth First Search) scan method to examine the FP-Tree. *Inverted DFS* starts to explore the tree from the leave nodes (bottom) and goes up to the (root node). The advantage to use *inverted DFS* respect to traditional DFS is related with the possibility to automatically prune (remove) the postfix part of a frequent pattern. All frequent pattern of a given item are mined following the links connecting all occurrences of the current item in the *FPTree* and computing the weighted support related with each path (frequent patterns), producing a new tree called β-Tree, used to mine rules. Postfix part of a frequent pattern is defined respect to a given item or itemset *I*. For example, taking into account the frequent pattern *FP=(a:5, b:4, x:3, t:1, z:1)* the postfix part of the item *x* is *Post(x)={t:1, z:1}*, while the prefix part is *Pre(x)={a:5, b:4}*. All frequent patterns of a given item are mined following the links connecting all occurrences of the current item in the *FP-Tree* and computing the support related with each path (frequent patterns). Each path is a set of ancestors of a given item called *Conditional Pattern Base (CPB)*, $CPB = \{Pre(I) \cup I \cup Post(I),$ *where Pre(I)=∅ or Post(I)=∅}*.

Starting from the leaf nodes, the Prefix part of a path can be used to mine rules, in particular, each path is a new tree called β-Tree, from which it is possible to apply the methodology explained previously to mine the meaningful rules. In particular, from the new β-Tree we prune all nodes (items) for which the condition $wS(node) < wminSupp$ is verified. The process goes ahead until all items of the current β-Tree are analyzed and/or we have reached the root *prefix set* related with the current item, or it is empty.

Algorithm 1. Gene Ontology Based Weighted Association Rules Mining (GO-WAR)

Require: A weighted Transaction Database WTB, A weighted minimum support $wminSupp$.

Ensure: A set of weighted association rules $Rules$.

   for all $wi \in WTB$ **do**

      Calculation of weighted support

      $ws(wi) \leftarrow computesupport$

   end for

   $frequentItemsList \leftarrow compute(wS, wminSupp)$ {Creation of FP-Tree}

   $Rules \leftarrow FP - Tree$ {Creation of Rules}

4.2 GO-WAR Performance Analysis

The space cost of GO-WAR algorithm is the size of the FP-Tree. The size of the FP-Tree is related with the dimension of the input dataset. In particular, the FP-Tree grows less than the database because, during the pre-processing step, all the items for which $ws(x) \leq weigthedminSupport$ are pruned (filtered), thus, after it has reached a particular dimension the tree remains constant, varying only the count of nodes. Even with huge dataset, the use of FP-Tree to represent data allows to save considerable amount of memory for storing the transactions.

Time Complexity Analysis. The time complexity of the algorithm varies on the scanning of the database and mining rules. Calculation of weighted support requires a linear time proportional with the dimension of the input dataset.

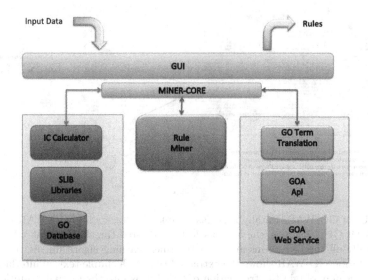

Fig. 3. Architecture of GO-WAR

Furthermore, in the same scan it is possible to locate frequent items, building, thus, the `FrequentItemsList`. Although the average time complexity of this composite step is: $O(n)$, `FP-Tree` creation is done in a linear time proportional to the number of frequent items identified. The time complexity to mine rules, is comparable with scanning a `n-ary tree`. In general, for a `n-ary tree` with height h, the upper bound for the maximum number of leaves is n^h. In our solution, sorting the elements in a descending order and using an inverted DFS scan strategy that allow us to obtain a time complexity equal to $O(n^2)$. Finally, adding all single times contribution, we can get the total complexity as: $3n + n^2$, where $3n$ is related with the computing of weighted support, `FrequentItemsList` and `FP-Tree` building. Complexity can be rewritten without loose generality as $n + n^2$, thus for huge value of n the complexity turns out to be $O(n^2)$.

4.3   GO-WAR Architecture and Implementation

GO-WAR has a layered architecture as depicted in Figure 3, that is composed of five main modules. The MINER-CORE receives user request and acts as a

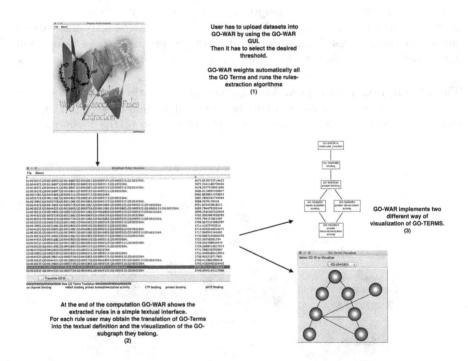

Fig. 4. GUI of GO-WAR. Initally user has to upload datasets into GO-WAR by using the GO-WAR GUI. Then he/she has to select the desired threshold. Go-WAR weights automatically all the GO-Terms and runs the rules learning algorithm. At the end of the computation GO-WAR shows the extracted rules in a simple textual interface. For each rule user may obtain the translation of GO-Terms into the textual definition and the visualization of the GO-graph they belong.

controller of other modules. Each submodule is controlled by using a master-slave approach that is internally realized through Java Threads in order to achieve efficient computation. The Rule Miner module is responsible for the calculation of frequent itemsets and for the extraction of final rules. The IC Calculator Module provides the calculation of IC for each GO Term. It is based on Semantic Measures Library and Toolkit libraries provided at (http://www.semantic-measures-library.org/) and on a local copy of Gene Ontology; The GO Term Translation module provides the complete description for each GO Term by invoking the Gene Ontology Annotation Database through Web Services. The GUI module based on Java Swing Technology provides to the user transparent access to all the implemented functionalities as depicted in Figure 4.

GO-WAR has been fully implemented by using Java Programming Language and it is available for download at `https://sites.google.com/site/weightedrules/`. Users may extract rules from the input dataset in an easy way as depicted in Figure 4.

5   Results

We tested GO-WAR on three different case studies: (i) comparison with respect to unweighted association rule learning; (ii) comparison with respect to Faria et al. approach, (iii) comparison with respect to Association Rule Learning in Weka.

The aims of these case studies are respectively: (i) to show that the weighted rule learning improves the quality of rules, (ii) to demonstrate the effectiveness of GO-WAR in a existing practical problem, and (iii) to demonstrate the efficency of GO-WAR.

5.1   Comparison with Respect to Pfam

In this case we compare GO-WAR with respect to a classical association rule algorithm ARule implemented into the Arules package of Bioconductor [1] [26] on an existing dataset of proteins in order to show as proof-of-concept that GO-WAR is able to mine more specific rules.

We used the Pfam database (protein families database) [27], that is a structured database of protein domains and families. The current release of Pfam (22.0) contains 9318 protein families. Each family of Pfam is, therefore, composed by proteins that have a similar structure and then may have a close function. Consequently, proteins in the same Pfam class should have similar annotations. Trivially, protein in the same Pfam classes should present statistically significant co-occurrence of annotations.

For each protein, corresponding GO annotations and the related IC have been determined forming the input dataset. Then *GO-WAR* has been used to mine weighted association rules. The software extracted 256 Rules that have a

---

[1] `www.bioconductor.org`

weighted support greater than 9.10. Finally, a post-processing phase filtered rules that include possibly redundant annotations. An annotation is redundant if it is implied by another more specific annotation of the same protein. For instance, the term `type I interferon signaling pathway` is parent of `interferon-gamma-mediated signaling`, then we filtered rules containing implication of such terms. Differently from other approaches, the use of weighted support discards a high number of rules with redundant annotations.

Top ranked rules are:

- IF (MHC class I protein complex AND plasma membrane AND Golgi membrane AND integral component of lumenal side of endoplasmic reticulum membrane) THEN interferon-gamma-mediated signaling pathway
- IF (MHC class I protein complex AND plasma membrane AND Golgi membran) THEN interferon-gamma-mediated signaling pathway
- IF (MHC class I protein complex AND plasma membrane) THEN interferon-gamma-mediated signaling pathway

The website of the project stores more example of rules. Results obtained by GO-WAR were compared to existing association rules approaches implemented in the Arules package of Bioconductor `www.bioconductor.org` [26]. After that non-weighted rules have been determined with these tools, the average weighted support for these rules was computed. The average value was 5.3. This value is lower than the average support of rules extracted by using GO-WAR confirming the effectiveness of the presented method.

A semantic-similarity based assessment was also performed [1,28]. A semantic similarity in biological scenario is a function defined over two terms extracted from an ontology (i.e. GO), which quantifies the similarity among sets of terms. The following hypothesis H_0 has been tested: given a rule extracted using GO-WAR, the average similarity among all the pairs of terms is significantly higher than random. For each rule, 1000 random rules were generated, all of them with the same structure (in terms of number of items), from the GO Database. Then the average semantic similarity has been calculated, i.e. we built all the possible pairs of items, then we calculated the pairwise semantic similarity. Finally we averaged all the obtained values. Since proteins within the same pathway/module/family play the same role, they are likely to have high semantic similarity. Finally, a non-parametric test was used to confirm that the average semantic similarity within the rule is higher than the random expectation. SimGIC [8] semantic similarity has been used because it does not use intrinsic IC, thus avoiding a trivial data circularity problem. Results confirmed that rules extracted by using GO-WAR have an average semantic similarity higher than random expectation.

5.2 Comparison between State of the Art: Approaches

This section shows how GO-WAR improves the current state of the art approaches for improving annotation consistency. We compare GO-WAR to the

Table 2. Comparison of Top Ranked Rules mined by Faria and extended by GO-WAR

Faria et al		GO-WAR		
IF	THEN	IF	THEN	INTERPRETATION
GO:0003924	GO:0005525	GO:0003924	GO:0005525	**IF** GTPase activity **THEN** GTP binding
GO:0030272	GO:0005524	GO:0030272 GO:0042803 GO:0046982	GO:0005524	**IF** 5-formyltetrahydrofolate cyclo-ligase activity, protein homodimerization activity, protein heterodimerization activity **THEN** ATP binding
GO:0004814	GO:0005524	GO:0004814 GO:0000049	GO:0005524	**IF** arginine-tRNA ligase activity, tRNA binding **THEN** ATP binding
GO:0008060	GO:0008270	GO:0008060 GO:0017137	GO:0008270	**IF** GTPase activator activity, Rab GTPase binding **THEN** zinc ion binding

paper of Faria et al. [8]. Authors analyze proteins stored in UniprotKB database focusing only on molecular function annotations. They define the meaning of incomplete and inconsistent annotations in a formal way and then they propose an association rule algorithm to improve annotation consistency aiming to help GO curators in developing the ontology. Authors do not consider weighted support but the classical support and confidence. They are aware of the specificity problem, thus they manually filter GO Terms with low specificity and redundant annotations before applying association rules. Conversely, GO-WAR approach is more flexible and avoids this manual intervention. Moreover, authors are interested in capturing implicit relationships between aspects of a single function (e.g. ATPase activity→ATP binding), while we explore the complete search space with the goal to highlight unknown relationships among biological functions, thus investigating a more broad perspective. Table 2 compares the top-ranked rules mined by Faria et al to those mined by GO-WAR.

5.3 Comparison with Weka

In order to analyze the performance of DMET-Miner against *Weka* we used a synthetic DMET dataset. All the experiments were performed on a MacBookPro with a Pentium i7 2.3 Ghz CPU, 16GB RAM and a 512GB SSH disk.

We have used a synthetic dataset containing the same number of transaction as the Pfam dataset. The dimension of the analyzed dataset is around 10 *MB* approximately. The reported execution times refer to average times, each value being computed by repeating the measure 10 times with the same settings.

Table 3. GO-WAR execution time and number of mined rules when varying the confidence (support=40%).

MinS%	Conf%	#Rules	ΔTime(ms)
40	20	20	6140.33
40	30	20	6400.00
40	40	20	6700.22
40	50	20	6156.33

Table 4. Weka execution time and number of mined rules when varying the confidence (support=40%).

MinS%	Conf%	#Rules	ΔTime(ms)
40	30	40	1,6410.33
40	50	40	1,6420.00
40	70	35	1,6504.66
40	80	0	1,9408.00
40	90	0	1,9310.00
40	95	0	1,9340.50
40	97	0	1,9010.33
40	99	0	1,9450.66

Tables 3 and, respectively, 4, show the computation time and the number of mined rules using GO-WAR and WEKA without the use of weights, when varying *minimum support* and *confidence*.

Moreover GO-WAR produces less rules than Weka (Apriori), but in the same time rules produced by GO-WAR are more informative than Weka rules. GO-WAR rules are presented in an IF-THEN form, making rules more easy to understand and interpret, and it is possible to directly visualize rules in the GO-graph.

6   Conclusion

Classical AR algorithms are not able to deal with different sources of production of GO annotations. Consequently, when used on annotated data they produce candidate rules with low IC. We here presented GO-WAR that is able to extract association rules with a high level of IC without losing Support and Confidence during the rule discovery phase and without the use of post-processing strategies for pruning uninteresting rules. We used publicly available GO annotation data to demonstrate our methods. Future works will regard testing of GO-WAR on larger datasets for improving annotation consistency.

References

1. Guzzi, P.H., Mina, M., Guerra, C., Cannataro, M.: Semantic similarity analysis of protein data: assessment with biological features and issues. Briefings in Bioinformatics 13(5), 569–585 (2012)
2. Harris, M.A., Clark, J., Ireland, A., Lomax, J., Ashburner, M., et al.: The gene ontology (go) database and informatics resource. Nucleic Acids Res. 32(Database issue), 258–261 (2004)
3. Camon, E., Magrane, M., Barrell, D., Lee, V., Dimmer, E., Maslen, J., Binns, D., Harte, N., Lopez, R., Apweiler, R.: The gene ontology annotation (goa) database: sharing knowledge in uniprot with gene ontology. Nucl. Acids Res. 32(suppl_1), D262–D266 (2004)
4. Hipp, J., Güntzer, U., Nakhaeizadeh, G.: Algorithms for association rule mining a general survey and comparison. ACM Sigkdd Explorations Newsletter 2(1), 58–64 (2000)
5. Guzzi, P.H., Milano, M., Cannataro, M.: Mining association rules from gene ontology and protein networks: Promises and challenges. Procedia Computer Science 29, 1970–1980 (2014)
6. Zaki, M.J., Parthasarathy, S., Ogihara, M., Li, W., et al.: New algorithms for fast discovery of association rules. In: KDD, vol. 97, pp. 283–286 (1997)
7. Cannataro, M., Guzzi, P.H., Sarica, A.: Data mining and life sciences applications on the grid. Wiley Interdisc. Rew.: Data Mining and Knowledge Discovery 3(3), 216–238 (2013)
8. Faria, D., Schlicker, A., Pesquita, C., Bastos, H., Ferreira, A.E.N., Albrecht, M., Falco, A.O.: Mining go annotations for improving annotation consistency. PLoS One 7(7), e40519 (2012)
9. Carmona-Saez, P., Chagoyen, M., Rodriguez, A., Trelles, O., Carazo, J.M., Pascual-Montano, A.: Integrated analysis of gene expression by association rules discovery. BMC Bioinformatics 7(1), 54 (2006)
10. Ponzoni, I., Nueda, M.J., Tarazona, S., Götz, S., Montaner, D., Dussaut, J.S., Dopazo, J., Conesa, A.: Pathway network inference from gene expression data. BMC Systems Biology 8(2), 1–17 (2014)
11. Tew, C., Giraud-Carrier, C., Tanner, K., Burton, S.: Behavior-based clustering and analysis of interestingness measures for association rule mining. Data Mining and Knowledge Discovery 28(4), 1004–1045 (2014)
12. Benites, F., Simon, S., Sapozhnikova, E.: Mining rare associations between biological ontologies. PLoS One 9(1), e84475 (2014)
13. Manda, P., Ozkan, S., Wang, H., McCarthy, F., Bridges, S.M.: Cross-ontology multi-level association rule mining in the gene ontology. PLoS One 7(10), e47411 (2012)
14. Nguyen, C.D., Gardiner, K.J., Cios, K.J.: Protein annotation from protein interaction networks and gene ontology. Journal of Biomedical Informatics 44(5), 824–829 (2011)
15. Manda, P., McCarthy, F., Bridges, S.M.: Interestingness measures and strategies for mining multi-ontology multi-level association rules from gene ontology annotations for the discovery of new go relationships. Journal of Biomedical Informatics 46(5), 849–856 (2013)
16. Naulaerts, S., Meysman, P., Bittremieux, W., Vu, T.N., Vanden Berghe, W., Goethals, B.: Kris Laukens. A primer to frequent itemset mining for bioinformatics. Briefings in Bioinformatics (2013)

17. Huttenhower, C., Hibbs, M.A., Myers, C.L., Caudy, A.A., Hess, D.C., Troyanskaya, O.G.: The impact of incomplete knowledge on evaluation: an experimental benchmark for protein function prediction. Bioinformatics 25(18), 2404–2410 (2009)
18. Alterovitz, G., Xiang, M., Hill, D.P., Lomax, J., Liu, J., Cherkassky, M., Dreyfuss, J., Mungall, C., Harris, M.A., Dolan, M.E., et al.: Ontology engineering. Nature Biotechnology 28(2), 128–130 (2010)
19. Harispe, S., Sánchez, D., Ranwez, S., Janaqi, S., Montmain, J.: A framework for unifying ontology-based semantic similarity measures: A study in the biomedical domain. Journal of Biomedical Informatics 48, 38–53 (2014)
20. Wang, W., Yang, J., Yu, P.S.: Efficient mining of weighted association rules (war). In: Proceedings of the Sixth ACM SIGKDD International Conference on Knowledge Discovery and Data Mining, pp. 270–274. ACM (2000)
21. Han, J., Pei, J., Yin, Y.: Mining frequent patterns without candidate generation. In: Chen, W., Naughton, J., Bernstein, P.A. (eds.) 2000 ACM SIGMOD Intl. Conference on Management of Data, pp. 1–12. ACM Press, May 2000
22. Borgelt, C.: Efficient implementations of apriori and eclat. In: Proc. 1st IEEE ICDM Workshop on Frequent Item Set Mining Implementations (FIMI 2003, Melbourne, FL). CEUR Workshop Proceedings 90 (2003)
23. du Plessis, L., Skunca, N., Dessimoz, C.: The what, where, how and why of gene ontology–a primer for bioinformaticians. Briefings in Bioinformatics 12(6), 723–735 (2011)
24. Sánchez, D., Batet, M., Isern, D.: Ontology-based information content computation. Knowledge-Based Systems 24(2), 297–303 (2011)
25. Han, J., Pei, J., Yin, Y.: Mining frequent patterns without candidate generation. SIGMOD Rec. 29(2), 1–12 (2000)
26. Hahsler, M., Grün, B., Hornik, K.: arules: Mining association rules and frequent itemsets (2006). http://cran.r-project.org/, r package version. SIGKDD Explorations 2, 0–4 (2007)
27. Finn, R.D., Tate, J., Mistry, J., Coggill, P.C., Sammut, S.J.J., Hotz, H.-R.R., Ceric, G., Forslund, K., Eddy, S.R., Sonnhammer, E.L., Bateman, A.: The pfam protein families database. Nucleic Acids Research 36(database issue), D281–D288 (2008)
28. Cho, Y.-R., Mina, M., Lu, Y., Kwon, N., Guzzi, P.H.: M-finder: Uncovering functionally associated proteins from interactome data integrated with go annotations. Proteome Sci. 11(suppl. 1), S3 (2013)

Extended Spearman and Kendall Coefficients for Gene Annotation List Correlation

Davide Chicco[1,2], Eleonora Ciceri[1], and Marco Masseroli[1]

[1] Dipartimento di Elettronica Informazione e Bioingegneria,
Politecnico di Milano, Milan, Italy
[2] Princess Margaret Cancer Centre,
University of Toronto, Toronto, Canada
davide.chicco@gmail.com,
eleonora.ciceri@polimi.it,
masseroli@elet.polimi.it

Abstract. Gene annotations are a key concept in bioinformatics and computational methods able to predict them are a fundamental contribution to the field. Several machine learning algorithms are available in this domain; they include relevant parameters that might influence the output list of predicted gene annotations. The amount that the variation of these key parameters affect the output gene annotation lists remains an open aspect to be evaluated. Here, we provide support for such evaluation by introducing two list correlation measures; they are based on and extend the Spearman ρ correlation coefficient and Kendall τ distance, respectively. The application of these measures to some gene annotation lists, predicted from Gene Ontology annotation datasets of different organisms' genes, showed interesting patterns between the predicted lists. Additionally, they allowed expressing some useful considerations about the prediction parameters and algorithms used.

Keywords: Biomolecular annotations, Spearman coefficient, Kendall distance, top-K queries.

1 Introduction

In molecular biology and bioinformatics, a *controlled biolomolecular annotation* is an association of a biomolecular entity (mainly a gene, or gene product) with a concept, described by a term of a controlled vocabulary (in this case part of an ontology), which represents a biomedical feature. This association states that the biomolecular entity has such feature. For instance, the association $\langle Entrez\,Gene\,ID\,1080,\,GO{:}0055085\rangle$ is a typical annotation of the human *CFTR* gene (*Cystic fibrosis transmembrane conductance regulator (ATP-binding cassette sub-family C, member 7)*), which has Entrez Gene ID 1080, with the concept represented by the *transmembrane transport* term of the Gene Ontology, which has ID GO:0055085. Thus, such annotation states that the human *CFTR* gene is involved in the *transmembrane transport*.

© Springer International Publishing Switzerland 2015
C. di Serio et al. (Eds.): CIBB 2014, LNCS 8623, pp. 19–32, 2015.
DOI: 10.1007/978-3-319-24462-4_2

Despite their biological significance, there are some issues concerning available biomolecular annotations [1]. In particular, they are incomplete: only a subset of the biomolecular entities of the sequenced organisms is known, and among those entities only a small part has been annotated by researchers so far. In addition, they may be erroneously annotated and not yet revisited, prior to their being stored into online data banks. Within this context, computational methods and software tools able to produce lists of available or new predicted annotations ranked based on their likelihood of being correct are an excellent contribution to the field [2].

For this reason, starting from a state-of-the-art algorithm [3] based on truncated Singular Value Decomposition (SVD) [4], we developed some enhanced variants that take advantage of available Gene Ontology (GO) [5] annotation data to predict new gene annotations of different organisms, including *Homo sapiens*. Specifically, we designed an automated algorithm that chooses the best the truncation level [6] for the truncated SVD method and developed some alternatives to the SVD, based on gene clustering [7] and Resnik's [8] term-term similarity metrics [9]. Similar to Khatri and colleagues papers [10] [11], we additionally implemented another version of this method with the enhancement of frequency and probability weights [12]. To this end, we also applied some *topic modeling* algorithms, such as Probabilistic Latent Semantic Analysis (pLSA) by Hofmann et al. [13], and Latent Dirichlet Allocation (LDA) by Blei et al. [14], obtaining relevant results, respectively, in [15] and [16]. Additionally, one of the authors recently took advantage of a *deep learning* algorithm, built on a multilayer autoencoder neural network, that lead to interesting prediction results in reference [17].

All these methods involve key parameters that strongly influence their output. To understand how the resulting annotation lists vary when these key parameters change, a similarity measure that compares different output annotation lists is required. Currently, several metrics are available to compare ranked lists of elements. A good example is the Goodman-Kruskal's γ [18], which measures the difference between rank-concordant or rank-discordant pairs of objects in two lists. However, Fagin and colleagues [19] state that the most useful and consistent measures are the Spearman ρ correlation coefficient [20] and Kendall τ distance [21]. In recent years, many variants were proposed to meet new needs that came with some state-of-the-art applications, e.g., top-K queries [22]. For example, the work proposed in [23] by Kumar and colleagues adapts the original formulation to measure weighted correlations, by placing more emphasis on items with high rankings. Applications have been shown in music signal prediction [24], recommendation systems [25] and computer vision [26].

In this work, we depart from a recent work by Ciceri et al. [27] to develop new weighted correlation metrics able to better compare biomolecular annotation lists. We adapt the weighted Kendall τ distance (proposed in [23]) and the Spearman ρ rank correlation coefficient (proposed in [20]) to the case in which multiple lists (initially containing the same items in different orders) are truncated up to a level K, thus resulting in sub-lists whose sets of contained items may not coincide.

The remainder of this chapter is organized as follows. Section 2 illustrates aspects related to the prediction of gene annotations. Section 3 introduces the Spearman ρ correlation coefficient and Kendall τ distance variants for the comparison of annotation lists. Section 4 describes some significant test results of the proposed measure variants. Finally, we conclude in Section 5.

2 Prediction of Gene Ontology Annotations

Let $\mathbf{A} = [a_{ij}]$ be an $m \times n$ matrix, where each row corresponds to a gene and each column corresponds to a Gene Ontology feature term ($a_{ij} = 1$ if gene i is annotated to feature term j, $a_{ij} = 0$ otherwise). Moreover, let θ be a fixed threshold value. The prediction algorithm elaborates the matrix $\mathbf{A}$ to produce an output matrix $\widetilde{\mathbf{A}}$, with the same dimensions of $\mathbf{A}$, where each likelihood value $\widetilde{a}_{ij}$ is used to categorize an annotation: $\langle \mathbf{gene}_i, \mathbf{feature}_j, \widetilde{a}_{ij} \rangle$. A high $\widetilde{a}_{ij}$ value indicates that the probability for $\mathbf{gene}_i$ to be associated with the feature $\mathbf{feature}_j$ is high. Each annotation $\langle \mathbf{gene}_i, \mathbf{feature}_j, \widetilde{a}_{ij} \rangle$ can be classified in four categories:

- *Annotation Predicted* (AP): $a_{ij} = 0 \wedge \widetilde{a}_{ij} > \theta$ (similar to False Positive);
- *Annotation Confirmed* (AC): $a_{ij} = 1 \wedge \widetilde{a}_{ij} > \theta$ (similar to True Positive);
- *Non-Annotation Confirmed* (NAC): $a_{ij} = 0 \wedge \widetilde{a}_{ij} \leq \theta$ (similar to True Negative);
- *Annotation to be Reviewed* (AR): $a_{ij} = 1 \wedge \widetilde{a}_{ij} \leq \theta$ (similar to False Negative).

Since APs and ARs can be considered as *presumed errors* with respect to the available annotations, we chose the value of θ as the one that minimizes their sum (APs + ARs), as Khatri et al. did in [3]. After a *ten fold cross validation* phase, in which the 10% of annotations are randomly set to zero (as explained in reference [9]), the software compares each input annotation with its corresponding output prediction. Based on the value of the threshold θ, a list of annotations with $\widetilde{a}_{ij} > \theta$ is created; it is subdivided in two sections: an APlist, i.e., *Annotation Predicted* list, and a NAClist, i.e., *Non-Annotation Confirmed* list, respectively containing the annotations from the original list that were classified as belonging to the AP or NAC category. Moreover, the defined categories are used to create a Receiver Operating Characteristic (ROC) curve, a graphical plot depicting the performance of a binary classifier system for different discrimination threshold values [28]. Similar to its original definition, which uses TPrate and FPrate, our ROC curve depicts the trade-off between the ACrate and the APrate, where:

$$\text{ACrate} = \frac{\text{AC}}{\text{AC} + \text{AR}} \qquad \text{APrate} = \frac{\text{AP}}{\text{AP} + \text{NAC}} \qquad (1)$$

for all possible values of θ. Notice that, in statistical terms, ACrate = *Sensitivity* and APrate = $1 - Specificity$. The output ROC space is thus defined by ACrate and APrate as x and y axes, respectively. In all of our test results, reported in Section 4, we consider only the APrate in the normalized interval $[0, 1]\%$, in order to evaluate the best predicted annotations (APs) having the highest likelihood score. Furthermore, only if the obtained ROC Area Under the Curve (AUC) percentage is greater than a fixed threshold $\theta_1 = 2/3 = 66.67\%$, we consider to have good reliability of reconstruction. We use the annotation category labels we just introduced to define our measures in the following sections.

3   Annotation List Correlation Measures

Each annotation prediction algorithm has some key parameters; changing their values usually leads to different output predicted annotation lists, i.e. APlists. For instance, the truncated SVD (tSVD) algorithm may produce quite different results when its truncation level k varies. In the two variants of tSVD that we enhanced with gene clustering and term-term similarity metrics (named Semantically IMproved tSVD variants, i.e., SIM1 and SIM2), different results may be obtained when varying the number of gene clusters C [9]. In Probabilistic Latent Semantic Analysis [13], a topic modeling method, a key role is played by the number of topics T selected before performing the evaluation [15]. To understand the amount by which the selected parameter values are able to influence the output results, it is important to define similarity metric that compares two output APlists resulting from different algorithm parameterizations. To this end, we present novel variants to two well-known similarity metrics:

- the *Spearman rank correlation coefficient* (ρ) [20]
- the *Kendall rank distance* (τ) [21]

which are respectively described in Subsection 3.1, and in Subsection 3.2.

3.1   Spearman Rank Correlation Coefficient

The *Spearman rank correlation coefficient* (ρ, sometimes also called *foot-rule*) [20] measures the statistical dependence between two variables X and Y. The measure expresses either *positive* correlation, i.e., Y increases when X increases, or *negative* correlation, i.e., Y decreases when X increases. A similar definition can be applied to pair of ranked lists. Let l_a and l_b be two ranked lists of biomolecular annotations. A maximum positive correlation $\rho = +1$ is returned when l_a and l_b are identical (i.e. having the same elements in the same order), while the maximum negative correlation $\rho = -1$ is returned when l_a and l_b contain the same elements, but in reverse order. The minimum correlation $\rho = 0$ (i.e., maximum diversity) is instead detected when the element order in l_a and l_b strongly diverges. Suppose l_a and l_b have the same length n. Given an element i, let x_i denote its position in l_a, y_i its position in l_b and $d_i = |x_i - y_i|$. The final normalized Spearman ρ value is then computed as:

$$\rho = 1 - \frac{6 \cdot \sum d_i^2}{n(n^2 - 1)} \qquad (2)$$

Table 1. Example of the application of the Spearman rank correlation coefficient; l_a and l_b are, respectively, the first and second compared lists; x_i is the position of the ith element of l_a in l_a, and y_i is the position of the ith element of l_a in l_b; d_i is the difference between positions.

l_a	x_i	l_b	y_i	d_i	d_i^2
a	1	c	4	3	9
b	2	b	2	0	0
c	3	e	1	2	4
d	4	a	5	1	1
e	5	d	3	2	4

For example, the Spearman rank coefficient computed for the l_a and l_b in Table 1 is: $\rho = 1 - \frac{6 \cdot 18}{5 \cdot 24} = 1 - \frac{18}{20} = 1 - 0.9 = 0.1$. The low value of ρ indicates that l_a and l_b have a low correlation, as shown in Table 1.

Weighted Spearman Rank Coefficient. As mentioned earlier, two lists l_a and l_b may contain different elements or have different length; in this case, the lists would not be properly handled by the classical Spearman rank correlation coefficient. In order to additionally consider this case, based on the work by Ciceri et al. [27], we introduce a new *Weighted Spearman rank coefficient* featuring penalty distance weights w_i for each element i absent from one list.

Let $q = |l_a \cup l_b|$. Thus, the penalty weight w_{si} for an object i in the lists l_a or l_b is computed as follows:

$$w_{si} = \begin{cases} 1 - \frac{1}{|x_i - y_i| + 1}, & i \in l_a \wedge i \in l_b \\ 1, & \text{otherwise} \end{cases} \tag{3}$$

The Weighted Spearman rank coefficient value is then computed as:

$$\rho_w = \frac{\sum_{i=1}^{q} w_{si}}{q} \tag{4}$$

High correlation is found when $\rho_w \simeq 0$ (i.e., very few penalties are assigned), while low correlation is found when $\rho_w \simeq 1$ (i.e., many penalties are assigned). If the two lists have no common elements, i.e., $q = |l_a| + |l_b|$, $\rho_w = 1$.

Extended Spearman Rank Coefficient. The Weighted Spearman rank coefficient shows a flaw in our biomolecular annotation prediction context: all elements $\{i : i \notin l_a \vee i \notin l_b\}$ are weighted equally.

As an example, let l_a (l_b) be a ranked list containing an APlist l_a^{AP} (l_b^{AP}) and a NAClist l_a^{NAC} (l_b^{NAC}), and let a' and a'' be two biomolecular annotations. If an annotation is not present in l_a^{AP} (l_b^{AP}), then it is likely present in the related NAClist l_a^{NAC} (l_b^{NAC}) (see Section 2 for details). However, the ρ_w coefficient would assign both lists the annotations $\{a' : a' \notin l_a^{AP}, a' \in l_a^{NAC}\}$ and $\{a'' : a'' \notin l_a^{AP}, a'' \notin l_a^{NAC}\}$ a maximum (equal) penalty (i.e., $w_{a'} = w_{a''} = 1.0$). The same holds for l_b. Thus, we designed a new coefficient more well-suited to our domain, where a' gets a lower penalty than a''. To do so, we first modified the position weight of each element i in the NAClist:

$$\hat{z}_i = z_i + 2 \cdot m \tag{5}$$

where z_i is the position of i in the NAClist l_b^{NAC}, m is the length of the associated APlist l_b^{AP} and 2 is a penalty factor for i not to be in l_b^{AP} and being in l_b^{NAC} (the value 2 keeps the penalty proportional to the position of i in the list). Then, we expressed the new penalty weight as follows:

$$v_{si} = \begin{cases} 1, & i \in l_a^{AP} \notin l_b^{AP} \notin l_b^{NAC} \\ 1 - \frac{1}{|x_i - \hat{z}_i| + 1}, & i \in l_a^{AP} \notin l_b^{AP} \in l_b^{NAC} \\ 1 - \frac{1}{|x_i - y_i| + 1}, & i \in l_a^{AP} \in l_b^{AP} \end{cases} \tag{6}$$

where x_i is the position of the i element in l_a^{AP}, y_i is its position in l_b^{AP} and $\hat{z}_i$ is its position in l_a^{NAC} (l_b^{NAC}).

We selected this function to reduce the penalties of those elements found in both the first AP lists (l_a^{AP}) and the second NAC lists (l_b^{NAC}), with respect to those only found in the first AP list (l_a^{AP}) but absent from the second one.

The Extended Spearman rank coefficient is thus computed as:

$$\rho_e = \frac{\sum_{i=1}^{q} v_{si}}{q} \tag{7}$$

As for the Weighted Spearman rank coefficient, high ρ_e values lead to low correlation, while $\rho_e \simeq 0$ suggests high correlation.

3.2   Kendall Rank Distance

The *Kendall rank distance* (τ) [21] counts the normalized number of pairwise disagreements between two ranked lists l_a and l_b, i.e., the number of bubble-sort swaps needed to sort l_a in the same order of l_b. Obviously, when the two lists are identical, $\tau = 0$. Conversely, if l_a is obtained by reversing the order of l_b, then $\tau = 1$. Let l_a and l_b be two lists of length n containing the same elements; given an element i, x_i is its position in l_a, while y_i is its position in l_b. Thus, the set $\mathcal{K}$ of required swaps between elements in lists l_a and l_b is computed as follows:

$$\mathcal{K}(l_a, l_b) = \{(i, j) : (x_i < y_i \wedge x_j > y_j) \vee (x_i > y_i \wedge x_j < y_j)\} \tag{8}$$

The normalized Kendall rank distance is given by:

$$\tau = \frac{|\mathcal{K}(l_a, l_b)|}{n(n-1)/2} \tag{9}$$

Notice that the Kendall rank distance does not express negative correlation between lists. Moreover, while the Spearman rank coefficient is focused on the

Table 2. Example of the application of the Kendall rank correlation metrics. For each pair of elements in the set (1st column), the ranks in l_a and l_b are provided (2nd and 3rd columns), along with the necessity of performing a bubble-sort swap (4th column).

Pair	l_a ranks	l_b ranks	Bubble-sort swap	Pair	l_a ranks	l_b ranks	Bubble-sort swap
(a, b)	1 < 2	4 > 2	✓	(b, d)	2 < 4	2 < 5	
(a, c)	1 < 3	4 > 1	✓	(b, e)	2 < 5	2 < 3	
(a, d)	1 < 4	4 < 5		(c, d)	3 < 4	1 < 5	
(a, e)	1 < 5	4 > 3	✓	(c, e)	3 < 5	1 < 3	
(b, c)	2 < 3	2 > 1	✓	(d, e)	4 < 5	5 > 3	✓

distance between the ranks of each element in the lists, the Kendall rank distance considers only the number of swaps in the element rank.

For example, consider the lists l_a and l_b in Table 1, whose rankings are summarized in Table 2. The number of bubble-sort swaps needed to give l_a and l_b the same ranking is $|\mathcal{K}(l_a, l_b)| = 5$. Thus, the Kendall rank distance between l_a and l_b is: $\tau = \frac{5}{(5\cdot4)/2} = \frac{5}{10} = 0.5$, i.e., the lists have medium correlation. This is discordant with the result obtained by applying the Spearman coefficient (see Section 3.1), which states that the lists have low correlation. Thus, the proposed example highlights the different nature of the two metrics.

Weighted Kendall Rank Distance. Like the Spearman rank coefficient (Subsection 3.1), a flaw of the classical normalized Kendall rank distance is that it works properly only when the two lists have the same size and contain the same elements. Analogous to what we did for the Spearman rank coefficient, we introduce weights to penalize elements that are present in one list, but absent from the other [27]. In this case, penalties are added so as to give more weight to bubble-sort swaps concerning elements in earlier positions and less weight to the those in later positions. Moreover, elements that appear only in one list are penalized further .

Let x_i and y_i be the positions of the element i in the lists l_a and l_b, respectively. In case $i \notin l_a$, $x_i = |l_a| + 1$; else if $i \notin l_b$, then $y_i = |l_b| + 1$. Then, the penalty for i is computed as:

$$w_{ki} = \begin{cases} \frac{1}{\log(x_i+2)} - \frac{1}{\log(x_i+3)} & i \in l_a \wedge i \in l_b \\ 0.5 & \text{otherwise} \end{cases} \tag{10}$$

We chose this weight function to make the weight larger if the element is in the earliest positions of the list and lower if it is in the later ones; we consider bubble-sort swaps in the first positions more important. The Weighted Kendall rank distance is then computed as:

$$\tau_w = \frac{\sum_{(i,j)\in\mathcal{K}(l_a,l_b)} w_{ki}w_{kj}}{|\mathcal{K}(l_a, l_b)|} \tag{11}$$

Thus, low correlations correspond to high distance values. On the other hand, if the two lists are identical, $\tau_w = 0$ (since no swaps occur).

Extended Kendall Rank Distance. Similar to what we did with the Extended Spearman rank coefficient (in Subsection 3.1), we introduce the Extended Kendall rank distance to furhter handle cases where an element i may be absent from an APlist, but present in its NAClist. If such an element i has a likelihood value h (that is, $\widetilde{a}_{ij}$ is in the output matrix and in the APlist), we set its weight to $v_{ki} = 0.5 - h$. Notice that annotations in the APlist have likelihood values ranked from the maximum to the minimum, in the interval $[1, \theta)$, while annotations in the NAClist are in the interval $[\theta, 0]$, i.e., the prediction likelihood real value h decreases along the rank. Accordingly, it is used to define the weight for the element i as follows:

$$
v_{ki} = \begin{cases} 0.5 & i \in l_a^{AP} \notin l_b^{AP} \notin l_b^{NAC} \\ 0.5 - h & i \in l_a^{AP} \notin l_b^{AP} \in l_b^{NAC} \\ \frac{1}{\log(x_i+2)} - \frac{1}{\log(x_i+3)} & i \in l_a^{AP} \in l_b^{AP} \end{cases} \tag{12}
$$

where x_i is the position of the i element in l_a^{AP}. In particular, the penalty reduction h can be defined as: $h = 0.5 - z_i/(m \cdot 2)$, where z_i is the position of i in l_a^{NAC} (l_b^{NAC}) and m is the length of l_a^{NAC} (l_b^{NAC}).

We selected the weight function in Equation 12 such that it decreases when the element gets a lower rank. Consequently:

$$
\tau_e = \sum_{(i,j) \in \mathcal{K}(l_a, l_b)} v_{ki} v_{kj} \tag{13}
$$

As for the Weighted Kendall rank coefficient (in Subsection 3.2), τ_e is high when l_a and l_b are very different and $\tau_e \simeq 0$ when the two lists are very similar.

4   Results and Discussion

In this Section we evaluate our proposed Extended Spearman rank coefficient and Extended Kendall rank distance measures on lists of annotations produced via the common truncated SVD method, explained in Figure 1. Specifically, we measure the ir prediction performance, while varying the SVD truncation level k. The measures are tested on a small dataset of GO Cellular Component (CC) and Molecular Function (MF) annotations of *Homo sapiens*, *Gallus gallus* (red junglefowl) and *Bos taurus* (cattle) genes, which were obtained from the Genomic and Proteomic Data Warehouse (GPDW) [29] (*H. sapiens* CC: number of genes: 7,868; number of CC feature terms: 684; number of annotations: 14,381. *G. gallus* MF: number of genes: 309; number of MF feature terms: 225; number of annotations: 509. *B. taurus* MF: number of genes: 543; number of MF feature terms: 422; number of annotations: 934).

Table 3a shows the quantitative amounts of AP and NAC annotations for different truncation levels k, provided by our best truncation level algorithm (described in [6]). In the case of *Homo sapiens* CC, the best chosen value is $k = 378$;

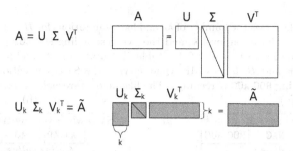

Fig. 1. An illustration of the Singular Value Decomposition (upper white image) and the Truncated SVD reconstruction (lower gray image) of the A matrix. In the classical SVD decomposition, $A \in \{0,1\}^{m \times n}$, $U \in \mathbb{R}^{m \times m}$, $\Sigma \in \mathbb{R}^{m \times n}$ $V^T \in \mathbb{R}^{n \times n}$. In the Truncated decomposition, where $k \in \mathbb{N}$ is the truncation level, $U_k \in \mathbb{R}^{m \times k}$, $\Sigma_k \in \mathbb{R}^{k \times k}$, $V_k^T \in \mathbb{R}^{k \times n}$, and the output matrix $\tilde{A} \in \mathbb{R}^{m \times n}$. The choice of k strongly influences the content of the output $\tilde{A}$ matrix, during reconstruction, even if it does not change the matrix dimensions

the application of the tSVD algorithm with this truncation level produces the *List0*. The values in Table 3a show that a fairly small change in the truncation level k may lead to different numbers of APs and NACs.

Table 4 shows the values of the Extended Kendall rank distances and the Extended Spearman rank coefficients between the nine lists in Table 3a. By analyzing the Extended Spearman rank coefficients (gray cells), we do not notice any interesting trends or patterns, since the values seem to vary stochastically. Conversely, the Extended Kendall rank distance values increase as the list difference increases monotonically, getting high values for almost all the list pairs, except for the pairs: $\langle List0, List1 \rangle$, $\langle List0, List2 \rangle$ and $\langle List1, List2 \rangle$.

The analysis of the Spearman ρ metrics, whose computation is based on the difference between list element ranks, suggests that the ROC AUC size does not directly influence the rank dissimilarity in the list. On the contrary, the Kendall τ values, whose computation is based on the number of bubble-sort swaps needed to make two lists identical, increase as the AUC difference increases; their analysis suggests the existence of a relationship between the ROC AUC values and the rankings of the APs in the lists from the tSVD generating high AUC percentages and very low AP numbers.

4.1   SVD Truncation Patterns

Another interesting result is revealed by sorting the lists on the basis of the truncation level k, as done in Table 5. While the Extended Spearman rank coefficients give no additional clues on specific trends, the Extended Kendall rank distances show that the higher the SVD truncation level difference between two lists is, and the less similar the lists are. Apart from the comparison between $\langle List0, List1 \rangle$, all other lists show value trends that increase when the distance between truncation levels increase. *List8*, for example, shows near maximum dissimilarity with all the other lists.

Table 3. Table 3a shows the number of AP and NAC annotations for *H. sapiens* CC when varying the SVD truncation level k, and their corresponding ROC AUC percentage.The likelihood threshold is fixed at $\theta = 0.49$. Table 3b Numbers of AP and NAC annotations for *G. gallus* MF and *B. taurus* MF when varying the SVD truncation level k, and their corresponding ROC AUC percentage. The likelihood threshold is fixed at $\theta = 0.50$

(a)

SVD when varying truncation k			
k	AP	NAC	ROC AUC
Homo sapiens CC			
List0	**378**	8	4,458,751	83.49%
List1	**402**	2	4,458,757	53.64%
List2	**390**	7	4,458,752	53.58%
List3	**291**	19	4,458,740	53.14%
List4	**349**	8	4,458,751	52.97%
List5	**233**	48	4,458,711	51.82%
List6	**175**	78	4,458,681	48.80%
List7	**117**	86	4,458,673	45.11%
List8	**59**	95	4,458,664	38.82%

(b)

SVD when varying truncation k			
k	AP	NAC	ROC AUC
Gallus gallus MF			
List0	**40**	9	39,340	75.38%
List1	**53**	5	39,344	74.33%
List2	**27**	10	39,339	73.69%
List3	**14**	11	39,338	67.25%
List4	**1**	164	39,185	35.98%
Bos taurus MF			
List0	**70**	11	120,318	74.74%
List1	**93**	8	120,321	74.59%
List2	**47**	11	120,318	73.57%
List3	**24**	32	120,297	68.59%
List4	**1**	369	119,960	35.21%

Table 4. Extended Spearman rank coefficient values (gray cells) and Extended Kendall rank distance values (white cells) from the comparison of the nine annotation lists in Table 3a, generated by the truncated SVD method applied to the *Homo sapiens* CC dataset with varying truncation level k. Intervals: MaxCorrelation = 0; MinCorrelation = 1. All lists are ordered from the one corresponding to the maximum ROC AUC percentage (*List0*, AUC = 83.49%) to the one corresponding to the minimum AUC (*List8*, AUC = 38.82%).

		Extended Spearman values								
		List0	List1	List2	List3	List4	List5	List6	List7	List8
E.	List0		0.129	0.07	0.301	0.224	0.593	0.841	0.443	0.501
	List1	0.621		0.01	0.535	0.512	0.708	0.838	0.585	0.539
K	List2	0.491	0.558		0.324	0.245	0.656	0.872	0.499	0.509
e	List3	0.984	0.962	0.978		0.57	0.158	0.362	0.249	0.772
n	List4	0.868	0.894	0.913	0.935		0.416	0.302	0.306	0.885
d	List5	0.995	0.995	0.995	0.952	0.984		0.283	0.069	0.805
a	List6	0.999	0.999	0.999	0.988	0.998	0.936		0.427	0.628
l	List7	0.996	0.998	0.997	0.997	0.998	0.990	0.946		0.522
l	List8	0.998	0.998	0.999	0.999	0.999	0.996	0.984	0.982	

Table 5. Extended Spearman rank coefficient values (gray cells) and Extended Kendall rank distance values (white cells) from the comparison of the nine annotation lists in Table 3a, generated by the truncated SVD method applied to the *Homo sapiens* CC dataset with varying truncation level k. Intervals: MaxCorrelation = 0; MinCorrelation = 1. All lists are ordered from the one generated with the greatest **SVD truncation level** (*List1*, $k = 402$) to the one generated with the lowest level (*List8*, $k = 59$).

		Extended Spearman values								
		List1	List0	List2	List4	List3	List5	List6	List7	List8
E.	List1		0.129	0.01	0.512	0.535	0.708	0.838	0.585	0.539
	List0	0.621		0.07	0.224	0.301	0.593	0.841	0.443	0.501
K	List2	0.558	0.491		0.245	0.324	0.656	0.872	0.499	0.509
e	List4	0.894	0.868	0.913		0.935	0.416	0.302	0.306	0.885
n	List3	0.962	0.984	0.978	0.57		0.158	0.362	0.249	0.772
d	List5	0.995	0.995	0.995	0.984	0.952		0.283	0.069	0.805
a	List6	0.999	0.999	0.999	0.998	0.988	0.936		0.427	0.628
l	List7	0.998	0.996	0.997	0.998	0.997	0.99	0.946		0.522
l	List8	0.998	0.998	0.999	0.999	0.999	0.996	0.984	0.982	

4.2 ROC AUC Patterns

Beyond its importance in providing new knowledge about the variation of the annotation lists, varying the SVD truncation level k has no utility for our validation process. Conversely, interesting trends can be found by comparing the ROC AUC percentages, the Extended Kendall rank distance and the Extended Spearman rank coefficient. As general examples, we show the cases of the *Bos taurus* MF and *G. gallus* MF datasets, in Table 3b and Table 6. As one may notice, these lists corresponding to similarly AUCs (*List0*, *List1*, *List2*), have similar low Extended Spearman rank coefficients. This means that the elements present in these lists have similar rankings.

Since there is a correlation between ROC AUC and list similarity, this may be helpful in finding the best predicted annotations. In fact, our prediction methods are able to produce ROC with maximum AUC, and the AUCs have an oscillatory trend. Since these oscillations produce low changes to the ROC AUC percentages, these will slightly influence the final prediction results. Thus, our methods, based on the optimization of the ROC AUCs, are quite robust. There is no need to find the overall best SVD truncation since it is sufficient to find truncation values that are close to the best ones; the algorithm can then select the best predicted annotations accordingly.

Table 6. Extended Spearman rank coefficient values (gray cells) and Extended Kendall rank distance values (white cells) from the comparison of the annotation lists in Table 3b, generated by the truncated SVD method applied to the *G. gallus* MF dataset with varying truncation level k. Intervals: `MaxCorrelation = 0`; `MinCorrelation = 1`. All lists are ordered from the one generated with the greatest SVD truncation level k to the one generated with the lowest level. We report in **bold** the cases showing low Spearman or Kendall values for lists with similar `AUCs`.

		Extended Spearman values					Extended Spearman values				
		List0	List1	List2	List3	List4	List0	List1	List2	List3	List4
		Gallus gallus MF					*Bos taurus* MF				
E	List0		**0.370**	**0.200**	0.620	0.85		**0.192**	**0.166**	0.364	0.794
x	List1	0.790		**0.330**	0.780	0.94	0.687		**0.279**	0.425	0.888
t.	List2	0.640	0.790		0.180	0.74	0.812	0.822		0.128	0.772
	List3	0.980	0.960	0.950		0.78	0.980	0.980	0.950		0.773
K.	List4	1.000	1.000	1.000	1.000		0.999	0.999	0.999	1.000	

5   Conclusions

Both our Extended Spearman rank coefficient and Extended Kendall rank distance measures resulted effective and useful to compute the level of "similarity" between two gene annotation lists, by focusing on either the list element position difference (Spearman) or on the number of list elements having different rankings (Kendall). Using them, we discovered a negative correlation between two generated annotation lists: the more the truncation level difference increases, the more dissimilar (in terms of bubble-sort swaps needed to make them identical) the annotation lists are. We also observed a positive correlation between the `ROC` `AUCs` and the similarity between two lists: the closer two `AUC` percentages are, the more similar the related predicted annotation lists are. In this case the similarity is expressed through the Extended Kendall rank distance, which measures the difference between ranked position of an element present in both analyzed lists.

In general, we can state that the Spearman coefficient is the most useful one when the user wants to take advantage of the global order of the items in the two compared lists; on the contrary, the Kendall distance is the best choice when the user wants to highlight the relative raking among items in the lists.

In the future, we plan to apply our metrics to annotation lists produced through different algorithms and study their variation when changing algorithm's key parameter values. For example, we will explore the similarity between annotation lists from the topic modeling algorithms (pLSA [15] and LDA [16]) when the number of topics changes.

Acknowledgments. This work is partially funded by the CUbRIK project (www.CubrikProject.eu).

References

[1] Karp, P.D.: What we do not know about sequence analysis and sequence databases. Bioinformatics 14(9), 753–754 (1998)

[2] Pandey, G., Kumar, V., Steinbach, M.: Computational approaches for protein function prediction: A survey. Twin Cities: Department of Computer Science and Engineering, University of Minnesota (2006)

[3] Khatri, P., Done, B., Rao, A., Done, A., Draghici, S.: A semantic analysis of the annotations of the human genome. Bioinformatics 21(16), 3416–3421 (2005)

[4] Golub, G.H., Reinsch, C.: Singular value decomposition and least squares solutions. Numerische Mathematik 14(5), 403–420 (1970)

[5] Consortium, G.O., et al.: Creating the gene ontology resource: design and implementation. Genome Research 11(8), 1425–1433 (2001)

[6] Chicco, D., Masseroli, M.: A discrete optimization approach for svd best truncation choice based on roc curves. In: 2013 IEEE 13th International Conference on Bioinformatics and Bioengineering (BIBE), pp. 1–4. IEEE (2013)

[7] Drineas, P., Frieze, A., Kannan, R., Vempala, S., Vinay, V.: Clustering large graphs via the singular value decomposition. Machine Learning 56(1-3), 9–33 (2004)

[8] Resnik, P.: Using information content to evaluate semantic similarity in a taxonomy. arXiv preprint cmp-lg/9511007 (1995)

[9] Chicco, D., Tagliasacchi, M., Masseroli, M.: Genomic annotation prediction based on integrated information. In: Biganzoli, E., Vellido, A., Ambrogi, F., Tagliaferri, R. (eds.) CIBB 2011. LNCS, vol. 7548, pp. 238–252. Springer, Heidelberg (2012)

[10] Done, B., Khatri, P., Done, A., Draghici, S.: Semantic analysis of genome annotations using weighting schemes. In: IEEE Symposium on Computational Intelligence and Bioinformatics and Computational Biology, CIBCB 2007, pp. 212–218. IET (2007)

[11] Done, B., Khatri, P., Done, A., Draghici, S.: Predicting novel human gene ontology annotations using semantic analysis. IEEE/ACM Transactions on Computational Biology and Bioinformatics (TCBB) 7(1), 91–99 (2010)

[12] Pinoli, P., Chicco, D., Masseroli, M.: Enhanced probabilistic latent semantic analysis with weighting schemes to predict genomic annotations. In: 2013 IEEE 13th International Conference on Bioinformatics and Bioengineering (BIBE), pp. 1–4. IEEE (2013)

[13] Hofmann, T.: Probabilistic latent semantic indexing. In: Proceedings of the 22nd Annual International ACM SIGIR Conference on Research and Development in Information Retrieval, pp. 50–57. ACM (1999)

[14] Blei, D.M., Ng, A.Y., Jordan, M.I.: Latent dirichlet allocation. the Journal of Machine Learning Research 3, 993–1022 (2003)

[15] Masseroli, M., Chicco, D., Pinoli, P.: Probabilistic latent semantic analysis for prediction of gene ontology annotations. In: The 2012 International Joint Conference on eural Networks (IJCNN), pp. 1–8. IEEE (2012)

[16] Pinoli, P., Chicco, D., Masseroli, M.: Latent dirichlet allocation based on gibbs sampling for gene function prediction. In: 2014 IEEE Conference on Computational Intelligence in Bioinformatics and Computational Biology, pp. 1–8. IEEE (2014)

[17] Chicco, D., Sadowski, P., Baldi, P.: Deep autoencoder neural networks for gene ontology annotation predictions. In: Proceedings of the 5th ACM Conference on Bioinformatics, Computational Biology, and Health Informatics, pp. 533–540. ACM (2014)

[18] Goodman, L.A., Kruskal, W.H.: Measures of association for cross classifications*. Journal of the American Statistical Association 49(268), 732–764 (1954)

[19] Fagin, R., Kumar, R., Sivakumar, D.: Comparing top k lists. SIAM Journal on Discrete Mathematics 17(1), 134–160 (2003)

[20] Spearman, C.: The proof and measurement of association between two things. The American Journal of Psychology 15(1), 72–101 (1904)

[21] Kendall, M.G.: A new measure of rank correlation. Biometrika, 81–93 (1938)

[22] Ilyas, I.F., Beskales, G., Soliman, M.A.: A survey of top-k query processing techniques in relational database systems. ACM Computing Surveys (CSUR) 40(4), 11 (2008)

[23] Kumar, R., Vassilvitskii, S.: Generalized distances between rankings. In: Proceedings of the 19th International Conference on World Wide Web, pp. 571–580. ACM (2010)

[24] Bertin-Mahieux, T., Eck, D., Maillet, F., Lamere, P.: Autotagger: A model for predicting social tags from acoustic features on large music databases. Journal of New Music Research 37(2), 115–135 (2008)

[25] Chen, Q., Aickelin, U.: Movie recommendation systems using an artificial immune system. arXiv preprint arXiv:0801.4287 (2008)

[26] Payne, J.S., Stonbam, T.J.: Can texture and image content retrieval methods match human perception?. In: Proceedings of 2001 International Symposium on Intelligent Multimedia, Video and Speech Processing, pp. 154–157. IEEE (2001)

[27] Ciceri, E., Fraternali, P., Martinenghi, D., Tagliasacchi, M.: Crowdsourcing for Top-K Query Processing over Uncertain Data. IEEE Transactions on Knowledge and Data Engineering (TKDE), 1–14 (preprint, 2015)

[28] Fawcett, T.: Roc graphs: Notes and practical considerations for researchers. Machine Learning 31, 1–38 (2004)

[29] Canakoglu, A., Masseroli, M., Ceri, S., Tettamanti, L., Ghisalberti, G., Campi, A.: Integrative warehousing of biomolecular information to support complex multi-topic queries for biomedical knowledge discovery. In: 2013 IEEE 13th International Conference on Bioinformatics and Bioengineering (BIBE), pp. 1–4. IEEE (2013)

Statistical Analysis of Protein Structural Features: Relationships and PCA Grouping

E. Del Prete[1], S. Dotolo[1], A. Marabotti[1,2], and A. Facchiano[1]

[1] Istituto di Scienze dell'Alimentazione, CNR, Via Roma 64, 83100 Avellino, Italy
{eugenio.delprete,angelo.facchiano}@isa.cnr.it,
serenadotolo@hotmail.it
[2] Dipartimento di Chimica e Biologia, Università degli Studi di Salerno,
Via Giovanni Paolo II 132, 84084 Fisciano (SA), Italy
amarabotti@unisa.it

Abstract. Subtle structural differences among homologous proteins may be responsible of the modulation of their functional properties. Therefore, we are exploring novel and strengthened methods to investigate in deep protein structure, and to analyze conformational features, in order to highlight relationships to functional properties. We selected some protein families based on their different structural class from CATH database, and studied in detail many structural parameters for these proteins. Some valuable results from Pearson's correlation matrix have been validated with a Student's t-distribution test at a significance level of 5% (p-value). We investigated in detail the best relationships among parameters, by using partial correlation. Moreover, PCA technique has been used for both single family and all families, in order to demonstrate how to find outliers for a family and extract new combined features. The correctness of this approach was borne out by the agreement of our results with geometric and structural properties, known or expected. In addition, we found unknown relationships, which will be object of further studies, in order to consider them as putative markers related to the peculiar structure-function relationships for each family.

Keywords: protein structure, global features, correlation, PCA.

1   Background

Proteins are biological macromolecules characterized by complex structural organizations, determined by subtle balance of energetic factors and able to confer different functionalities. Within different organisms, there are proteins with similar function and structure, so that these proteins are considered to descend from a common ancestor, and therefore belonging to the same family. These homologous proteins are similar at level of global organization, although differences exist and modulate functional properties such as activity and thermostability.

We are interested in exploring novel methods to investigate protein structure and to analyze conformational features and their differences within protein families, or among them, in order to find relationships that could be related to functional properties. In this study, we decided to investigate ten protein families, chosen to represent

© Springer International Publishing Switzerland 2015
C. di Serio et al. (Eds.): CIBB 2014, LNCS 8623, pp. 33–43, 2015.
DOI: 10.1007/978-3-319-24462-4_3

different CATH database classes [1] and having a suitable number of PDB structures from different organisms: *beta-lactamase* (BLA), *cathepsin B* (CTS), *ferritin* (FTL), *glycosyltransferase* (GTF), *hemoglobin* (HGB), *lipocalin 2* (LCN), *lysozyme* (LYS), *proliferating cell nuclear antigen* (PCNA), *purine nucleoside phosphorylase* (PNP), *superoxide dismutase* (SOD).

2   Materials and Methods

2.1   Analysis Workflow

We retrieved 153 crystallographic structures (*Table 1*) from **PDBe database** [2]. In order to analyze similar structures within each family, we used only one chain (A, where available), length difference of 50 residues at most; in addition, to have similar contributions from each family in the global analysis, we selected no more than 19 proteins for family.

Table 1. Protein families and PDB structures used in this study

Protein family (with CATH code)	PDB files
Beta-Lactamase (3.30.450)	1D1J, 1EW0, 1F2K, 1N9L, 1P0Z, 2V9A, 2VK3, 2VV6, 2ZOH, 3BW6, 3BY8, 3CI6, 3CWF, 3EEH
Cathepsin B (3.90.70)	1AEC, 1B5F, 1S4V, 2B1M, 2BDZ, 2DC6, 2P7U, 2WBF, 3AI8, 2BCN, 3CH2, 3LXS, 3P5U
Ferritin (1.20.1260)	1JI4, 1QGH, 1R03, 1S2Z, 1TJO, 2FKZ, 2XJM, 2YW6, 3AK8, 3E1J, 3KA3, 3MPS, 3R2H, 3RAV
Glycosyltransferase (1.50.10)	1GAH, 1HVX, 1KRF, 1KS8, 1NXC, 1R76, 1X9D, 2NVP, 2P0V, 2XFG, 2ZZR, 3P2C, 3QRY
Hemoglobin (1.10.490)	1CG5, 1FLP, 1GCW, 1HLM, 1RQA, 1UVX, 2C0K, 2QSP, 2VYW, 3BJ1, 3NG6, 3QQR, 3WCT, 4IRO, 4NK1
Lipocalin 2 (2.40.128)	1AQB, 1BEB, 1CBI, 1CBS, 1GGL, 1GM6, 1IIU, 1JYD, 1KQW, 1KT6, 1LPJ, 1OPB, 1QWD, 2CBR, 2NND, 2RCQ, 2XST, 3S26, 4TLJ
Lysozyme (1.10.530)	1BB6, 1FKV, 1GD6, 1GHL, 1HHL, 1IIZ, 1JUG, 1QQY, 1REX, 1TEW, 2EQL, 2GV0, 2IHL, 2Z2F, 3QY4
Proliferating Cell Nuclear Antigen (3.70.10)	1AXC, 1B77, 1CZD, 1DML, 1T6L, 1UD9, 1UL1, 1HII, 1IJX, 2OD8, 3HI8, 3LX1, 3P83, 3P91, 4CS5
Purine Nucleoside Phosphorylase (3.40.50)	1A9O, 1JP7, 1M73, 1ODK, 1PK9, 1QE5, 1TCU, 1V4N, 1VMK, 1XE3, 1Z33, 2P4S, 3KHS, 3OZE, 3SCZ, 3TL6, 3UAV, 4D98
Superoxide Dismutase (1.10.287 – 2.60.40)	1BSM, 1IDS, 1JCV, 1MA1, 1MMM, 1MY6, 1Q0E, 1WB8, 2ADP, 2JLP, 2W7W, 3BFR, 3ECU, 3EVK, 3LIO, 3QVN, 3SDP

Different online and local tools have been used to extract from PDB files protein structural properties of our interest: *Vadar* [3] for secondary structures (also confirmed with *DSSP* [4]), hydrogen bonds, accessible surface areas, torsion angles, packing defects, charged residue numbers, free energy of folding; *McVol* [5] for volume, with a mean difference of 4% from that calculated with Vadar, but using a more robust algorithm; an in-house-developed **Perl-script** for automatic search of salt bridge conditions [6].

Moreover, they have been normalized in a z-score form for a better stability relative to using a statistical approach.

Finally, features have been transformed into simple and partial correlation matrices, to extract useful relationships. Then, they have been used as features for a family-specific and a general PCA, in order to verify the existence of common information (*Fig. 1*).

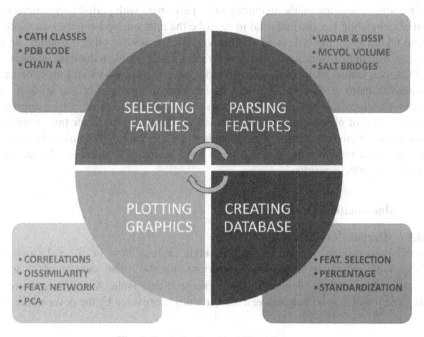

Fig. 1. Analysis Graphical Workflow

R packages in *RStudio IDE* [7] have been used for computations, some of them integrated, others available online:

— *stringr*: rearrange file names [8];
— *RCurl*: manage connection for downloading [9]
— *bio3d*: compute DSSP inside R and read PDB files [10];
— *corrplot*: plot graphical correlation matrix [11];

— *hmisc*: calculate correlation matrix with p-value [12];
— *ppcor*: calculate partial and semi-partial correlations with p-value [13];
— *ggplot2*: plot PCA clustering [14];
— *GeneNet*: plot features network [15].

2.2   Statistical Methods

Starting from Ding and coworker analysis [16], two robust statistical procedures have been chosen for our work: correlation and principal component analysis (PCA).

Simple correlation is in a matrix form, where every element is a Pearson's correlation coefficient between pairwise variables. The application of this matrix is supposed to be well-placed, because we have chosen structural features bound each other, in order to prevent possible spurious correlations: the interest is focused on a quantitative measure among them.

In addition, a check on the features can be performed with partial correlations help. Partial correlation is a method used to describe the relationship between two variables while taking away the effects of another variable, or several other variables, on this relationship. In this work, it has been used to detected possible redundant features.

PCA is a statistical method, used to reduce the number of variables or, better, to summarize them in new variables, to be labelled on the basis of the kind of data. Moreover, given the orthogonality of these new variables, PCA can be applied to obtain a kind of clustering (as clarified in [17]) depending on inner information derived from explained variance. This grouping helps in seeking possible outliers when executed on a protein family: it is a good habit searching for outliers, because they could polarize PCA results.

2.3   Mathematical Overview

2.3.1   Partial Correlation

Given two protein variables with discrete values $X = (x_1, \ldots, x_r)$ and $Y = (y_1, \ldots, y_r)$, where r is rows-observations number, the density $f(x_i, y_j)$ is represented by a single element in the normalized data table. A measure of strength and direction of association between the variables is provided by the covariance:

$$\sigma_{xy} = E[(X - \mu_x)(Y - \mu_y)] = E[XY] - \mu_x \mu_y$$

with

$$E[XY] = \sum_{i=1}^{r} \sum_{j=1}^{c} x_i y_j f(x_i, y_j)$$

where μ_x, μ_y are the expected values for a single variable. An index of covariation between X and Y is provided by the correlation coefficient $\rho_{xy} = \sigma_{xy}/\sigma_x \sigma_y$, where σ_x, σ_y are the standard deviations for a single variable. Given a third protein variable Z, the partial correlation coefficient between X and Y after removing the effect of Z is:

$$\rho_{yx \cdot z} = \frac{\rho_{yz} - \rho_{yx}\rho_{zx}}{\sqrt{1 - \rho_{yx}^2}\sqrt{1 - \rho_{zx}^2}}$$

and it is possible to extend the formula in case of removing the effect of all the variables but one, as shown in [18].

2.3.2 Principal Component Analysis

Given a data table in matrix form, with r observations and c variables V_k, it is possible to determine new PC_k variables composed by a linear combination of the old ones:

$$\begin{cases} PC_1 = a_{11}V_1 + a_{12}V_2 + \cdots + a_{1k}V_k \\ PC_2 = a_{21}V_1 + a_{22}V_2 + \cdots + a_{2k}V_k \\ \quad\quad \cdots \\ PC_l = a_{l1}V_1 + a_{l2}V_2 + \cdots + a_{lk}V_k \end{cases}$$

where $l \leq k$. Loading vectors a_m with $m = 1, 2, \ldots, l$, are determined by:

$$(\Sigma - \lambda_m I)a_m = 0$$

where Σ is the covariance matrix of the original data, λ_m are eigenvalues in descending order associated with a_m eigenvectors and I is the identity matrix. This is an eigenvalues-eigenvectors problem, computationally resolvable with Singular Value Decomposition factorization. After calculating the contribution of every eigenvalue $\lambda_m / \sum_{n=1}^l \lambda_n$, it is possible to choose the first several λ_m that cover a preset quantity of explained variability. In other words, new data table composed by PC_k, always in matrix form, represent the old one with a reduced dimensionality [18] [19].

The challenge with this method is the PC_k interpretation in the reality: that is, new variables are not so intuitive and their interpretation is delegated to investigators' experience.

3 Results

Lower triangular correlation matrix can be used as a basic analysis for a protein family; acronyms are explained in *Fig. 2*. Some correlations are family-specific and their pattern may be considered as a signature, as shown in *Fig. 3*. The numbers in the matrices are the correlation kicked off because of their low statistical significance: in particular, a Student's t distribution test has been used to validate them, considering as good the correlations with a p-value ≤ 0.05 and as strong those with a threshold ≥ 0.65, value deducted from data. For example, in *Fig. 4*, there is a "four-relationship" between percentage of secondary structure (%A, %B), percentage of buried charged residues (%BC) and free energy of folding (FEF).

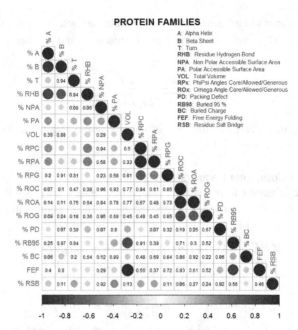

Fig. 2. Circular Triangular Correlation Matrix for the whole protein dataset. Features legend is reported in the inset.

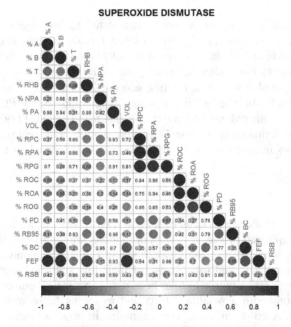

Fig. 3. Circular Triangular Correlation Matrix for Superoxide Dismutase. Features legend is the same for Fig. 2.

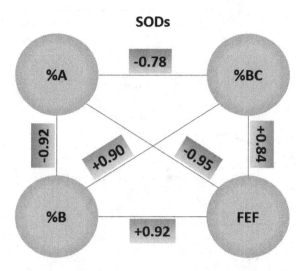

Fig. 4. Family-specific "Four-Relationship" for Superoxide Dismutase. Features legend is the same for Fig. 2.

Some common patterns in torsion angles (%RPx, %ROx with x = C, A, G) between overall and SODs graphical correlation matrices suggest to investigate in details a possible presence of redundant features. Indeed, other correlations are obviously family-independent, because they depend on intrinsic physical-conformational relationships present in every protein. Trying to link them each other could lead to ignore multicollinearity problem. Therefore, it is possible to simplify the work using partial correlations, in order to create with them a features network and to obtain two hints: seeking for spurious correlations and pruning the excessive ones.

Features network in Fig. 5, plotted by means of partial correlation and graphical Gaussian model [20], shows in the squares how torsion angles information results peripheral in the network and, at the same time, unnecessary for our purposes.

PCA has been performed for a double purpose. At first, for a single family it could be useful in underlining possible outliers: the lowest point with code 3SDPA (Pseudomonas putida SOD A chain) in Fig. 6 is an evident example.

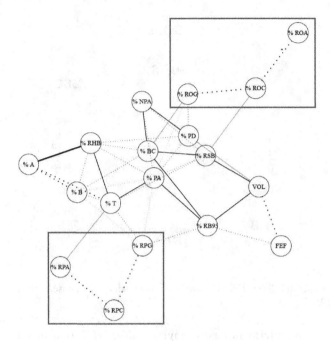

Fig. 5. Features Network

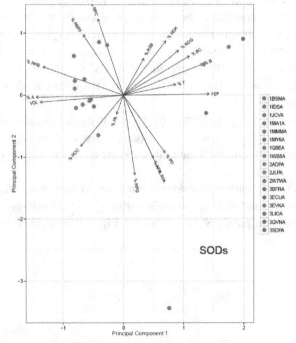

Fig. 6. Superoxide Dismutase Principal Component Analysis. The legend of the right refers to PDB codes (see Table 1) with the addition of the chain identifier.

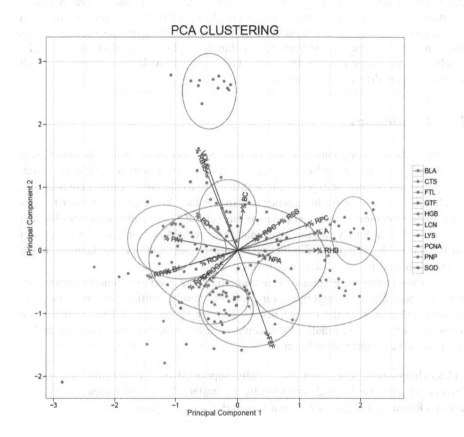

Fig. 7. Overall Principal Component Analysis. The legend of the right indicates the protein family, accordingly with acronyms shown in Background section.

An overall PCA on the whole dataset is more informative, it allows to extract the real important features for every protein family with a sort of clustering and to avoid well-known mathematical-statistical issues, such as overfitting or singular matrix inversion, usually coming from datasets with a small number of observations. A quantitative analysis has shown a cumulative proportion of explained variance of around 40%, 23.6% for the first component and 16.7% for the second one. This not so high value depends most likely on three main factors:

- many features distributions have an high skewness;
- many features distributions are strongly multimodal;
- many pairwise correlations are weak.

With reference to the scree plot (not shown), the first elbow is subsequent to the second component, so that the biplot in *Fig.7* can be used in order to extract information. In our example, with a loadings threshold of 0.4 in absolute value, first principal component is composed by structural features (%A, %RHB) and second principal component by energy-geometrical ones (VOL, FEF, %RB95). It is relevant to touch on omega torsion angle (%ROx) in the third principal component (despite of previous result in features network). Moreover, FTL family is placed near positive PC1, GTF family is placed near positive PC2 and SOD family is wide-open, expected result because of its multi-structural choice.

4   Conclusions

Best interactions among parameters have been investigated in detail. The correctness of this analysis is borne out by the agreement of our results with geometric and structural properties, known or expected. In addition, we found novel features connections, which will be object of further studies to validate their consistence. In order to consider some structural features as putative markers of the peculiar structure-function properties of each family, we are investigating more protein families, with the aim of verifying the existence of peculiarities and compare them within super-families, in the view of a more deep interpretation of results.

About used methods, graphical correlation may be a good tool to make exploratory analysis and taken as a fingerprint for a protein family. Partial correlation helps in features selection, thus the dataset becomes easier to manage and the results more robust and explanatory. Finally, PCA could be used in outliers identification and in family-specific connection to features.

Acknowledgments. This work is partially supported by the *Flagship InterOmics Project* (PB.P05, funded and supported by the Italian Ministry of Education, University and Research and Italian National Research Council organizations).

References

1. Sillitoe, I., Cuff, A.L., Dessailly, B.H., Dawson, N.L., Furnham, N., Lee, D., Lees, J.G., Lewis, T.E., Studer, R.A., Rentzsch, R., Yeats, C., Thornton, J.M., Orengo, C.A.: New functional families (FunFams) in CATH to improve the mapping of conserved functional sites to 3D struc-tures. Nucleic Acids Res. 41(database issue), D490–D498 (2013). http://www.cathdb.info/
2. Velankar, S., Best, C., Beuth, B., Boutselakis, C.H., Cobley, N., Sousa Da Silva, A.W., Dimitropoulos, D., Golovin, A., Hirshberg, M., John, M., Krissinel, E.B., Newman, R., Oldfield, T., Pajon, A., Penkett, C.J., Pineda-Castillo, J., Sahni, G., Sen, S., Slowley, R., Suarez-Uruena, A., Swaminathan, J., van Ginkel, G., Vranken, W.F., Henrick, K., Kley-wegt, G.J.: PDBe: Protein Data Bank in Europe. Nucleic Acids Res. 38(suppl. 1), D308–D317 (2010). http://www.ebi.ac.uk/pdbe/
3. Willard, L., Ranjan, A., Zhang, H., Monzavi, H., Boyko, R.F., Sykes, B.D., Wishart, D.S.: VADAR: a web server for quantitative evaluation of protein structure quality. Nucleic Acids Res. 31(13), 3316–3319 (2003). http://vadar.wishartlab.com/
4. Kabsch, W., Sander, C.: Dictionary of protein secondary structure: pattern recognition of hydrogen-bonded and geometrical features. Biopolymers 22(12), 2577–2637 (1983). http://swift.cmbi.ru.nl/gv/dssp/

5. Till, M.S., Ullmann, G.: McVol - A program for calculating protein volumes and identifying cavities by a Monte Carlo algorithm. J. Mol. Mod. 16, 419–429 (2010). http://www.bisb.uni-bayreuth.de/index.php?page=data/mcvol/mcvol
6. Costantini, S., Colonna, G., Facchiano, A.M.: ESBRI: A web server for evaluating salt bridges in proteins. Bioinformation 3(3), 137–138 (2008). http://bioinformatica.isa.cnr.it/ESBRI/
7. R Core Team, R: A language and environment for statistical computing. R Foundation for Statistical Computing, Vienna, Austria (2014). http://www.R-project.org/
8. Wickham, H.: Stringr: Make it easier to work with strings. R package version 0.6.2 (2012). http://CRAN.R-project.org/package=stringr
9. Temple Lang, D.: RCurl: General network (HTTP/FTP/..) client interface for R. R package version 1.95-4.3 (2014). http://CRAN.R-project.org/package=RCurl
10. Grant, B.J., Rodrigues, A.P., ElSawy, K.M., McCammon, J.A., Caves, L.S.: Bio3d: an R package for the comparative analysis of protein structures. Bioinformatics 22(21), 2695–2696 (2006). http://thegrantlab.org/bio3d/index.php
11. Wei, T.: corrplot: Visualization of a correlation matrix. R package version 0.73 (2014). http://CRAN.R-project.org/package=corrplot
12. Harrell Jr., F.E., Dupont, C., et al.: Hmisc: Harrell Miscellaneous. R package version 3.14-5 (2014). http://CRAN.R-project.org/package=Hmisc
13. Kim, S.: ppcor: Patial and Semi-partial (Part) correlation. R package version 1.0 (2012). http://CRAN.R-project.org/package=ppcor
14. Wickham, H.: A layered grammar of graphics. J. of Comput. Graph. Stat. 19(1), 3–28 (2010). http://ggplot2.org/
15. Schaefer, J., Opgen-Rhein, R., Strimmer, K.: GeneNet: Modeling and Inferring Gene Networks. R package version 1.2.10 (2014). http://CRAN.R-project.org/package=GeneNet
16. Ding, Y., Cai, Y., Han, Y., Zhao, B., Zhu, L.: Application of principal component analysis to determine the key structural features contributing to iron superoxide dismutase thermostability. Biopolymers 97(11), 864–872 (2012). doi:10.1002/bip.22093.
17. Ding, C., He, X.: K-means Clustering via Principal Component Analysis. In: Proceedings of the 21st International Conference on Machine Learning, Banff, Canada (2004)
18. Jobson, J.D.: Applied Multivariate Data Analysis. Volume I: Regression and Experimental Design, 4th edn. Springer Texts in Statistics. Springer, NY (1999)
19. Jolliffe, I.T.: Principal Component Analysis, 2nd edn. Springer Series in Statistics. Springer, Heidelberg (2002)
20. Schaefer, J., Strimmer, K.: An empirical Bayes approach to inferring large-scale gene association networks. Bioinformatics 21, 754–764 (2005)

Exploring the Relatedness of Gene Sets

Nicoletta Dessì, Stefania Dessì, Emanuele Pascariello, and Barbara Pes

Dipartimento di Matematica e Informatica,
Università degli Studi di Cagliari, Via Ospedale 72, 09124 Cagliari, Italy
{dessi,dessistefania,pes}@unica.it,
emanuele.pascariello@gmail.com

Abstract. A key activity for life scientists is the exploration of the relatedness of a set of genes in order to differentiate genes performing coherently related functions from random grouped genes. This paper considers exploring the relatedness within two popular bio-organizations, namely gene families and pathways. This exploration is carried out by integrating different resources (ontologies, texts, expert classifications) and aims to suggest patterns that facilitate the biologists in obtaining a more comprehensive vision of differences in gene behaviour. Our approach is based on the annotation of a specialized corpus of texts (the gene summaries) that condense the description of functions/processes in which genes are involved. By annotating these summaries with different ontologies a set of descriptor terms is derived and compared in order to obtain a measure of relatedness within the bio-organizations we considered. Finally, the most important annotations within each family are extracted using a text categorization method.

Keywords: Gene relatedness, Semantic similarity, Ontology annotation, Text mining.

1   Introduction

Recognizing putative interactions and detecting common functions in a set of genes is a key activity for life scientists in order to assess the significance of experimentally derived gene sets and prioritizing those sets that deserve follow-up. This interest is shifting the focus on data analysis from individual genes to families of genes that are supposed to interact each other in determining a pathological state or influencing the outcome of a single trait (i.e. a phenotype). Because of the large number of genes and their multiple functions, discovering computational methods to detect a set of functionally coherent genes is still a critical issue in bioinformatics [1]. Biologists have dealt with these challenges in part by leveraging the biological principle commonly referred to as "guilt by association" (GBA) [2]. GBA states that genes with related functions tend to share properties such as genetic or physical interactions. For example, if two genes interact, they can be inferred to play roles in a common process leading to the phenotype.

The relatedness of a set of genes is a measure which differentiates a set of genes performing coherently related functions from ones consisting of random grouped

© Springer International Publishing Switzerland 2015
C. di Serio et al. (Eds.): CIBB 2014, LNCS 8623, pp. 44–56, 2015.
DOI: 10.1007/978-3-319-24462-4_4

genes [1]. Specifically, the relatedness considers two aspects. First, whether genes share similar functions or whether they participate in the same biological process. Second, whether the distinct functions are related.

The semantic similarity is a popular approach to evaluate the relatedness of two genes. Basically, the semantic similarity is a numerical representation of how close two concepts are from each other in some ontology. In discovering gene interactions, different works [3] have focused on the definition of semantic similarity measures tailored to the characteristics of GO [4] which is the "de facto" standard for knowledge representation about gene products. Within GO, the level of interaction of two genes is measured by the distance between the terms that describe their semantic content. For example, if a gene is annotated with the term "protein phosphatase activity" and another one is labeled with the term "phosphatase", they are considered as sharing the same functions because both are annotated by terms which are semantically alike.

It should be noted that GO covers three aspects (cellular component (CC), molecular function (MF), biological process (BP)) which are entirely orthogonal, being they represented by disconnected sub-graphs (i.e. three sub-ontologies). This reflects the notion that these aspects are independent, when, in reality, they are strongly correlated in all biological processes. Being "is-a" and "part-of" relationships extensively used in GO to express that two concepts are alike, the semantic similarity does not account for the existence of other relationships such as "has-part" and "is-a-way-of-doing", which typically are present in a group of genes belonging to the same pathway. As stressed in [5] the concept of relatedness is more specific than that of similarity. As well, the relatedness within two concepts is estimated by counting the overlapping words in their definitions [5, 6].

In this paper we describe an approach to explore the relatedness within a set of genes belonging to popular bio-organizations, namely gene families and gene pathways. The relatedness exploration is carried out using three modules.

The first module identifies entities of interest by annotating a corpus of gene summaries. Compiled by expert curators and freely available on the Internet [7], a gene summary is a short text about a single gene which describes functions and processes related to that gene. Given these annotations, the second module, namely the exploration module, evaluates and compares the relatedness within each single bio-organization and suggests patterns that facilitate a more comprehensive vision of differences in gene behavior. Finally, the third module uses a data mining method to extract the most important annotations within a family.

Our proposal is based on the integration of different resources (i.e. ontologies, texts, expert classifications) and aims to capture and present additional information about how genes work together in manner that biologists of all levels can rapidly understand.

The paper is organized as follows. Section 2 details the composition of the corpus of summaries. Section 3 presents the annotation module. The exploration module and experimental results are presented in section 4. Section 5 illustrates the third module i.e. the categorization process for extracting the most significant concepts which characterize each family. Related work is exposed in the section 6. Section 7 gives concluding remarks and presents future work.

2   The Corpus of Summaries

The published literature contains virtually every important biological development (including the necessary information for assessing whether a group of genes represents a common biological function) and a number of approaches have been proposed which exploit article abstracts about genes in order to automatically extract biomedical event from text. The above extraction typically takes place in two phases: first, entities of interest are recognized, next relations between recognized entities are determined using a number of approaches of which we mention simple co-occurrence of entities [8], rule-based [9], and machine learning based techniques [10].

The extraction of useful information from biomedical texts is an essential process, useful in itself, and necessary to help scientist research activity, both to understand experimental results and to design new ones. Only when biomedical named entities are correctly identified could other more complex tasks, such as discovering inherent interrelation between bio-entities, be performed effectively. Recognizing coherent gene groups from the literature is a difficult problem because some genes have been extensively studied, whereas others have only been recently discovered. In addition, a given gene may have many relevant documents or none, and the documents about it may cover a wide spectrum of functions. Consequently, the available text can skew performance of text analysis algorithms.

In order to deal with this complexity of analyzing biomedical literature, we conduct experiments on the corpus of gene summaries from [7]. As previously mentioned, we were interested in exploring the relatedness between genes belonging to two bio-organizations: families and pathways.

In more detail, a family of genes defines structural domains where genes carry-on similar functions. A family consists in a list of unique gene symbols but sometimes not enough is known about a gene to assign it to an established family. In other cases, genes may fit into more than one family. No formal guidelines define the criteria for grouping genes together.

HUGO Gene Nomenclature Committee [11] (a worldwide consortium that is responsible for approving unique symbols and names for human loci, including protein coding genes, RNA genes and pseudo-genes) makes available an online repository of gene families.

Represented by a semantic network, a pathway exhibits a large-scale structure where genes perform a variety of functions and have complex interactions with other genes. Information about pathways is limited as it strongly depends on the advances in current knowledge of molecular and cellular biology.

Our study considers 5 families and 2 pathways. In particular, to compose our corpus, we have chosen 10 summaries of genes within each of these bio-organizations and 10 summaries of genes randomly selected in manner that they do not belong to these bio-organizations.

3   The Annotation Module

Within this module, the corpus of summaries is annotated using existing knowledge resources i.e. domain ontologies which centralize and organize valuable knowledge in a structured form ready to be exploited. We used NCBO annotator [12] as the basis of our annotation module. The NCBO Annotator (formerly referred to as the Open Biomedical Annotator (OBA)) is an ontology-based Web service that annotates public datasets with biomedical ontology concepts based on their textual metadata. The biomedical community can use the annotator service to tag their data automatically with ontology concepts. These concepts come from the Unified Medical Language System (UMLS) Meta-thesaurus [13] and the National Center for Biomedical Ontology (NCBO) BioPortal ontologies [14]. Currently, annotation is based on a highly efficient syntactic concept recognition (using concept names and synonyms) engine developed in collaboration with National Center for Integrative Biomedical Informatics [15] and on a set of semantic expansion algorithms that leverage the semantics in ontologies (e.g., is_a relations and mappings). First, direct annotations are created from raw text according to a dictionary that uses terms from a set of ontologies. Second, different components expand the first set of annotations using ontology semantics.

Table 1. Example of gene summary and its annotations.

GENE HOXA5 SUMMARY	
In vertebrates, the genes encoding the class of transcription factors called homeobox genes are found in clusters named A, B, C, and D on four separate chromosomes. Expression of these proteins is spatially and temporally regulated during embryonic development. This gene is part of the A cluster on chromosome 7 and encodes a DNA-binding transcription factor which may regulate gene expression, morphogenesis, and differentiation. Methylation of this gene may result in the loss of its expression and, since the encoded protein upregulates the tumor suppressor p53, this protein may play an important role in tumorigenesis.	
Annotations	**Ontology**
transcription	Gene Regulation Ontology
LBX1	Regulation of Transcription Ontology
cluster	Regulation of Transcription Ontology
binding	Gene Ontology Extension
transcription factor	Gene Regulation Ontology
gene expression	Computer Retrieval of Information on Scientific Projects Thesaurus
anatomical structure morphogenesis	Gene Ontology Extension
histogenesis	Computer Retrieval of Information on Scientific Projects Thesaurus
methylation	Gene Regulation Ontology
methylation reaction	Regulation of Transcription Ontology
cellular tumor antigen p53	Protein Ontology
neoplasm/cancer	Computer Retrieval of Information on Scientific Projects Thesaurus
TP53	Regulation of Transcription Ontology
neoplastic growth	Computer Retrieval of Information on Scientific Projects Thesaurus
carcinogenesis	Computer Retrieval of Information on Scientific Projects Thesaurus

Within NCBO, we rely on 20 different ontologies because a single ontology may be incomplete and is not expected to contain all the domain-specific instances.

For each summary, the annotation module outputs a set of terms, namely the summary Bag Of Words (BOW). Table 1 shows an example of gene summary and its annotations from NCBO. Note that annotations differ from the terms of summary and concentrate the most important biological concepts within the summary in order to reflect the semantic content of the text and constrain the potential interpretation of terms.

4   The Exploration Module

Starting from BOWs of summaries, the exploration module evaluates and compares the relatedness within each single bio-organization we considered (i.e. the 5 families and the 2 pathways). Broadly, the interaction between the genes can be classified according to the following basic types of behavior:

- *Complementary*. Genes cooperate to make a product. Both need to be present for either to work.
- *Supplementary* (Epistasis). One gene alters the outcome of another gene or the second gene adds more to the first.
- *Collaborative*. Two genes interact to make a product different to that which either could make independently.
- *Polygenetics*. Many genes control a single phenotype.

Here we are interested in evaluating the relatedness of genes belonging to each single family and each single pathway and highlighting the differences in the behavior between genes belonging to different organizations. Toward this end, for each group of genes belonging to the same bio-organization, we considered all the possible pair-wise combinations of their BOWs. In practice, the BOW of each gene is compared with the BOW of all other genes belonging to the same organization.

For comparing two BOWs, we adopt the Dice Coefficient [16], a popular measure used in information retrieval for comparing the semantic similarity of two sets of terms. Based on the co-occurrence of terms of interest, this measure assumes that if two terms are mentioned together in two different texts, these texts are probably related.

Like the Jaccard similarity index [17], the Dice coefficient also measures set agreement using the following metric:

$$D(S,T) = 2 \; |S \text{ and } T| \; / \; (|S|+|T|)$$

where S and T are the two sets of annotated terms. More simply, this formula represents the size of the overlapping terms in the two sets divided by the sum of their size. A value of 0 indicates no overlap; a value of 1 indicates perfect agreement.

Accordingly, given two BOWs, namely B1 and B2, their similarity Sim (B1, B2) is defined as follows:

$$Sim (B1, B2) = 2*Cterms / (N1+N2)$$

where Cterms is the number of terms that B1 and B2 have in common, N1 is the number of terms in B1 and N2 is the number of terms in B2.

For each bio-organization, the exploration module outputs a sequence of similarity values i.e. a discrete variable which features the relatedness of genes belonging to that organization.

Fig.1 and Fig. 2 show the frequency distribution of values that score the relatedness for the family MAPK and for the Pi3AKT pathway, respectively.

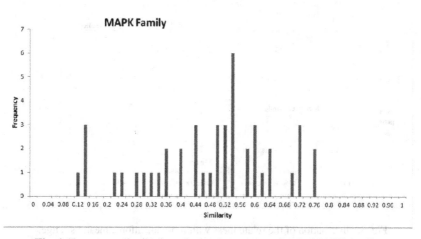

Fig. 1. Frequency distribution of relatedness scores for the MAPK family.

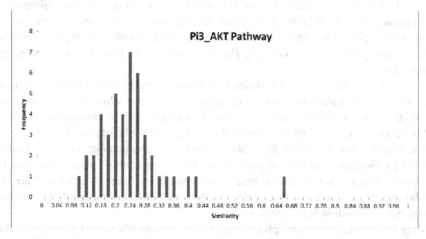

Fig. 2. Frequency distribution of relatedness scores for the Pi3AKT pathway.

We observe that these distributions are not informative about the diversity between families and pathways as we cannot derive from them a representative interaction profile which takes account for the type of behavior in gene interactions. Following a popular approach in biostatistics, we highlighted this behavior using the scatter plot of the statistical parameters that score the relatedness (i.e. mean, range, standard deviation etc).

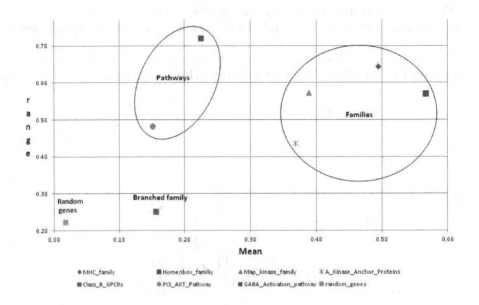

Fig. 3. Distribution of the relatedness within organizations: mean vs range.

Fig. 3 shows the scatter plot of the mean vs the range for the relatedness distribution within the organizations. Here, families are clustered on the right corner of the scatter plot. The mean of the relatedness within families is greater than the corresponding parameter in pathways. The relatedness of the randomly selected genes is negligible.

According to what asserted by biomedical domain experts, results seem to confirm that the couple of parameters (mean, range) is an indicator which discriminates the organizations we consider. Indeed, a complementary type of interaction happens within genes belonging to the families in the right corner of the scatter plot.

The branched family "Class_B_GPRs" is a notable exception that validates what is asserted by biomedical domain experts: genes belonging to this family participate in the same process that is hierarchically structured i.e. their interactions are of supplementary and collaborative type. Finally, in pathways clustered on the left corner, gene interaction is polygenetic.

More interesting, the scatter plot in Fig. 4 shows the coefficient of variation vs the range. Expressed by the ratio of standard deviation to the mean, the coefficient of variation compares the degree of variation from one data series to another, even if the means are drastically different from each other. Here, the coefficient of variation measures the inverse of the relatedness because the higher the dispersion the lower is the functional coherence of the genes. Indeed, due to the similarity of processes that the genes perform, the dispersion in families is very low. Note that even the branched family "Class_B_GPRs" exhibits a low level of dispersion. In pathways, the functional coherence is lower than that of families. Finally, random selected genes are very dispersed as they don't interact.

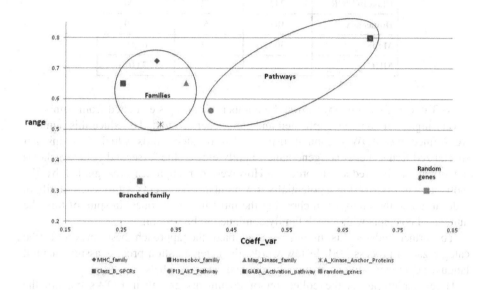

Fig. 4. Semantic similarity distribution within organizations: coefficient of variation vs range.

The above scatter plots enable the researcher to obtain a visual comparison of the coherent organizations that we considered, and help him to automatically discover what kind of differences there might be between families and pathways.

5 Extracting the Most Relevant Annotations within a Family

A gene family is an efficient and useful way for grouping a large number of related genes, and already works well for a number of established gene groupings. Genes are grouped into families when they perform similar functions and share often also a significant degree of resemblance in the sequences of the DNA-Building blocks encoding for proteins that derive from these genes. As such, biologists are interested in discovering which are the essential processes, namely concepts from now on, that characterize the gene interactions within a family. This is the purpose of our third module which explores the BOWs resulting from the first module in order to extract the above concepts.

Table 2. Number of concepts extracted for each family.

FAMILIES	Total Concepts	Refined Concepts	Percentage Reduction (%)
A Kinase	250	113	54,80
Class BGPCR	247	153	38,06
Homeobox	183	56	69,40
Mapk	307	117	61,89
MHC	215	58	73,02

As Table 2 (first column) shows, the number of concepts extracted from Annotator is very high as it contains duplicated and auto-generated text. To reduce this number, we refined the BOW's content using a list of stop-words which contains not specialized terms (i.e. gene, gene name, family etc). Table 2 (second column) shows the effect of this reduction process. However, our approach was guided by the motivation of discovering a potentially low number of concepts within each family in order to guide the biologist in choosing the most important ones. In spite of this, the number of concepts within each family continued to be very high.

To further reduce this number we adopted the approach best known as the categorization process [18]. In few words, the categorization process assigns natural language documents to one or more predefined category labels.

Here, our corpus is the collection of documents and their BOWs contain the corresponding concepts or *features*. The dominant approach to categorization process considers the employment of a general inductive process that automatically builds a classifier by learning, from a set of pre-classified documents, the characteristics of the categories [18].

However, these algorithms may be not completely suitable to our case, as our text collection (i.e. the corpus of 50 summaries) has a moderate size and hundreds of terms (i.e. the refined concepts) most of which are irrelevant for the classification task and some of them even introduce noise that may decrease the overall performance.

Leveraging on our previous work [19], we considered a single list which contains all the refined concepts (see Table 2, second column). After the elimination of duplicate concepts, the list resulted in 346 terms which were assumed to be the features of our categorization process.

First, we constructed a reference matrix M(50x346) where rows represent summaries, columns are the features. A generic term $M(i,j)$ is equal to 1 if the term in the j-column belongs to the BOW of the summary at the i-row or to 0 otherwise.

Then we applied the Information Gain to score features according to their discriminative power, i.e. their capacity of separating the five families. It resulted in an ordered list where features appear in descending order of relevance.

Because we were interested in obtaining an average number of 8 features per family, we considered only the first 40 elements of the scoring list and we reduced the size of the matrix M by considering only the columns which represent these elements.

As previously mentioned, a classical approach to categorization considers to build a classifier on M and finds the best and unique classifier (i.e. the group of features which categorize each family with the maximum accuracy).

As a family can be categorized by different groups of features, we considered the use of a Genetic Algorithm (GA) [20] whose characteristic is to find optimal predictors i.e. the GA explores every possible solution and provides different best solutions. Additionally, it removes redundant terms and originates more accurate and small-sized subsets of terms for categorization.

Starting from the scoring list we constructed nested subsets of features where the first set contained the first 3 top-ranked features and the remaining sets were built by progressively adding to the previous set the next feature in the list until obtaining a set which contained all the 40 features.

Then we applied the Genetic Algorithm (GA) for exploring and discovering optimal predictors within each subset. Specifically, for each subset, the GA initializes a population of individuals randomly, each individual being codified by a binary vector whose dimension equals the size of the subset. In the binary vector, the value 1 means the selection of the respective term, otherwise the value is 0. A fitness function evaluates the individuals by means of a classifier and selects the individuals that maximize the classification accuracy. Then, the current population undergoes genetic operations (i.e. selection, mutation, and crossover) and a new population is generated and evaluated. This evolution process is repeated within a pre-defined number of generations and it outputs the best individual, i.e. the subset of terms that best categorizes the subset.

Leveraging on our previous studies about tuning GA parameters [21, 22], we set the following values: population size = 30, crossover probability = 1, mutation probability = 0.02, number of generations = 50. Since the GA performs a stochastic search, we considered the results over different trials. We used a Naïve Bayes classifier [23] for evaluating the fitness. Using a 10-fold cross validation, we evaluate and compare solutions from each subset using the F-measure, a popular metric which rates the harmonic mean between precision and recall. The overall analysis was implemented using the Weka data mining environment [24].

Table 3. Number of predictors and annotations selected for each family.

FAMILIES	Perfect Predictors	Total Annotations	Distinct Annotations	Expert Selections
A Kinase family	13	56	19	9
Class BGPCR	10	55	20	11
Homeobox	18	76	26	8
MapK	3	11	7	1
MHC	2	6	6	3

Table 3 resumes our results. Specifically, the first column shows the number of perfect predictors within each family. A perfect predictor is a set of features (i.e. concepts) whereby the classifier reaches an F-measure equal to 1. The second column shows the total number of concepts belonging to these perfect predictors. As a single concept can belong to different predictors, we eliminate duplicates. The third column depicts the number of distinct concepts within the perfect predictors.

A comparison of Table 2 with Table 3 shows a drastic reduction in the number of concepts that best represent a specific family. In particular, in MapK and MHC families only few concepts are enough to characterize the family. As well the number of concepts increases when the collaboration is more articulate, as it happens in the branched family Class_B_GPCR.

According to a common practice in bioinformatics, concepts were further examined and refined by a domain expert. The last column in Table 3 shows the result of this refinement and Table 4 details the concepts within each family.

Table 4. Final list of annotations for each family.

Family	List of Annotations
A Kinase family	akap1, camp, extracellular adherence protein, flagellum, mitochondrion, protein kinase, protein kinase a, receptor, sperm
Class BGPCR	adenylate cyclase, adrenocorticotropic hormone, corticotropin releasing factor, cyclase, glucagon, homeostasis, hormone, microtubule associated protein, receptor, secretion, vasoactive intestinal peptide
Homeobox	anatomical structure morphogenesis, dwarfism, hindbrain, histogenesis, homeobox gene, lbx1, rhombencephalon, transcription factor
MapK	mitogen-activated protein kinase
MHC	peptide binding, receptor, tnfsf14

6   Related Work

Related to our approach, [25] conducts a detailed study of gene functional similarity on randomly selected gene pairs from the mouse genome and their annotations with GO. The paper compares several measures of similarity with term overlap methods, including Dice measure. Experiments suggest that term overlap can serve as a simple and fast alternative to other approaches which use other measures or require complex pre-calculations. The paper recommends Dice measure as an alternative to other more complex semantic similarity measures.

[26] evaluates the similarity of two genes from the same family and from different families using different similarity measures on GO annotations. Results compare well with our experiments as they confirm that overlap methods perform well and Dice score is significantly stronger for genes belonging to the same family. However, experiments consider only the similarity between two genes.

Towards evaluating the functional coherence of a gene set, [1] presents complex metrics that are based on the topological properties of graphs comprised of genes and their annotations. Using a real world microarray dataset, the paper demonstrates that, according to the expert assessment, the proposed metric correctly classifies biologically coherent clusters.

7   Conclusions

The work presented in this paper is a preliminary approach towards exploring the functional relatedness of a gene list in order to improve the selection of functional related genes from an experimental list of genes. We have investigated the annotation of gene summaries to extract gene events and the use of Dice measure to evaluate the relatedness within gene organizations. According to a common practice in bioinformatics, our results were further examined and refined by a domain expert.

Being preliminary, our approach has several limitations. It currently does not try to distinguish between different types of interactions. However, it has been observed [1] that it remains to be developed a method for unifying all the aspects of the relatedness. A second limitation is that the source of gene summaries and the UMLS Meta-thesaurus do not provide extensive coverage of genes. The incorporation of relations from other sources of knowledge may remedy this drawback. Another related limitation is the use of a few number of summaries.

As future research we plan to scale up our experiments to much large gene organizations. As well we will investigate about highly discriminative measures of the relatedness that make possible differentiating coherent gene sets from random ones.

Acknowledgments. This research was supported by RAS, Regione Autonoma della Sardegna (Legge regionale 7 agosto 2007, n. 7), in the project *"DENIS: Dataspaces Enhancing the Next Internet in Sardinia"*.

Stefania Dessì gratefully acknowledges Sardinia Regional Government for the financial support of her PhD scholarship (P.O.R. Sardegna F.S.E. Operational Programme of RAS).

References

1. Richards, A.J., Muller, B., Shotwell, M., Cowart, L.A., Rohrer, B., Lu, X.: Assessing the functional coherence of gene sets with metrics based on the Gene Ontology graph. Bioinformatics 26(12), i79–i87 (2010)
2. Oliver, S.: Guilt-by-association goes global. Nature 403, 601–603 (2000)
3. Guzzi, P.H., Mina, M., Guerra, C., Cannataro, M.: Semantic similarity analysis of protein data: assessment with biological features and issues. Briefings in Bioinformatics 13(5), 569–585 (2012)
4. http://www.geneontology.org/GO.cite.shtml
5. Pedersen, T., Pakhomov, S.V., Patwardhan, S., Chute, C.G.: Measures of semantic similarity and relatedness in the biomedical domain. Journal of Biomedical Informatics 40(3), 288–299 (2007)

6. Patwardhan, S.: Using WordNet-based context vectors to estimate the semantic relatedness of concepts. In: Proceedings of the EACL, pp. 1–8 (2006)
7. http://ghr.nml.nih.gov/geneFamily
8. Kandula, S., Zeng-Treitler, Q.: Exploring relations among semantic groups: a comparison of concept co-occurrence in biomedical sources. Stud. Health Technol. Inform. 160, 995–999 (2010)
9. Jang, H., Lim, J., Lim, J.H., Park, S.J., Lee, K.C., Park, S.H.: Finding the evidence for protein-protein interactions from PubMed abstracts. Bioinformatics 22, e220–e226 (2006)
10. Kang, N., Van Mulligen, E.M., Kors, J.A.: Comparing and combining chunkers of biomedical text. J. Biomed. Inform. 44, 354–360 (2011)
11. Gray, K.A., Daugherty, L.C., Gordon, S.M., Seal, R.L., Wright, M.W., Bruford, E.A.: Genenames.org: the HGNC resources in 2013. Nucleic Acids Res. 41(database issue), D545–D5452 (2013)
12. Jonquet, C., Shah, N.H., Musen, M.A.: The open biomedical annotator. Summit on Translat Bioinforma 2009, 56–60 (2009)
13. http://www.nlm.nih.gov/research/umls/knowledge_sources/metat hesaurus/
14. Whetzel, P.L.: NCBO Team: NCBO Technology: Powering semantically aware applications. J. Biomed. Semantics 4(suppl. 1), S8 (2013)
15. http://portal.ncibi.org/gateway/
16. Dice, L.: Measures of the Amount of Ecologic Association Between Species. Ecology 26, 297–302 (1945)
17. http://en.wikipedia.org/wiki/Jaccard_index
18. Sebastiani, F.: Machine learning in automated text categorization. ACM Computing Surveys 34(1), 1–47 (2002)
19. Cannas, L.M., Dessì, N., Dessì, S.: A Model for term selection in text categorization problems. In: DEXA Workshops 2012, pp. 169–173 (2012)
20. Goldberg, D.E.: Genetic algorithms in search, optimization and machine learning. Addison-Wesley (1989)
21. Cannas, L.M., Dessì, N., Pes, B.: Tuning evolutionary algorithms in high dimensional classification problems (extended abstract). In: Proceedings of the 18th Italian Symposium on Advanced Database Systems (SEBD 2010), pp. 142–149 (2010)
22. Cannas, L.M., Dessì, N., Pes, B.: A filter-based evolutionary approach for selecting features in high-dimensional micro-array data. In: Shi, Z., Vadera, S., Aamodt, A., Leake, D. (eds.) IIP 2010. IFIP, vol. 340, pp. 297–307. Springer, Heidelberg (2010)
23. Mccallum, A., Nigam, K.: A comparison of event models for naive bayes text classification. In: AAAI 1998 Workshop on 'Learning for Text Categorization' (1998)
24. Bouckaert, R.R., Frank, E., Hall, M.A., et al.: WEKA - Experiences with a Java Open-Source Project. Journal of Machine Learning Research 11, 2533–2541 (2010)
25. Mistry, M., Pavlidis, P.: Gene Ontology term overlap as a measure of gene functional similarity. BMC Bioinformatics 9, 237 (2008)
26. Popescu, M., Keller, J.M., Mitchell, J.A.: Fuzzy Measures on the Gene Ontology for Gene Products Similarity. IEEE/ACM Transactions on Computational Biology and Bioinformatics 3(3) (2006)

Consensus Clustering in Gene Expression

Paola Galdi[1], Francesco Napolitano[2], and Roberto Tagliaferri[1]

[1] NeuRoNe Lab, Department of Informatics, University of Salerno,
via Giovanni Paolo II 132, 84084, Fisciano (SA), Italy
[2] Systems and Synthetic Biology Lab, Telethon Institute of Genetics and Medicine
(TIGEM), via Pietro Castellino 111, 80131, Naples, Italy
{pgaldi,rtagliaferri}@unisa.it, f.napolitano@tigem.it

Abstract. In data analysis, clustering is the process of finding groups in
unlabelled data according to similarities among them in such a way that
data items belonging to the same group are more similar between each
other than items in different groups. Consensus clustering is a methodology for combining different clustering solutions from the same data set in
a new clustering, in order to obtain a more accurate and stable solution.
In this work we compared different consensus approaches in combination with different clustering algorithms and ran several experiments on
gene expression data sets. We show that consensus techniques lead to an
improvement in clustering accuracy and give evidence of the stability of
the solutions obtained with these methods.

Keywords: clustering, consensus clustering, cluster ensembles, affinity
propagation, k-means, dynamic tree cut, random projections.

1 Background

Clustering is an unsupervised learning technique used in exploratory data analysis to study the underlying natural structure of data, in order to uncover latent
relationships between the objects of interest. The clustering process involves
grouping unlabelled data according to similarities among them in such a way
that data items belonging to the same group (or cluster) are more similar between each other than items in different clusters.

It has been applied in various field, ranging from image segmentation to text
mining, and has also been applied in bioinformatics, for example in gene expression data analysis to identify functional gene modules and in proteomics to
identify protein subfamilies. Another usual application in biomedical studies is
that of identifying patient subclasses or cancer subtypes that were not known a
priori.

Choosing the best algorithm to analyse a specific data set is a difficult task.
Probably no algorithm can obtain the optimal result on all data sets, as in
the case of supervised learning, where this result is proved by the No-Free-
Lunch Theorem [1]. Data analysts need also to face the problem of properly
selecting model parameters (e.g. the number of groupings to be searched for),
since assumptions made on data might not be reflected in their natural structure.

© Springer International Publishing Switzerland 2015
C. di Serio et al. (Eds.): CIBB 2014, LNCS 8623, pp. 57–67, 2015.
DOI: 10.1007/978-3-319-24462-4_5

In order to compensate for the lack of prior knowledge, some clustering algorithms have been developed to automatically detect the number of clusters. Alternatively, assessment techniques can be used in combination with algorithms that require to specify the number of clusters beforehand in order to choose the best clustering with respect to a given quality criterion [2].

Once the number of clusters has been selected, a desirable characteristic of the final solution is that of being stable with respect to data set perturbations. In this context, consensus clustering is a methodology that allows to quantify the consensus between different clustering solutions in order to evaluate the stability of the detected clusters.

In this work we compare four consensus techniques combined with five different clustering algorithms to show that consensus clustering can improve the clustering accuracy and provide more reliable results identifying more robust solutions.

2   Materials and Methods

Consensus clustering consists in combining different clusterings from the same data set into a final one in order to improve the quality of individual data groupings [3]. Consensus methodologies differ in two main aspects, that are how base clusterings are combined and how the concordance between different solutions is quantified. For instance, multiple clustering solutions may be obtained by subsampling a proportion of patterns and features (rows and columns) from a data matrix and clustering each subsample with a chosen algorithm, or, where appropriate, by randomly initializing the model parameters [4]. Once the set of base clusterings (also known as cluster ensemble) has been generated, many algorithms proceed with computing pairwise similarities between data points based on the consensus between the base clusterings. In the following we briefly describe the clustering algorithms and the different approaches to consensus clustering that we compare in this study and present the data sets used in our experiments.

2.1   Clustering Algorithms

We used five different algorithms to obtain the base clustering solutions to be combined by a consensus method. Affinity Propagation (AP) is a clustering algorithm based on message passing proposed in [5]. It takes as input pairwise similarities between data points and finds out the most representative items of the data set (the so called exemplars) to build clusters around them. It operates by simultaneously considering all data points as candidate exemplars and exchanging real-valued messages between data points until a good set of exemplars and clusters emerges. The algorithm also requires in input a vector of real values called *preferences* which, according to the authors, can be used to indicate a set of potential exemplars, but should be set to a common value (the *preference value*) when no prior knowledge is available.

It is not required to specify in advance the number of clusters to find, nevertheless the number of clusters found by the algorithm is close to being monotonically related to the preference value [5].

We also used two modifications of AP that take as input the number of clusters to find: APclusterK (APK) [6], a bisection method that runs AP several times with different preference values searching for exactly K clusters; K-AP [7], which introduces a constraint to limit the number of clusters to be K in the message passing process.

Dynamic Tree Cut (DTC), described in [8], is a hierarchical clustering based algorithm which implements a dynamic branch cutting method in order to detect the proper number of clusters in a dendrogram depending on the clusters shape.

Finally, we used the classical k-means algorithm (KM), which tries to find a partition of the data set that minimizes the distance between cluster points and centroids (distortion error). Since this algorithm is conditioned by local minima, it is executed 1000 times in order to select the run that achieves the minimum distortion error.

2.2   Subsampling and Consensus Matrix

In the approach described in [4,9], different solutions are obtained by clustering randomly generated subsamples of the data set. Then a Consensus matrix $\mathcal{M}$ is built that stores, for each pair of data points, the proportion of clustering runs in which the two items are grouped together.

Formally, let $D^{(1)}, D^{(2)}, \ldots, D^{(H)}$ be the H subsamples of the original data set of N data points and let $M^{(h)}$ be the $N \times N$ connectivity matrix corresponding to subsample $D^{(h)}$, for $h = 1, \ldots, H$, defined as follows:

$$M^{(h)}(i,j) = \begin{cases} 1 & \text{if items } i \text{ and } j \text{ belong to the same cluster} \\ 0 & \text{otherwise.} \end{cases} \tag{1}$$

Furthermore, let $I^{(h)}$ be a $N \times N$ indicator matrix such that the (i,j)-th entry is equal to 1 if both items i and j are present in the subsample $D^{(h)}$, and 0 otherwise. Then, the Consensus matrix $\mathcal{M}$ can be defined as:

$$\mathcal{M}(i,j) = \frac{\sum_h M^{(h)}(i,j)}{\sum_h I^{(h)}(i,j)} \ . \tag{2}$$

Finally, a consensus clustering is computed applying an agglomerative hierarchical clustering algorithm to the dissimilarity matrix $1 - \mathcal{M}$.

2.3   Random Projections

In [10] a method is proposed for generating multiple clustering solutions using random projections.

The rationale behind random projections is justified by the Johnson-Lindenstrauss lemma [11], that states that a set of n points in high-dimensional space can be mapped down onto an $O(\log n/\epsilon^2)$ dimensional subspace such that the distances between the points are not distorted more than a factor of $1 \pm \epsilon$ (see [12,13] for further details).

The main idea in [10] is to project the original d−dimensional data set to different k−dimensional ($k << d$) subspaces using random matrices whose elements are Gaussian distributed; a clustering is then executed on each subspace and a Consensus matrix is built in the same way as described above.

2.4   Link-Based Cluster Ensemble

In [14] two methods are proposed that try to refine the approach based on the Consensus matrix. The main idea is that of taking into account the similarity between the clusters to which a pair of objects is assigned, instead of just counting for co-occurrences: two objects of the data set are similar if they belong to similar clusters of the cluster ensemble. Two different similarity matrices are presented.

The Connected-Triple Similarity (CTS) matrix assesses the similarity between two clusters of the ensemble computing the proportion of items that they share, that is:

$$w_{ij} = \frac{|X_{C_i} \cap X_{C_j}|}{|X_{C_i} \cup X_{C_j}|} \tag{3}$$

where X_{C_i} denotes the set of data points belonging to cluster C_i.

The Approximated Sim-Rank Similarity (ASRS) matrix is based on a bipartite graph representation of the cluster ensemble in which vertices represent both clusters and data points and edges connect data points to the clusters to which they belong. The idea is that two data objects are similar if their neighbourhoods (the set of clusters to which they are assigned in the cluster ensemble) are similar.

Formally, the similarity between two data points i and j, is defined as:

$$ASRS(i,j) = \frac{1}{|N_i| |N_j|} \sum_{i' \in N_i} \sum_{j' \in N_j} Sim^{Clus}(i', j') \tag{4}$$

where N_x denotes the set of clusters to which x belongs and $Sim^{Clus}(y, z)$ is the similarity value between clusters y and z, once again quantified by the proportion of items that they share.

As for the previous approach, the base clusterings are generated by random projections of the feature space, while the final consensus clustering is computed by a hierarchical clustering algorithm.

2.5   Data Sets

We ran our experiments on three data sets of global mRNA expression profiling:

- the Oxford Breast Cancer data set (ID: *oxf*), presented in [15], available at Gene Expression Omnibus (GEO) [16], superSeries GSE22219, consisting of 201 cases of early-invasive breast cancers divided into four subtypes (luminal A, luminal B, HER-2 Overexpression and basal);

- the TCGA Breast Cancer data set (ID: *tcga*), publicly available from the Tcga Genome Atlas [17], consisting of 151 cases of invasive breast cancers divided into four subtypes (luminal A, luminal B, HER-2 Overexpression and basal);
- the TCGA Glioblastoma data set (ID: *glio*), publicly available from the Tcga Genome Atlas, consisting of 167 cases of glioblastoma cancer divided into four classes (classical, mesechymal, neural and proneural).

To run multiple experiments in a reasonable time, data were preprocessed in order to reduce data dimensionality following the feature selection methodology proposed in [18]. Both Pearson correlation coefficient and negative Euclidean distance were used as similarity measures between data items.

In the following every experiment is identified by a code, composed by: the data set ID (*oxf*, *tcga* or *glio*); a letter indicating the similarity measure employed (E for Euclidean distance and P for Pearson correlation coefficient); a number indicating the number of clusters sought[1]. For instance, the experiment *glioE7* ran on the TCGA Glioblastoma data set, using the negative Euclidean distance and searching for 7 clusters.

3   Results and Discussion

In a previous work [19] we adopted internal validation criteria to select the best clustering using information intrinsic to data alone, with no use of prior knowledge. In this work, since we are interested in finding new data subclasses, we adopted the Adjusted Rand Index (ARI) [20] as a criterion to evaluate the results comparing against the known classes of data points. More specifically, ARI compares the clustering outcome with class labels counting the number of pair-wise co-assignments of data items; if two partitions are independent then the index assumes values close to 0, while the maximum value 1 is achieved in case of perfect agreement.

As we mentioned above, clustering can be a difficult task when a priori assumptions on data are not reflected in their natural structure. In our experiments ARI values tend indeed to be small, but our purpose is to observe if the application of consensus techniques can lead to an improvement in the results.

Table 1 shows for each clustering algorithm and for each experiment which consensus method obtained the best result with respect to ARI (while the corresponding ARI scores are shown in Table 2). In addition, using the Borda count as an election method, we show which consensus method obtains the best performance for each experiment, i.e. the winner is selected combining the rankings expressed by each clustering algorithm. In most cases the best solution is identified by a consensus method; in particular the most frequent winners are the methods based on subsampling and the Consensus matrix and on random projections and the Consensus matrix, which also ranks first in five experiments out

---

[1] Note that in a simple execution of AP and DTC the number of clusters is detected automatically.

Table 1. Best methods w.r.t. Adjusted Rand Index (where *simple* refers to a simple execution of the clustering algorithm without a consensus technique; *consensus* refers to the method based on subsampling and Consensus matrix; *randproj* refers to the method based on random projections and Consensus matrix and *CTS* and *ASRS* refer to the methods based on CTS and ASRS matrices respectively). In bold the method that obtained the best score for each experiment.

	AP	DTC	APK	KAP	KM	BORDA
glioE6	ASRS	**consensus**	randproj	ASRS	consensus	consensus
glioE7	ASRS	**consensus**	ASRS	ASRS	consensus	ASRS
glioP6	randproj	**simple**	randproj	consensus	consensus	randproj
glioP7	randproj	**simple**	CTS	ASRS	randproj	randproj
oxfE6	CTS	CTS	CTS	CTS	**simple**	CTS RP
oxfP6	randproj	consensus	CTS	randproj	**simple**	randproj
tgcaE6	randproj	randproj	**CTS**	CTS	consensus	randproj
tgcaP6	consensus	consensus	randproj	randproj	**consensus**	randproj

Table 2. Best ARI scores obtained by each clustering algorithm (in combination with the consensus method indicated in Table 1). In bold the best score for each experiment.

	AP	DTC	APK	KAP	KM
glioE6	0,5451	**0,4390**	0,5475	0,5354	0,5912
glioE7	0,5379	**0,4315**	0,5379	0,5379	0,4906
glioP6	0,5464	**0,4640**	0,4778	0,4791	0,5400
glioP7	0,5504	**0,4640**	0,5245	0,5129	0,5202
oxfE6	0,4688	0,4299	0,4644	0,4675	**0,3513**
oxfP6	0,4808	0,4489	0,4170	0,4491	**0,3513**
tgcaE6	0,4009	0,3613	**0,3581**	0,3628	0,4185
tgcaP6	0,4348	0,3599	0,3827	0,3918	**0,3465**

of eight in the Borda ranking. When looking at the best overall score for each experiment (in bold in Table 2), we can observe that every time the simple execution of the clustering algorithm performs better than consensus methods, it also achieves the best overall score for that experiment. We can hence conclude that in general consensus clustering can improve the accuracy of the results, except when the simple execution of a clustering algorithm obtains a good solution that is not further improved by consensus techniques.

To assess the stability of the solutions and extending the method described in [21], a technique that produces a non-partitive clustering solution calculating the intersection between different clusterings, we developed a method based on a majority voting scheme that, starting from n clustering solutions, assigns two objects to the same clusters if they were assigned to the same cluster in at least $\lfloor \frac{n}{2} \rfloor + 1$ clustering solutions.[2] In particular, we selected, for each of the five clustering algorithms, the solution produced by the method that performed better with respect to ARI, therefore in the final solution a cluster is formed if at

---

[2] In the following we will refer to this method as *intersection method* for brevity.

least three clusterings out of five agree. We thought of this method as a simple yet effective way to quickly verify the clustering stability, following the idea that different solutions are more likely to agree on stable clusters (i.e. groups that reflect the natural structure of data), but we did not intend it to be an alternative consensus technique.

Figure 1 shows the best ARI scores obtained by each clustering algorithm (in combination with the consensus method indicated in Table 1) and the ARI score obtained by the intersection method, while Fig. 2 compares the mean ARI scores obtained for each experiment with the ARI score obtained by the intersection method. In all but one experiment the clustering obtained by the intersection method has a better ARI score, indicating that the clusters that proved to be stable with respect to the change of the clustering algorithm are consistent with the known classes.

Note that since the intersection method produces a non partitive clustering, several data points may be excluded and hence we need to consider how many points are preserved. Therefore, in order to have an insight of the stability of the original clusterings we also observed the percentage of data points selected by the intersection method. Figure 3 compares the percentage of data points preserved when we consider the agreement between at least 3, 4 and 5 (all) clusterings, respectively. As we can observe only in two cases out of eight the percentage of points selected by the intersection method (black bar) is below 60% and in half of the cases the percentage is above 80%, hence we can conclude that the original clusters detected by consensus methods are rather stable. Nevertheless, when considering the percentage of points on which 4 algorithms agree (grey bar) in three cases the percentage stays above 60%, while in the other cases, as well as when considering the agreement on all algorithms (white bar), the values lower drastically.

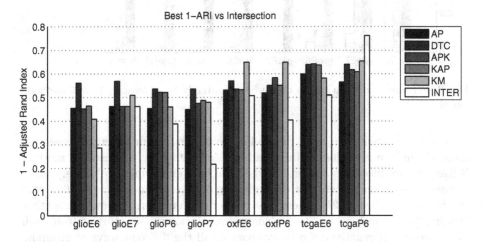

Fig. 1. Best ARI score for each algorithm vs. ARI score obtained by intersection method (the lower the better).

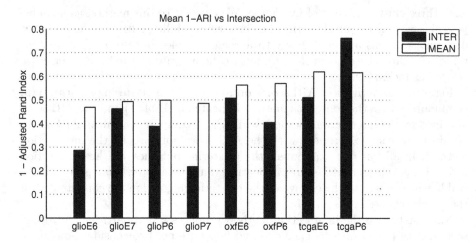

Fig. 2. Mean ARI score obtained for each experiment vs. ARI score obtained by intersection method (the lower the better).

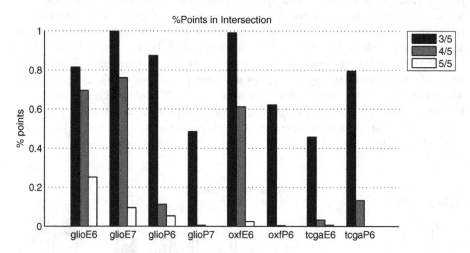

Fig. 3. Percentage of data points clustered accordingly by at least 3, 4, or 5 methods out of 5.

When clusterings are very dissimilar, combining them by an intersection method causes a great loss of information. More generally, a consensus method is likely to produce a poor solution when the base clusterings are unmergeable. A possible solution is then that of combining only solutions that are sufficiently similar, as proposed in [21], where base clustering solutions are grouped together forming "meta-clusters" and then a consensus solution is built starting from each meta-cluster. Alternatively one may consider all the different ways of grouping the data simultaneously through membership functions and fuzzy sets, allowing each data point to belong to more than one group.

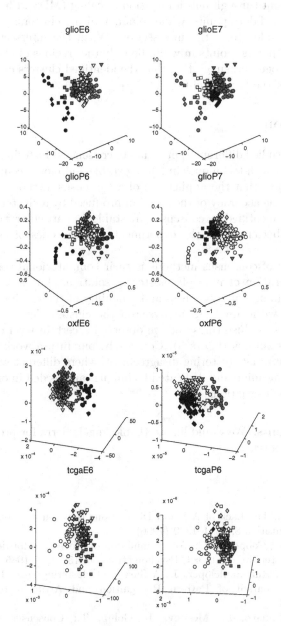

Fig. 4. MDS plots of clusterings obtained by intersection method. Shapes indicates classes while colours indicates clusters; white points are excluded by intersection (see online version for colours).

Plots in figure 4 represent the clusterings obtained by the intersection method for each experiment through multidimensional scaling (MDS). Shapes represent the known classes of data points, while colours indicate clusters; points coloured in white are excluded from the intersection. When the intersection preserves most of the data points, points in white lie in border regions of classes or where two classes mix together, suggesting that the identified clusters can be regarded as core regions of the original clusters.

4   Conclusion

In this work we studied different approaches to consensus clustering and analysed the advantages of such techniques in finding stable and more accurate solutions.

Our results show that the application of a consensus method can lead to an improvement in the accuracy of the solution produced by a clustering algorithm and that clusters identified by consensus are stable. In particular, the approaches that performed better across the experiments are the ones based on random projections.

The drawback of consensus methods is their computational cost, since they require the generation of multiple clustering solutions, but it is a reasonable trade-off when aiming for robustness and reliability of the results.

In this study we focused our analysis on the clusters derived from the intersection of various clusterings and we observed a certain level of consistency across different solutions (figure 3). Conversely, our future work will focus on the cases of substantial clustering disagreement, where different solutions could convey different meaning and dedicated techniques are needed in order to detect and deal with such situations.

Acknowledgments. We would like to thank Angela Serra for providing us the preprocessed data sets.

References

1. Wolpert, D.H.: The Lack of A Priori Distinctions Between Learning Algorithms. Neural Computation 8, 1341–1390 (1996)
2. Milligan, G.W., Cooper, M.C.: An examination of procedures for determining the number of clusters in a data set. Psychometrika 50, 159–179 (1985)
3. Vega-Pons, S., Ruiz-Shulcloper, J.: A Survey of clustering ensemble algorithms. International Journal of Pattern Recognition and Artificial Intelligence 25, 337–372 (2011)
4. Monti, S., Tamayo, P., Mesirov, J., Golub, T.: Consensus Clustering: A Resampling-Based Method for Class Discovery and Visualization of Gene Expression Microarray Data. Machine Learning 52, 91–118 (2003)
5. Frey, B.J., Dueck, D.: Clustering by Passing Messages Between Data Points. Science 315, 972–976 (2007)
6. Frey Lab, Probabilistic and Statistical Inference Group, University of Toronto. http://www.psi.toronto.edu/affinitypropagation

7. Zhang, X., Wang, W., Nørvag, K., Sebag, M.: K-AP: Generating Specified K Clusters by efficient affinity propagation. In: 2010 IEEE 10th International Conference on Data Mining (ICDM), pp. 1187–1192 (2010)
8. Langfelder, P., Zhang, B., Horvath, S.: Defining clusters from a hierarchical cluster tree: the Dynamic Tree Cut package for R. Bioinformatics 24, 719–720 (2008)
9. Wilkerson, M.D., Hayes, D.N.: ConsensusClusterPlus: a class discovery tool with confidence assessments and item tracking. Bioinformatics 26, 1572–1573 (2010)
10. Fern, X.Z., Brodley, C.E.: Random projection for high dimensional data clustering: A cluster ensemble approach. In: ICML, vol. 3, pp. 186–193 (2003)
11. Johnson, W.B., Lindenstrauss, J.: Extensions of Lipschitz mappings into a Hilbert space. Contemporary Mathematics 26(1), 189–206 (1984)
12. Bertoni, A., Valentini, G.: Ensembles based on random projections to improve the accuracy of clustering algorithms. In: Apolloni, B., Marinaro, M., Nicosia, G., Tagliaferri, R. (eds.) WIRN/NAIS 2005. LNCS, vol. 3931, pp. 31–37. Springer, Heidelberg (2006)
13. Bingham, E., Mannila, H.: Random projection in dimensionality reduction: applications to image and text data. In: Proceedings of the Seventh ACM SIGKDD International Conference on Knowledge Discovery and Data Mining, pp. 245–250. ACM (2001)
14. Iam-on, N., Garrett, S.: LinkCluE: A MATLAB package for link-based cluster ensembles. J. Stat. Software 36(9), 1–36 (2010)
15. Buffa, F.M., Camps, C., Winchester, L., Snell, C.E., Gee, H.E., Sheldon, H., Taylor, M., Harris, A.L., Ragoussis, J.: microRNA-Associated Progression Pathways and Potential Therapeutic Targets Identified by Integrated mRNA and microRNA Expression Profiling in Breast Cancer. Cancer Res. 71, 5635–5645 (2011)
16. Gene Expression Omnibus (GEO). http://www.ncbi.nlm.nih.gov/geo/
17. Tcga Genome Atlas. https://tcga-data.ncl.nih.gov/tcga/
18. Serra, A., Fratello, M., Fortino, V., Raiconi, G., Tagliaferri R., Greco, D.: MVDA: a multi-view genomic data integration methodology. BMC Bioinformatics 2015 16, 261 (2015)
19. Galdi, P., Napolitano, F., Tagliaferri, R.: A comparison between Affinity Propagation and assessment based methods in finding the best number of clusters. In: Di Serio, C., Li, P., Richardson, S., Tagliaferri, R. (eds.) Proceedings of Eleventh International Meeting on Computational Intelligence Methods for Bioinformatics and Biostatistics (CIBB 2014), Cambridge, pp. 978–988, June 2014. ISBN: 978-88-906437-4-3
20. Hubert, L., Arabie, P.: Comparing partitions. Journal of Classification 2, 193–218 (1985)
21. Bifulco, I., Fedullo, C., Napolitano, F., Raiconi, G., Tagliaferri, R.: Robust clustering by aggregation and intersection methods. In: Lovrek, I., Howlett, R.J., Jain, L.C. (eds.) KES 2008, Part III. LNCS (LNAI), vol. 5179, pp. 732–739. Springer, Heidelberg (2008)

Automated Detection of Fluorescent Probes in Molecular Imaging

Fiona Kathryn Hamey[1], Yoli Shavit[2], Valdone Maciulyte[3], Christopher Town[2],
Pietro Liò[2], and Sabrina Tosi[3]

[1] Wellcome Trust - Medical Research Council Stem Cell Institute
University of Cambridge, Cambridge, UK
fkh23@cam.ac.uk
[2] Computer Laboratory University of Cambridge, Cambridge, CB3 0FD, UK
{ys388,cpt23,pl219}@cam.ac.uk
[3] Brunel University London, Middlesex UB8 3PH, UK
{bb11vvm1,sabrina.tosi}@brunel.ac.uk

Abstract. Complex biological features at the molecular, organelle and cellular levels, which were traditionally evaluated and quantified visually by a trained expert, are now subjected to computational analytics. The use of machine learning techniques allows one to extend the computational imaging approach by considering various markers based on DNA, mRNA, microRNA (miRNA) and proteins that could be used for classification of disease taxonomy, response to therapy and patient outcome. One method employed to investigate these markers is Fluorescent *In Situ* Hybridization (FISH). FISH employs probes designed to hybridise to specific sequences of DNA in order to display the locations of regions of interest. We have developed a method to identify individual interphase nuclei and record the positions of different coloured probes attached to chromatin regions within these nuclei. Our method could be used for obtaining information such as pairwise distances between probes and inferring properties of chromatin structure.

Keywords: FISH, Nuclear Architecture, Image Analysis.

1 Introduction

Chromatin positioning plays an important role in biological processes such as gene-gene interactions and gene regulation. One way of investigating chromatin locations is by using the method of Fluorescent *In Situ* Hybridization (FISH). This method is described in [1–3]. Fluorescent probes are designed to hybridize to specific genomic regions thus allowing us to view the positions of these regions in images. Different coloured probes can be used together and the separate fluorescent images superimposed in order to compare positions of several regions. From such images it is possible to obtain information on the relative positions of homologous chromosomes and the distances between specific genomic regions in order to study features such as gene-gene interactions.

© Springer International Publishing Switzerland 2015
C. di Serio et al. (Eds.): CIBB 2014, LNCS 8623, pp. 68–75, 2015.
DOI: 10.1007/978-3-319-24462-4_6

Although FISH images can provide important information simply from observation (for example providing evidence for rearrangement of genetic material in diseases including leukaemia) knowing the exact positions of probes can be useful for extracting information such as distance measurements from images. The quantitative analysis of distances and positions is a key step in validation studies and in supporting the translational process of taking innovation into the clinical environment.

We have developed a method for automatic detection of cell nuclei and the position of probes in FISH images. We implemented our algorithm in the MathWorks MATLAB programming environment to first segment the image into separate nuclei and to then detect probes of different colour in each nucleus. Probe positions are recorded for each nucleus allowing for distances between probes to be calculated as well as information to be obtained about positioning of homologous chromosomes. We provide results of applying our algorithm to a set of FISH images of lymphoblastoid cell lines with 3 different coloured probes, showing that our method is useful for detecting probes and nuclei. Existing methods for analysis of FISH images are limited and so our algorithm provides a valuable contribution to this field. Further advantages of our method are discussed in the Results and Discussion section.

2 Algorithm and Data Analysis

Images for FISH studies are formed from several multicolour images superimposed to create a false colour image that shows the positions of different coloured probes. Based on this our method of detecting probe positions was designed to analyse a set of $n + 1$ images (such as those shown in figures 1a-1d). These images are stored in $p \times q$ matrices $P_1, \ldots, P_n$ (showing the probe locations) and B (showing background chromatin staining). Each element of a matrix represents a pixel and can take values in a set range, which in our data was $[0, 255]$. In order to develop our method we considered multicoloured FISH images from lymphoblastoid cell lines GM12878, GM17208 and CRL-2630 with background chromatin staining and three different probes coloured red, green and purple. Each image was provided in a set of 5 consisting of 4 greyscale images (taken under different fluorescent lights to show the background chromatin staining and the positions of the different probes) along with a false colour image which was a superposition of the other 4.

Algorithm 1 explains our method of obtaining probe positions in each nucleus. The purpose of lines 1-29 is to segment the image to identify all of the interphase nuclei. Iterative thresholding in line 1 was implemented using a previously described method [2] to detect regions corresponding to either background or cell nucleus. We then performed a dilation of this image in order to smooth the edges of the detected regions. The aim of lines 3-7 is to remove components with an area less than A_k from the image as these correspond to either noise or mitotic chromosomes. The set $\hat{C}$ corresponds to the set of components detected in the image. If this set is empty then we have detected no nuclei and so no probes

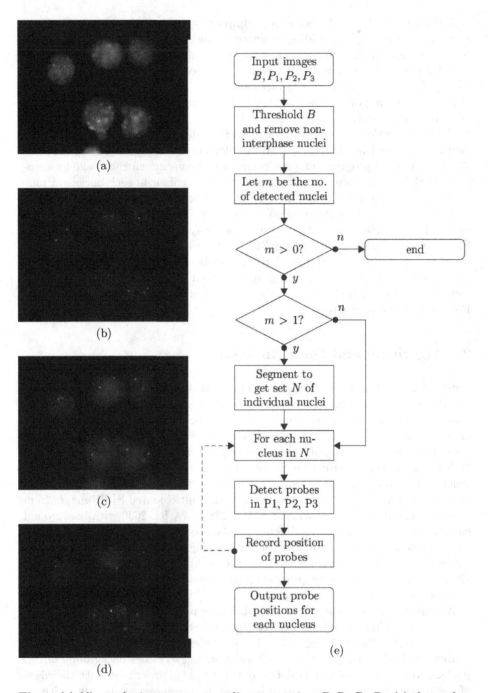

Fig. 1. (a)-(d) are the images corresponding to matrices B, P_1, P_2, P_3. (a) shows the background chromatin staining and (b), (c), (d) show the location of the green, red and purple probes respectively. (e) is a flowchart explaining the image processing procedure.

---

input : Set of $p \times q$ matrices $\{B, P_1, \ldots, P_n\}$
output: Positions of probes in each nucleus

1 Perform iterative thresholding to set elements of B to 0 or 1;
2 Let C be the set of connected components in B;
3 **for** c *in* C **do**
4   **if** *size of* $c < A_k$ **then**
5 | Set $b_{ij} = 0$ $\forall b_{ij} \in c$
6   **end**
7 **end**
8 Let $\hat{C}$ be the set of connected components in B;
9 **if** $|\hat{C}| = 0$ **then**
10 | Terminate algorithm
11 **end**
12 **if** $|\hat{C}| > 1$ *or* eccentricity$(\hat{c}_i) > 0.5$ *for some* $\hat{c}_i \in \hat{C}$ **then**
13   Calculate $B^\dagger$ = Euclidean distance transform of B;
14   Set $\hat{B} = -B^\dagger$ with $\hat{b}_{ij} = -\inf$ for $b_{ij} = 0$;
15   Calculate B^* = watershed transform of $\hat{B}$;
16   Let C^* be the set s.t. for $c_k^* \in C$, c_k^* is the set of all $b_{ij} = a_k$ for some constant a_k ;
17   **for** c_k^* *in* C^* **do**
18   **if** Min bounding box$(c_k^*) \leq 5$ *and*
        Max bounding box$(c_k^*) \geq \min(p,q) - 5$ **then**
19 | set $b_{ij}^* = 0$ $\forall b_{ij}^* \in c_k^*$;
20   **end**
21   **else**
22   **for** $l > k$, $l \leq |C^*|$ **do**
23   **if** $|\text{centroid}(c_k^*) - \text{centroid}(c_l^*)| < R$ **then**
24 | Set $b_{ij} = a_k$ $\forall b_{ij} \in c_l^*$;
25   **end**
26   **end**
27   **end**
28   **end**
29 **end**
30 Let S be the set s.t. for $s_k \in S$, s_k is the set of all $b_{ij} = a_k$ for some constant a ;
31 **for** s_k *in* S **do**
32   **for** m *in* $1 : n$ **do**
33   Let $T = (t_{ij})$ s.t. $t_{ij} = (P_m)_{ij}$ if $b_{ij} \in s_k$ else $t_{ij} = 0$;
34   Compute extended maxima transform of T_{ij} to find set of connected components ;
35   Record position of centroid of connected components with size $< D$ alongside k and m ;
36   **end**
37 **end**

Algorithm 1. Probe detection

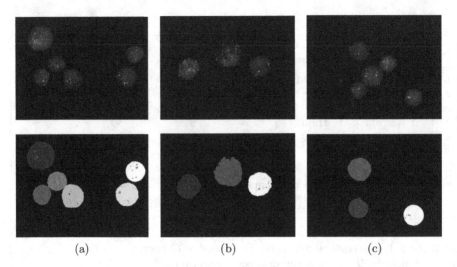

(a)   (b)   (c)

Fig. 2. The top panel shows the 3 false colour FISH images. The bottom panel displays the results of applying our algorithm to these images. Separate nuclei identified in the segmentation process are shown in different shades of grey and the detected positions of the red, green and purple probes are indicated in red, green and blue respectively. (a) shows an example of successful segmentation and probe detection by the algorithm. (b) shows a successful segmentation but not all probes are detected. (c) has all probes detected in the nuclei which were successfully segmented but only 3 out of 5 nuclei were segmented.

can be found (lines 9-11). If we have detected more than one nucleus (line 12) then it is necessary to segment the thresholded image in order to identify individual nuclei. This is done by computing the watershed transform (lines 14-15) and merging components with centres of gravity less than R apart (lines 17-28). Then, for each nucleus we have identified (line 31), the images for each different probe (line 32) are masked so that only the area corresponding to the chosen nucleus is considered (line 33).

We determined appropriate levels for probe detection from experimentation. We recommend using a sample of images in order to optimise this for a set of images. Using these levels, the extended maxima transform for each image was calculated in order to identify bright regions which correspond to probes (lines 34-35). Regions of area greater than a constant D were excluded as these were too unreliable to measure positions or occurred due to having no probes of the relevant colour in the nucleus. The centre of mass of each detected probe was recorded alongside the colour of the probe and the nucleus that it came from. We implemented our algorithm in MATLAB version R2013b with the aid of some functions from the Image Processing Toolbox. This implementation is available upon request from the authors.

Table 1. Results of visual verification

	Nuclei	Green probes	Red probes	Purple probes
Percentage detected overall	88.3	87.9	93.2	64.0
Percentage of images with > 0.8 success rate	81.3	81.3	87.5	42.6
Percentage of images with > 0.5 success rate	97.9	100.0	97.9	63.8

3 Results and Discussion

We used our algorithm to find probe positions in FISH images of lymphoblastoid cell lines with 3 different coloured probes. In the majority of cases we were able to successfully identify a high proportion of the interphase nuclei and the probes within them. Figure 2 displays the segmentation of nuclei and detected positions of probes in 3 cases showing different levels of success. The positions of probes obtained can then easily be used to calculate properties such as pairwise distances in each nucleus.

Validation of the algorithm was carried out by comparing the false colour image (see top panel of figure 2) to an image visualising the output of the algorithm (bottom panel of figure 2). This image was generated by taking the greyscale matrix showing nuclei segmentation (the result of lines 1-29 in algorithm 1) and converting it to a $p \times q \times 3$ matrix before setting the elements corresponding to each detected probe to be red, green or blue for the red, green and purple probes respectively.

Figure 2 shows 3 examples of nuclear segmentation and probe detection by the algorithm. By comparing the top and bottom images we are able to see how well the segmentation worked. In figure 2a the algorithm was able to segment all of the nuclei (including ones positioned close together, which present more of a challenge) as well as detecting the probe positions in these nuclei. Figure 2b displays an image with successful segmentation of nuclei but where not all of the probes were detected. We can see that probes in the top right of the central nucleus were not detected. This is most likely related to the quality of the greyscale images displaying probe locations. From studying these images (data not shown) we were able to observe that the regions corresponding to the top right probes were of a lower intensity than for the central probes. In images such as this the method struggles to simultaneously detect all of the probes, which is a limitation. However, depending on the aim of the experiment, having partial probe detection in some nuclei can still provide valuable information, e.g. when calculating pairwise distances between probes. In other cases the method was not able to detect all interphase nuclei. This usually occurred in cases such as shown in figure 2c where nuclei were positioned too close together and were therefore difficult to segment using the watershed transform.

In order to provide some quantitative measure of the success of our algorithm, we considered a test set of 49 images. Our complete data set consisted of 7 subsets of images (each with a different combination of probes types and cell line) and so 7 images were randomly chosen from each set. The number of nuclei and each type of probe were established visually and recorded from the false colour images, as were the number of detected nuclei and probes from the algorithm. Probes in the false colour image were only counted in nuclear regions which had been successfully detected. The results are shown in table 1 which displays the overall percentages of nuclei and probes detected across all images. Some success rates in individual images are also included in the table.

The overall success rates are high for the detection of nuclei and green/red probes. The majority of purple probes were detected but not as successfully. This was most likely due to the quality of the images for the purple probes that, for some of the subsets in particular, had a poor distinction between probe and background. It can be seen that over half of purple probes were detected in 60% of the images, with the majority detected in almost all of the images for nuclei and red/green probes. It is possible that a different level for the probe detection would have produced a higher detection rate of purple probes in the images.

As noted previously, there are not many existing methods for analysis of FISH images. The automated iterative thresholding we employ provides an advantage over methods where only thresholding at a single user-specified level is used [4] by allowing identification of a greater proportion of nuclei. The light intensity of background chromatin staining can vary widely between images and so an automated thresholding approach is required to avoid losing information. Our approach also provides advantages over methods such as FISH finder [5] since we describe our algorithm in detail, thus enabling others to re-implement our methods using open source software image analysis tools such as R or OpenCV, whereas FISH finder requires purchase of the commercial MATLAB software.

A further benefit of our method is that the probe positions are returned for each nucleus, thereby allowing a range of nuclear properties to be calculated. These properties could include measurements between homologous chromosomes, or provide evidence of translocations or inversions in chromosomes. Complementary methods, such as Chromosome Conformation Capture, could also benefit from the statistics of the probe distances and positions, for example for calibration [6] and validation of predicted models.

However, our approach does have some limitations. The levels chosen for the probe detection were experimentally determined from a sample of images. It is possible that the choice of levels could be improved to obtain a higher probe detection rate by optimising for a larger sample of images. From the visual validation, we saw that two distinct cases were responsible for the majority of probe detection failures: either there were two probes very close together (caused by the replication of chromatin), in which case only one probe was detected, or probes from the same nucleus had different intensities, in which case only the brightest probe was detected. Failure to segment nuclei was most often caused by nuclei being too close together for the algorithm to separate them.

Other failure cases were caused by having some nuclei in an image much brighter than others, in which case the darker nuclei were not detected. Thus, possible extensions to our method could include supervised training with sample data sets for estimating parameters and noise, and incorporation of signal processing methods to improve the robustness of the thresholds used (specifically addressing issues of poor intensity and low signal-to-noise ratio) and the estimation of shape regularity.

4   Conclusion

We have developed a new algorithm for obtaining probe positions in individual nuclei from FISH images. These positions can be used to compute probe distances and the positioning of homologous chromosomes. Our algorithm successfully separates nuclei and detects multi-coloured probes, and is readily accessible through its MATLAB implementation and detailed description. While there are other programs available for FISH image analysis, we have addressed the automatic detection of specific probes for quantitative spatial analysis in translational medicine tasks.

References

1. Volpi, E.V., Bridger, J.M.: FISH glossary: an overview of the fluorescence in situ hybridization technique. Biotechniques 45, 385–409 (2008)
2. Liehr, T.: Fluorescence In Situ Hybridization (FISH) - Application Guide. Springer Protocols. Springer, Heidelberg (2009)
3. Yurov, Y.B., Vorsanova, S.G., Iourov, I. (eds.): Human Interphase Chromosomes: Biomedical Aspects. Springer Science+Business Media, LLC (2013)
4. Shopov, A., Williams, S.C., Verity, P.G.: Improvements in image analysis and fluorescence microscopy to discriminate and enumerate bacteria and viruses in aquatic samples. Aquatic Microbial Ecology 22, 103–110 (2000)
5. Ty, S., Shirley, J., Liu, X., Gilbert, D.M.: FISH FINDER: A high-throughput tool for analyzing FISH images. Bioinformatics 27, 933–938 (2011)
6. Shavit, Y., Hamey, F.K., Lio', P.: FisHiCal: an R package for iterative FISH-based calibration of Hi-C data. Bioinformatics 30, 3120–3122 (2014)

Applications of Network-based Survival Analysis Methods for Pathways Detection in Cancer

Antonella Iuliano[1,*], Annalisa Occhipinti[2,*], Claudia Angelini[1],
Italia De Feis[1], and Pietro Liò[2]

[1] Istituto per le Applicazioni del Calcolo "Mauro Picone",
[2] Consiglio Nazionale delle Ricerche, via Pietro Castellino 111, 80131 Napoli, Italy
{a.iuliano,c.angelini,i.defeis}@na.iac.cnr.it
[3] Computer Laboratory, University of Cambridge, CB3 0FD, UK
{ao356,pl219}@cam.ac.uk

Abstract. Gene expression data from high-throughput assays, such as microarray, are often used to predict cancer survival. Available datasets consist of a small number of samples (n patients) and a large number of genes (p predictors). Therefore, the main challenge is to cope with the high-dimensionality. Moreover, genes are co-regulated and their expression levels are expected to be highly correlated. In order to face these two issues, network based approaches can be applied. In our analysis, we compared the most recent network penalized Cox models for high-dimensional survival data aimed to determine pathway structures and biomarkers involved into cancer progression.

Using these network-based models, we show how to obtain a deeper understanding of the gene-regulatory networks and investigate the gene signatures related to prognosis and survival in different types of tumors. Comparisons are carried out on three real different cancer datasets.

Keywords: Cancer, comorbidity, Cox model, high-dimensional data, gene expression data, network analysis, regularization, survival data.

1 Introduction

Cancer is a *multi-factorial disease* since it is caused by a combination of genetic and environmental factors working together in a still unknown way. Genetic screening for mutations associated with multi-factorial diseases cannot predict exactly whether a patient is going to develop a disease, but only the risk to have the disease. Hence, a woman inheriting an alteration in the BRCA2 gene can develop breast cancer more likely than other women, although she may also remain disease-free. Genetic mutation is only one risk factor among many. Lifestyle, environment and other biological factors are also involved in the study of the disease development. The integration of all this supplementary information is the key point to stress the mechanism of disease progression and identify reliable biomarkers.

---

* These two authors contributed equally to this work.

© Springer International Publishing Switzerland 2015
C. di Serio et al. (Eds.): CIBB 2014, LNCS 8623, pp. 76–88, 2015.
DOI: 10.1007/978-3-319-24462-4_7

The advancement of recent biotechnology has increased our knowledge about the molecular mechanism involved into cancer progression. However, this biological knowledge is still not fully exploited since the integration of all those different types of data leads to the curse of dimensionality. Indeed, the number of covariates (molecular and clinical information) exceed the number of observations (patients). As a result, many classical statistical methods cannot be applied to analyse this kind of data and new techniques need to be proposed to cope with the high-dimensionality.

In cancer research is also important to study survival analysis, that can be used to investigate microarray gene expression data and evaluate cancer outcomes depending on time intervals. Those intervals start at a survival time and end when an event of interest occurs (a death or a relapse). The exploitation of the relationship between event distributions and gene expression profiles permits to achieve more accurate prognoses or diagnoses. The Cox regression [2] is the most popular method to analyse censored survival data. However, due to high-dimensionality, it cannot be directly applied to obtain the estimated parameters. Therefore, penalized techniques based on lasso type penalties [5,17,18] have been taken into account. Moreover, those methods perform estimation and variable selection by shrinking some parameters to zero. These methods solve the "$p \gg n$" issue but ignore the strong-correlation among variables (i.e. genes). For this reason, the elastic net method (an improved variant of the lasso for high-dimensional data, [13,21]) can be applied to achieve some grouping effects ([3,23]) and to incorporate pathway information of genes. A pathway is given by a group of genes that are involved in the same biological process and have similar biological functions. Those genes are co-regulated and their expression levels are expected to be highly correlated. The pathway structures play a biologically important role to understand the complex process of cancer progression.

The purpose of this paper is (i) to describe a systematic approach to compare the most recent methods based on the integration of pathway information into penalized-based Cox methods and (ii) to evaluate their performance. We considered three methods. *Net-Cox* [20] explores the co-expression and functional relation among gene expression features using an L_2-norm constrain plus a Laplacian penalty. The L_2-norm smooths the regression coefficients reducing their variability in the network; the Laplacian take into account the grouping effects. *Adaptive Laplacian net* [16] uses an L_1-penalty to enforce sparsity of the regression coefficients and a quadratic Laplacian penalty to encorage smoothness between the coefficients of neighboring variables on network. Finally, *Fastcox* method [7] is a new fast algorithm for computing the elastic net penalized Cox model. We compare three different types of cancer by using the penalized regression methods presented before in order to provide an interesting investigation from a biological, medical and computational point of view.

The paper is organized as follows. In Section 2, we introduce the network-based regularized methods for high-dimensional Cox regression analysed in our comparisons. Cross-validation and parameter tuning are discussed in Section 3.

Real data analysis is presented in Section 4, with the main results obtained in the analysis. We conclude with a brief discussion about future works in Section 5.

2   Methodology

In this section, we describe the three methods for Cox's proportional hazard model that we used for our analysis. We first review the Cox model and then, we introduce the three regularization methods.

2.1   The Cox Model

Prediction of cancer patients survival based on gene expression profiles is an important application of gene expression data analysis. Usually it is difficult to select the most significant genes (i.e. covariates) for prediction, as these may depend on each other in a still unknown way. Because of the large number of expression values, it is easy to find predictors that perform well on the fitted data, but fail in external validation, leading to poor prediction rules.

The problem can be formulated as a prediction problem where the response of interest is a possibly censored survival time and the predictor variables are the gene expression values. The Cox Proportional hazards model [2] is used to describe the relationship between survival times and predictor covariates.

Given a sample of n subjects, let T_i and C_i be the survival time and the censoring time respectively for subject $i = 1, \ldots, n$. Let $t_i = min\{T_i, C_i\}$ be the observed survival time and $\delta_i = I(T_i \leq C_i)$ the censoring indicator, where $I(\cdot)$ is the indicator function (i.e $\delta_i = 1$ if the survival time is observed and $\delta_i = 0$ if the survival time is censored) and $\mathbf{X}_i = (X_{i1}, \ldots, X_{ip})'$ be the p-variable vector for the ith subject (i.e. the gene expression profile of the ith patient over p genes). The survival time T_i and the censoring time C_i are assumed to be conditionally independent given $\mathbf{X}_i$. Furthermore, the censoring mechanism is assumed to be non-informative. The observed data can be represented by the triplets $\{(t_i, \delta_i, \mathbf{X}_i), i = 1, \ldots, n\}$. The Cox regression model assumes that the hazard function $h(t|\mathbf{X}_i)$, which means the risk of death at time t for the ith patient with gene expression profile $\mathbf{X}_i$, can be written as

$$h(t|\mathbf{X}_i) = h_0(t)exp\left(\sum_{i=1}^{p} \mathbf{X}_i'\boldsymbol{\beta}\right) = h_0(t)exp(\mathbf{X}'\boldsymbol{\beta})$$

where $h_0(t)$ is the baseline hazard and $\boldsymbol{\beta} = (\beta_1, \ldots, \beta_p)'$ is the column vector of the regression parameters.

Since the number of predictors p (genes) is much greater than the number of observations n (patients), the Cox model cannot be applied directly and a regularization approach needs to be used to select important variables from a large pool of candidates. For instance, a Lasso penalty ([17,18]), can be used to remove the not significant predictors by shrinking their regression coefficients exactly to zero. The lasso type approach solves the high dimensionality issue but don't take into account the functional relationships among genes. For this reason, in the last years, network-based regularization methods have been introduced in order to identify the functional relationships between genes and overcome the gap between genomic data analysis and biological mechanisms. By using these network-based models, it is possible to obtain a deeper understanding of the gene-regulatory networks and investigate the gene signatures related to the cancer survival time. In this context, the regression coefficients are estimated by maximizing the penalized Cox's log-partial likelihood function

$$l_{pen}(\boldsymbol{\beta}) = \sum_{i=1}^{n} \delta_i \left\{ \boldsymbol{X}_i' \boldsymbol{\beta} - log \left[\sum_{j \in R(t_i)} exp(\boldsymbol{X}_j' \boldsymbol{\beta}) \right] \right\} - P_\lambda(\boldsymbol{\beta}), \qquad (1)$$

where t_i is the survival time (observed or censored) for the ith patient, $R(t_i)$ is the risk set at time t_i (i.e., the set of all patients who still survived prior to time t_i) and $P_\lambda(\boldsymbol{\beta})$ is a network-constrained penalty function on the coefficients $\boldsymbol{\beta}$.

2.2 Network-regularized Cox Regression Models

We assume that the relationships among the covariates (genes) are specified by a network $G = (V, E, W)$ (weighted and undirected graph). Here $V = \{1, \ldots, p\}$ is the set of vertices (genes/covariates); an element (i, j) in the edge set $E \subset V \times V$ indicates a link between vertices i and j; $W = (w_{ij})$, $(i, j) \in E$ is the set of weights associated with the edges. Each edge in the network is weighted between [0,1] and indicates the functional relation between two genes [6]. For instance, in a gene regulatory network built from data, the weight may indicate the probability that two genes are functionally connected.

Net-Cox [20] integrates gene network information into the Cox's proportional hazard model by the following

$$P_{\lambda,\alpha}(\boldsymbol{\beta}) = \lambda \left[\alpha \|\boldsymbol{\beta}\|_2^2 + (1 - \alpha) \Phi(\boldsymbol{\beta}) \right], \qquad (2)$$

where $\lambda > 0$ and $\alpha \in (0, 1]$ are two regularization parameters in the network constraint and

$$\Phi(\boldsymbol{\beta}) = \sum_{(i,j) \in E} w_{i,j} (\beta_i - \beta_j)^2. \qquad (3)$$

The penalty (2) consists of two terms: the first one is an L_2-norm of β that regularizes the uncertainty in the network constraint; the second term is a network Laplacian penalty $\Phi(\boldsymbol{\beta}) = \boldsymbol{\beta}'[(1-\alpha)\boldsymbol{L}+\alpha\boldsymbol{I}]\boldsymbol{\beta}$ that encourages smoothness among correlated gene in the network and encode prior knowledge from a network. In the penalty, $\boldsymbol{L}$ is a positive semi-definite matrix derived from network information and $\boldsymbol{I}$ is an identity matrix. Given a normalized graph weight matrix $\boldsymbol{W}$, by using Eq.(3), *Net-Cox* assumes that co-expressed (related) genes should be assigned similar coefficients by defining the following cost term over the coefficients $\Phi(\boldsymbol{\beta}) = \boldsymbol{\beta}'(\boldsymbol{I} - \boldsymbol{W})\boldsymbol{\beta} = \boldsymbol{\beta}'\boldsymbol{L}\boldsymbol{\beta}$. More precisely, for any pair of genes connected by an high weight edge and with a large difference between their coefficients, the objective function will result in a significant cost in the network.

AdaLnet [16] (*Adaptive Laplacian net*) is a modified version of a network-constrained regularization procedure for fitting linear-regression models and for variable selction [10,11] where the predictors are genomic data with graphical structures. *AdaLnet* is based on prior gene regulatory network information, represented by an undirected graph for the analysis of gene expression data and survival outcomes. Denoting with $d_i = \sum_{i:(i,j)\in E} w_{ij}$ the degree of vertex i, *AdaLnet* defines the normalized Laplacian matrix $\mathbf{L} = (l_{ij})$ of the graph G by

$$l_{i,j} = \begin{cases} 1, & \text{if } i = j \text{ and } d_i \neq 0, \\ -w_{ij}/\sqrt{d_i d_j}, & \text{if}(i,j) \in E, \\ 0, & \text{otherwise.} \end{cases} \tag{4}$$

Note that $\mathbf{L}$ is positive semi definite. The network-constrained penalty in Eq. (1) is given by

$$P_{\lambda,\alpha}(\boldsymbol{\beta}) = \lambda \left[\alpha \|\boldsymbol{\beta}\|_1 + (1 - \alpha) \Psi(\boldsymbol{\beta})\right], \tag{5}$$

with

$$\Psi(\boldsymbol{\beta}) = \sum_{(i,j)\in E} w_{i,j} \left(sgn(\tilde{\beta}_i)\beta_i/\sqrt{d_i} - sgn(\tilde{\beta}_j)\beta_j/\sqrt{d_j}\right)^2. \tag{6}$$

Equation (5) is composed by two penalties. The first one is an L_1-penalty that induces a sparse solution, the second one is a quadratic Laplacian penalty $\Psi(\boldsymbol{\beta}) = \boldsymbol{\beta}'\tilde{\mathbf{L}}\boldsymbol{\beta}$ that imposes smoothness of the parameters β between neighboring vertices in the network. Note that $\tilde{\mathbf{L}} = \mathbf{S}'\mathbf{L}\mathbf{S}$ with $\mathbf{S} = diag(sgn(\tilde{\beta}_1),\ldots,sgn(\tilde{\beta}_p))$ and $\tilde{\boldsymbol{\beta}} = (\tilde{\beta}_1,\ldots,\tilde{\beta}_p)$ is obtained from a preliminary regression analysis. The scaling of the coefficients $\boldsymbol{\beta}$ respect to the degree allows the genes with more connections (i.e., the hub genes) to have larger coefficients. Hence, small changes of expression levels of these genes can lead to large changes in the response.

An advantage of using penalty (5) consists in representing the case when two neighboring variables have opposite regression coefficient signs, which is reasonable in network-based analysis of gene expression data. Indeed, when a transcription factor (TF) positively regulate gene i and negatively regulate gene j in a certain pathway, the corresponding coefficients will result with opposite sign.

Finally, *Fastcox* [7] computes the solution paths of the elastic net penalized Cox's proportional hazards model. In this method the penalty function in Eq. (1) is given by

$$P_{\lambda,\alpha}(\boldsymbol{\beta}) = \lambda \left[\alpha w\|\boldsymbol{\beta}\|_1 + \frac{1}{2}(1 - \alpha)\|\boldsymbol{\beta}\|_2^2\right],$$

where the non-negative weights w allows more flexible estimation.

3   Tuning Parameters by Cross-validation

All above described methods require to set two hyper-parameters: λ and α controlling the sparsity and the network influence, respectively. To determine the optimal tuning parameters λ and α to use in our study, we performed five-fold cross-validation following the procedure proposed by [22]. In the cross-validation, four folds of data are used to build a model for validation on the fifth fold, cycling through each of the five folds in turn. Then, the (λ,α) pair that minimizes the cross-validation log-partial likelihood (CVPL) are chosen as the optimal parameters. CVPL is defined as

$$CVPL(\lambda,\alpha) = -\frac{1}{n}\sum_{k=1}^{K}\{\ell(\hat{\beta}^{(-k)}(\lambda,\alpha)) - \ell^{(-k)}(\hat{\beta}^{(-k)}(\lambda,\alpha))\}, \qquad (7)$$

where $\hat{\beta}^{(-k)}(\cdot)$ is the estimate obtained from excluding the kth part of the data with a given pair of (λ,α), $\ell(\cdot)$ is the Cox log-partial likelihood on all the sample and $\ell^{(-k)}(\cdot)$ is the log-partial likelihood when the kth fold is left out.

4   Real Case Studies

In this section we describe the performances of the methods presented in Section 2 on three different types of cancer. In the following we first describe the datasets, then the results.

4.1   Datasets

We applied the three methods on three datasets containing large-scale microarray gene expression measurements from different type of cancer together with their (possible censored) survival informations (times and status). In particular, we used gene expression datasets downloaded from Gene Expression Omnibus as raw .CEL files. All the three datasets were generated by Affymetrix U133A. The raw files were processed and normalized individually by RMA package available in Bioconductor [4].

We consider the human gene functional linkage network [6] constructed by a regularized Bayesian integration system [6]. Such network contains maps of functional activity and interaction networks in over 200 areas of human cellular biology with information from 30.000 genome-scale experiments. The functional linkage network summarizes information from a variety of biologically informative perspectives: prediction of protein function and functional modules, crosstalk among biological processes, and association of novel genes and pathways with known genetic disorders [6]. The edges of the network are weighted between $[0,1]$ and express the functional relation between two genes. Thus, the functional linkage network plays an important role in our tests since it includes more information than Human protein-protein interaction, frequently used as the network prior knowledge. It is clear that taking into account such biological knowledge helps in identifying significant genes that are functionally related in order to obtain important results biologically interpretable.

We use HEFaIMp [6] tool to identify the edge's weight of between two genes on the network. After merging probes by gene symbols and removing probes with no gene symbol, we use KEGG pathways [8,9] in order to obtain a network consisting of a fixed number of unique genes derived from a large pool of probes and overlapped with the functional linkage network. The three datasets analysed are the following:

1. **Breast Cancer Microarray Data.** The first dataset is from Nagalla et al. [12] (accession number: GSE45255) and consist of $p = 2431$ gene expression measurements from $n = 93$ patients with breast cancer.
2. **Lung Cancer Microarray Data.** The second dataset is from Chen et al. [1] (accession number: GSE37745) and contains $p = 2259$ gene expression measurements from $n = 100$ patients with lung cancer.
3. **Ovarian Cancer Microarray Data.** The third dataset is from Zhang et al.[20] (accession number: GSE26712) and contains gene expression measurements from $N = 153$ patients with ovarian cancer. We use a list of $p = 2372$ genes.

4.2   Model Evaluation Criteria

In order to evaluate the three methods we first divided each dataset randomly into two parts: (*i*) *training set* consisting of about 2/3 of the patients used for estimation; (*ii*) *testing set* consisting of about 1/3 of the patients used for evaluate and test the prediction capability of the models. We denoted the parameter estimate from the training data for a given method by $\hat{\beta}_{train}$. This estimate is computed as described in Section 3 by using five-fold cross-validation to select the optimal tuning parameter values $(\hat{\lambda}_{train}, \hat{\alpha}_{train})$, and then by fitting the corresponding penalized function $P_{\hat{\lambda}_{train}, \hat{\alpha}_{train}}(\hat{\beta}_{train})$ on the training set. In particular, we first set α to a sufficiently fine grid of values on $[0, 1]$. For each fixed α, λ was chosen from $\{1e-5, 1e-4, 1e-3, 1e-2, 1e-1, 1\}$ for *Net-Cox*, while we set λ to a decreasing sequence of values λ_{max} to λ_{min} automatically choosen by *AdaLnet* and *Fastcox*. Note that, when $\alpha = 1$ all the three methods listed in Section 2.2 ignore the network information. The results are given in Table 1. Interestingly, the optimal α is often 0.1 and 0.5, indicating the optimal CVPL is a balance of the information from gene expressions and the network. These results highlight that the network information is useful for improving survival analysis.

The estimated $\hat{\beta}_{train}$ is used to calculate the prognostic index (PI) for each patient i in the training set, given by

$$PI_i^{train} = x_i' \hat{\beta}_{train}, \tag{8}$$

Table 1. Cross-validation parameters

Datasets	Net-Cox λ	Net-Cox α	AdaLnet λ	AdaLnet α	Fastcox λ	Fastcox α
Breast	0.001	0.5	0.16	0.5	0.22	0.5
Lung	0.0001	0.1	1.90	0.1	0.60	0.5
Ovarian	0.001	0.5	11.94	0.01	0.25	0.95

where x_i is the vector of gene expression value associated to the ith patient. By using the PI_i^{train}, it is possible to divide the patients in two subgroups, i.e., *high-risk* and *low-risk* prognosis groups. Thus, the patient i in the training set is assigned to the *high-risk* (or *low-risk*) group if its prognostic index PI_i^{train}, Eq. (8), is above (or below) the quantile selected on a grid of given values that spans from 30% to 70%. We select as PI^* the optimal cutoff in terms of PI^{test} corresponding to the lowest p-value in a log rank test. Then, we calculate the prognostic index PI_i^{test} by using $\hat{\beta}_{train}$. Each patient i in the testing set is assigned into the *high*-and-*low-risk* groups if its prognostic index PI_i^{test} is above (or below) threshold PI^* chosen as stated before. To evaluate the performance of rule, we applied a log rank test and used the p-value as an evaluation criterion (the significance level was set at 5%, i.e., $p < 0.05$). For each datasets, Kaplan-Meier survival curves are drawn and the log-rank test is performed to assess differences between groups. For instance, Fig.1 shows the survival probabilities for these two groups obtained for cancer ovarian patients selected in the testing set by using *AdaLnet* and *Net-Cox*, respectively. More precisely, first we look at survival time in the training set for patients in the top 45% (40%) compared to the lower 55% (60%) testing *Net-Cox* (*AdaLnet*), as described before. We determine the cutoff in terms of PI^*. Then, the prognostic PI_i^{test} is calculated and patients are assigned into the *high*-and-*low-risk* groups by comparing with the cutoff obtained from the training set.The log-rank test on the test-set gives a p-value of 0.0103 for *AdaLnet* (Fig.1(a)), which means the two groups can be separeted and the selected pathways and genes are significant. In Fig.1(b), even if the log-rank test gives a p-value of 0.0189 for *Net-Cox*, we observe that a patient (bottom-right) of the *high-risk* group falls in the *low-risk* group. In particular, we observed that in predicting the survival probabilities, *AdaLnet* and *Net-Cox* discriminate the risk groups better than *Fastcox*.

We performed the same analysis for *high*-and-*low risk* patients in the other two datasets. In the lung cancer dataset, we noticed that even though the Kaplan-Meier survival curves generated by the three methods are well separated, the p-value is not significant. On the other hand, in the breast cancer dataset, the survival probabilities for *high*-and-*low risk* patients result not separated.

To further understand the role of the network information in cross-validation and to overcome the drawbacks of investigating only one split, in future studies we will split the dataset using a cross-validation based method for estimating the survival distribution of two or more survival risk groups. All the patients classified as *low-risk* and *high-risk* in every loop of the cross-validation are grouped together and a single Kaplan-Meier curve is computed for each group [14].

4.3   Genes and Subnetworks Selected

As mentioned in the beginning, one of the aim of this paper is to find the pathways and the genes selected by the analyzed methods in different types of cancer (breast, ovarian and lung cancer). This study is expected to produce high-quality and well-curated data because of the structure of the different methods. We applied each penalized Cox regression method to the datasets described in

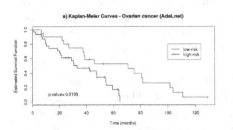
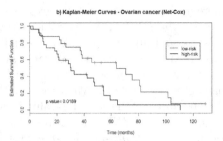

Fig. 1. Cross-ovarian dataset survival prediction. The patients are divided in *high-risk* and *low-risk* groups based on the selected pathways and genes. The survival probabilities of these two groups are compared using the log-rank test. a) By using *AdaLnet* the p-value means the two groups are well separated and the pathways and genes are significant; b) by using *Net-Cox* (functional linkage network) we note that even if the p-value is significant, one patient of the *high-risk* group falls in the *low-risk* group.

Table 2. Number of genes selected by the three methods.

Datasets	Net-Cox	AdaLnet	Fastcox
Breast	122	38	26
Lung	111	61	4
Ovarian	119	308	12

Section 4.1. Here, we present the KEGG networks associated to the non-isolated genes (subnetworks) simultaneously selected by the three methods (Fig.2). The number of genes selected by each method is shown in Table 2. In particular, since *Net-Cox* is a method based on ridge regression, the genes are only shrinkaged and it is necessary to fix a threshold to select the most relevant ones. We fixed the threshold at the 95*th* percentile of the regression coefficients to determine the number of genes showed in Table 2 for *Net-Cox*. We observed that *AdaLnet* identified many more genes and edges on the KEGG network than *Net-Cox* and *Fastcox* for the ovarian cancer dataset, while *Net-Cox* selected many more genes and edges than *AdaLnet* and *Fastcox* for the breast and lung cancer dataset.

In the breast cancer dataset, a subnetwork of the cancer pathway *M12868* was selected, including the *DAPK1* and *RALA* genes strictly involved in cell apoptosis and differentiation (Fig.2(a)). The other two subnetworks are part of the extracellular matrix (ECM) receptor interaction (*M7098*) and focal adhesion (*M7253*) pathways. Both of them are related to important biological processes including cell motility, cell proliferation, cell differentiation, regulation of gene expression and cell survival.

In the lung cancer dataset, *Fastcox* selected only four isolated genes (*CCL22, CSNK1D, HUWE1* and *SLC1A2*). Hence, Fig. 2(b), which reports the not isolated genes, represents the subnetworks selected only by *Net-Cox* and *AdaLnet* in the lung cancer dataset. The gene *IGF1R* appeared in *M12868* which is a

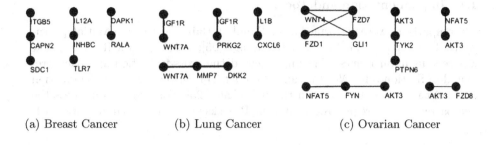

(a) Breast Cancer   (b) Lung Cancer   (c) Ovarian Cancer

Fig. 2. KEGG Subnetworks. The figure shows the subnetworks of the KEGG pathways simultaneously identified by the three algorithms. Only not isolated genes are shown. Figures (a) and (c) represent the subnetworks selected by all the three methods in the breast and ovarian cancer datasets respectively. Since *Fastcox* selected just 4 isolated genes in the lung cancer dataset, (b) shows the subnetworks simultaneously identified only by *Net-Cox* and *AdaLnet*.

well known pathway in cancer. Indeed, *IGF1R* plays an important role in cancer since it is highly overexpressed in most malignant tissues where it functions as an anti-apoptotic agent by enhancing cell survival. Gene *WNT7A*, encodes proteins that are implicated in oncogenesis [28]. The three-node subnetwork *WNT7A–MMP7–DKK2* is part of the WNT signaling pathway (*M19428*) and it is strictly related to the WNT proteins involved in cancer. Finally, the subnetwork *IL1B–CXCL6* is part of the cytokine-cytokine receptor interaction pathway (*M9809*) which is crucial for intercellular regulators and mobilizers of cells engaged in adaptive inflammatory host defenses, cell growth, differentiation, cell death and angiogenesis.

Applying the methods to the ovarian cancer dataset, 5 KEGG subnetworks were selected (Fig.2(c)). The largest connected component is part of the basal cell carcinoma pathway (*M17807*) which includes the *WNT4* gene. This gene is structurally related to genes encoding secreted signaling proteins and it has been implicated in oncogenesis and in several developmental processes, including regulation of cell fate and patterning during embryogenesis. *GLI1* is a gene that encodes a transcription activator involved in oncogene development [26]. The other two genes involved in this subpathway, *FZD1* and *FZD7*, are receptors for *WNT* signaling proteins. The most relevant subnetwork is the one including *AKT3*, *TYK2* and *PTPN6* genes and it is part of the Jak-STAT signaling pathway (*M17411*). This pathway is one of the core ones suggested by [27] and it is the principal signaling mechanism for a wide array of cytokines and growth factors. The subnetwork *AKT3–FZD8* is part of the cancer pathway *M12868* and both the genes are known to be regulators of cell signaling in response to growth factors. They are involved in a wide variety of biological processes including cell proliferation, differentiation, apoptosis, tumorigenesis. The other two subnetworks are related to the T and B cell receptors signaling pathway which are important components of adaptive immunity.

4.4 Implementation and Tools

All comparisons were performed using R and Matlab. *Net-Cox* is a Matlab package available at [24]; *Fastcox* is a R package [25]; *AdaLnet* is a R code and it
was sent us upon request. We implemented the cross-validation approach presented in Section 3 for *Net-Cox* and *AdaLnet*. For *Fastcox* we used the function
`cv.cocktail()` implemented in the R package [25]. For real data analysis the
microarray data were preprocessed using R packages, as described in Section 4.1.

5 Discussion and Conclusions

A central problem in genomic research is to identify genes and pathways involved in cancer in order to create a prediction model linking high-dimensional
genomic data and clinical outcomes. In cancer genomic, gene expression levels
provide important molecular signatures which can be useful to predict the survival of cancer patients. Since gene expression data are characterized by a small
set of samples and a large number of variables, the main challenge is to cope
with the high-dimensionality and the high-correlation among genes (genes are
not independent). To tackle this problem, various network penalized Cox proportional hazards models have been proposed. In this paper, we have compared
three methods for the analysis of microarray gene expression data in order to better understand the disease's mechanism. Moreover a grouped/network approach
[19] can help us to: (*i*) identify core pathways and significant genes within those
pathways related to cancer survival; (*ii*) build a predictive model for survival
of future patients based on the identification genetic signatures. Furthermore,
this kind of analysis is important to understand how patients' features (i.e., age,
gender and coexisting diseases-comorbidity [15]) can influence cancer treatment,
detection and outcome.

The Cox model has achieved widespread use in the analysis of time-to-event
data with censoring and covariates. The covariates, for example a treatment or
other exposure, may change their values over time. It seems natural and appropriate to use the covariate information that varies over time in an appropriate
statistical model. One method of doing this is the time-dependent Cox model.
The form of a time-dependent covariate is much more complex than in Cox
models with fixed (time-independent) covariates. It involves the use of a time
dependent function. However, the use of time-dependent covariates offers several opportunities for exploring associations and potentially causal cancer mechanisms. The evolutionary patterns of cancer disease trajectories across different
stages and cell heterogeneities provide an effective explanation of the remodula-
tion of disease markers, i.e., the emergence of new disease markers or the change
of weight of existing one inside a group of markers induced by changes in phase
of the disease or the presence of comorbidity states induced by drugs/therapies
or other diseases. We will investigate such problems in future studies.

Acknowledgements. This research was partially supported by BioforIU Project and by InterOmics Project. We would like to thank Prof. Hokeun Sun for sharing AdaLnet code.

References

1. Chen, R., Khatri, P., Mazur, P.K., Polin, M., Zheng, Y., Vaka, D., Hoang, C.D., Shrager, J., Xu, Y., Vicent, S., Butte, A., Sweet-Cordero, E.A.: A meta-analysis of lung cancer gene expression identifies PTK7 as a survival gene in lung adenocarcinoma. Cancer Res. 74, 2892–2902 (2014). Published OnlineFirst March 20, doi: 10.1158/0008-5472.CAN-13-2775
2. Cox, D.R.: Regression models and life-tables (with discussion). J. Roy. Stat. Soc. Ser. B 34, 187–220 (1972)
3. Engler, D., Li, Y.: Survival analysis with high-dimensional covariates: An application in microarray studies. Stat. Appl. Genet. Mol. Bio. 8, Article 14 (2009)
4. Gentleman, R.C., Carey, V.J., Bates, D.M., Bolstad, B., Dettling, M., Dudoit, S., Ellis, B., Gautier, L., Ge, Y., Gentry, J., Hornik, K., Hothorn, T., Huber, W., Iacus, S., Irizarry, R., Leisch, F., Li, C., Maechler, M., Rossini, A.J., Sawitzki, G., Smith, C., Smyth, G., Tierney, L., Yang, J.Y., Zhang, J.: Bioconductor: open software development for computational biology and bioinformatics. Genome Biol. 5(10), R80 (2004)
5. Gui, J., Li, H.: Penalized Cox regression analysis in the high-dimentional and low-sample size setting, with applications to microarray gene expression data. Bioinformatics 21, 3001–3005 (2005)
6. Huttenhower, C., Haley, E.M., Hibbs, M.A., Dumeaux, V., Barrett, D.R., Coller, H.A., Troyanskaya, O.G.: Exploring the human genome with functional maps. Genome Research 19(6), 1093–1106 (2009)
7. Yang, Y., Zou, H.: A cocktail algorithm for solving the elastic net penalized Cox's regression in high dimensions. Statistics and Its Interface 6, 167–173 (2013)
8. Kanehisa, M., Goto, S.: KEGG: Kyoto Encyclopedia of Genes and Genomes. Nucleic Acids Res. 28, 27–30 (2000)
9. Kanehisa, M., Goto, S., Sato, Y., Kawashima, M., Furumichi, M., Tanabe, M.: Data, information, knowledge and principle: back to metabolism in KEGG. Nucleic Acids Res. 42, D199–D205 (2014)
10. Li, C., Li, H.: Network-constrained regularization and variable selection for analysis of genomic data. Bioinformatics 24(9), 1175–1182 (2008)
11. Li, C., Li, H.: Variable selection and regression analysis for graph-structured covariates with an application to genomics. Ann. Appl. Stat. 4, 1498–1516 (2010)
12. Nagalla, S., Chou, J.W., Willingham, M.C., Ruiz, J., Vaughn, J.P., Dubey, P., Lash, T.L., Hamilton-Dutoit, S.J., Bergh, J., Sotiriou, C., Black, M.A., Miller, L.D.: Interactions between immunity, proliferation and molecular subtype in breast cancer prognosis. Genome Biology 14, R34 (2013)
13. Simon, N., Friedman, J., Hastie, T., Tibshirani, R.: Regularization Paths for Coxs Proportional Hazards Model via Coordinate Descent. J. Stat. Soft. 39, 1–13 (2011)
14. Simon, R.M., Subramanian, J., Li, M.C., Menezes, S.: Using cross-validation to evaluate predictive accuracy of survival risk classifiers based on high-dimensional data. Briefings in Bioinformatics 12, 203–214 (2011)
15. Sogaard, M., Thomsen, R.W., Bossen, K.S., Sorensen, H.T., Norgaard, M.: The impact of comorbidity on cancer survival: a review. Clinical Epidemiology 5, 3–29 (2013)

16. Sun, H., Lin, W., Feng, R., Li, H.: Network-Regularized high-dimensional cox regression for analysis of genomic data. Statistica Sinica 24, 1433–1459 (2014)
17. Tibshirani, R.: Regression shrinkage and selection via the lasso. J. Roy. Statist. Soc. Ser. B 58, 267–288 (1996)
18. Tibshirani, R.: The lasso method for variable selection in the Cox model. J. Roy. Stat. Med. 16, 385–395 (1997)
19. Wu, T.T., Wang, S.: Doubly regularized Cox regression for high-dimensional survival data with group structures. Statistics and Its Interface 6, 175–186 (2013)
20. Zhang, W., Ota, T., Shridhar, V., Chien, J., Wu, B., Kuang, R.: Network-based Survival Analysis Reveals Subnetwork Signatures for Predicting Outcomes of Ovarian Cancer Treatment. PLoS Comput. Bio. 9(3), e1002975 (2013). doi:10.1371/journal.pcbi.1002975
21. Zou, H., Hastie, T.: Regularization and variable selection via the elastic net. J. Roy. Stat. Soc. Ser. B 67, 301–320 (2005)
22. van Houwelingen, H.C., Bruinsma, T., Hart, A.A.M., van't Veer, L.J., Wessels, L.F.A.: Cross-validated Cox regression on microarray gene expression data. Stat. Med. 25, 3201–3216 (2006)
23. Wu, Y.: Elastic net for Cox's proportional hazards model with a solution path algorithm. Statist. Sinica 22, 271–294 (2012)
24. http://compbio.cs.umn.edu/Net-Cox/
25. http://code.google.com/p/fastcox/
26. Liu, C.Z., Yang, J.T., Yoon, J.W., Villavicencio, E., Pfendler, K., Walterhouse, D., Iannaccone, P.: Characterization of the promoter region and genomic organization of GLI, a member of the Sonic hedgehog-Patched signaling pathway. Gene 209(1-2), 1–11 (1998)
27. Jones, S., Zhang, X., Parsons, D.W., Lin, J.C., Leary, R.J., Angenendt, P., Mankoo, P., Carter, H., Kamiyama, H., Jimeno, A., Hong, S.M., Fu, B., Lin, M.T., Calhoun, E.S., Kamiyama, M., Walter, K., et al.: Core signaling pathways in human pancreatic cancers revealed by global genomic analyses. Science 321(5897), 1801–1806 (2008)
28. Ikegawa, S., Kumano, Y., Okui, K., Fujiwara, T., Takahashi, E., Nakamura, Y.: Isolation, characterization and chromosomal assignment of the human WNT7A gene. Cytogenetic and Genome Research 74(1-2), 149–152 (1996)

Improving Literature-Based Discovery with Advanced Text Mining

Anna Korhonen[1], Yufan Guo[1,2], Simon Baker[1], Meliha Yetisgen-Yildiz[2],
Ulla Stenius[3], Masashi Narita[4], and Pietro Liò[1]

[1] Computer Laboratory, University of Cambridge
15 JJ Thomson Avenue, Cambridge CB3 0FD, UK
alk23,yg244,sb895,pl219@cam.ac.uk
[2] Biomedical and Health Informatics, School of Medicine, University of Washington
Box 358047 Seattle, WA 98109, USA
melihay@uw.edu
[3] Institute of Environmental Medicine, Karolinska Institute
SE-171 77 Stockholm, Sweden
Ulla.Stenius@ki.se
[4] Cancer Research UK Cambridge Institute, University of Cambridge
Li Ka Shing Centre, Robinson Way, Cambridge, CB2 0RE, UK
masashi.narita@cruk.cam.ac.uk

Abstract. Automated Literature Based Discovery (LBD) generates new knowledge by combining what is already known in literature. Facilitating large-scale hypothesis testing and generation from huge collections of literature, LBD could significantly support research in biomedical sciences. However, the uptake of LBD by the scientific community has been limited. One of the key reasons for this is the limited nature of existing LBD methodology. Based on fairly shallow methods, current LBD captures only some of the information available in literature. We discuss how advanced Text Mining based on Information retrieval, Natural Language Processing and data mining could open the doors to much deeper, wider coverage and dynamic LBD better capable of evolving with science, in particular when combined with sophisticated, state-of-the-art knowledge discovery techniques.

1 Scientific Background

The volume of scientific literature has grown dramatically over the past decades, particularly in rapidly developing areas such as biomedicine. PubMed (the US National Library of Medicine's literature service) provides now access to more than 24M citations, adding thousands of records daily[1]. It is now impossible for scientists working in biomedical fields to read all the literature relevant to their field, let alone relevant adjacent fields. Critical hypothesis generating evidence is often discovered long after it was first published, leading to wasted research time and resources [20]. This hinders the progress on solving fundamental problems

---

[1] PubMed: http://www.ncbi.nlm.nih.gov/pubmed

© Springer International Publishing Switzerland 2015
C. di Serio et al. (Eds.): CIBB 2014, LNCS 8623, pp. 89–98, 2015.
DOI: 10.1007/978-3-319-24462-4_8

such as understanding the mechanisms underlying diseases and developing the means for their effective treatment and prevention.

Automated Literature Based Discovery (LBD) aims to address this problem. It generates new knowledge by combining what is already known in literature. It has been used to identify new connections between e.g. genes, drugs and diseases and it has resulted in new scientific discoveries, e.g. identification of candidate genes and treatments for illnesses [6,21].

Facilitating large-scale hypothesis testing and generation from huge collections of literature, LBD could significantly support scientific research [15]. However, based on fairly shallow techniques (e.g. dictionary matching) current LBD captures only some of the information available in literature. Enabling automatic analysis and understanding of biomedical texts via techniques such as Natural Language Processing (NLP), advanced Text Mining (TM) could open the doors to much deeper, wider coverage and dynamic LBD better capable of evolving with science. The last decade has seen massive application of such methodology to biomedicine and has produced tools for supporting important tasks such as literature curation and the development of semantic data-bases [20,19]. Although advanced TM could similarly support LBD, little work exists in this area, e.g. [25].

In this paper we discuss the state of the art of LBD and the benefits of an approach based on advanced TM. We describe how such an approach could greatly improve the capacity of LBD, in particular when combined with sophisticated knowledge discovery techniques. We illustrate our discussion by highlighting the potential benefit in the literature-intensive area of cancer biology. Since LBD is of wide interest and its potential applications are numerous, improved LBD could, in the future, support scientific discovery in a manner similar to widely employed retrieval and sequencing tools.

2   Materials and Methods

2.1   Literature-Based Discovery: The State of the Art

Literature-based discovery was pioneered by Swanson [22] who hypothesised that the combination of two separately published premises "A causes B" and "B causes C" indicates a relationship between A and C. He discovered fish oil as treatment for Raynaud's syndrome based on their shared connections to blood viscosity in literature. Since then, considerable follow-up research has been conducted on LBD (see [6] for a recent review).

LBD has been used for both closed and open discovery. Closed discovery (i.e. hypothesis testing) assumes a potential relationship between concepts A and C and searches for intermediate concepts B that can bridge the gap between A and C and support the hypothesis. It can help and find an explanation for a relationship between two concepts. Open discovery (i.e. hypothesis generation), in contrast, takes as input concept A and aims to identify a set of concepts C that are likely to be linked to A via an intermediate concept B. It can, for example, be used to find new treatments for a given disease or new applications for an existing drug.

The first step of LBD is to identify the concepts of interest (e.g. genes, diseases, drugs) in literature. Most current systems use dictionary-based matching for this. The MetaMap tool (http://metamap.nlm.nih.gov/) which identifies biomedical concepts by mapping text to the Unified Medical Language System (UMLS) Metathesaurus [7] is a popular choice. Unfortunately, the dictionary-based method suffers from poor coverage because it cannot find linguistically complex concepts (e.g. event-like concepts describing biomedical processes), concepts indicated by anaphoric expressions (e.g. pronouns or anaphoric expressions spanning sentences) or newly introduced concepts still missing in dictionaries.

The second step of LBD is to discover relations between concepts. This is typically done using co-occurrence statistics. However, since most co-occurring concepts are unrelated, this simple approach is error-prone and also fails to explain *how* two concepts might be related (e.g. that there is an *interaction* or *activation* relationship between them, or possibly a negative association). Semantic filtering based on relations in a thesaurus such as UMLS can help [6] but suffers from the limitations of dictionary-based approaches. While use of advanced text mining could enable the discovery of novel concepts and relations in context, it remains relatively unexplored in LBD [25].

For knowledge discovery, most systems use Swanson's ABC model or its extensions, e.g. concept chains [8], network analysis [16], and logical reasoning [24] (see [21] for a survey of such extensions). For a concept pair A and C, these models identify the most obvious B and return a ranking of pairs using measures such as average minimum weight, linking term count and and literature cohesiveness [26]. Based on partial B evidence only, these models are not optimally accurate and also do not produce data suitable for statistical hypothesis testing. The latter would be valuable for users of LBD as it could guide them towards highly confident hypotheses.

Evaluation of LBD is challenging as successful techniques discover knowledge that is not proven valuable at the time of discovery. Metrics for direct system comparisons are now available [26] and some existing techniques have been integrated in practical LBD tools which have been made freely available to scientists. Examples of such tools include Arrowsmith [23], BITOLA [5], Semantic MEDLINE [1], and FACTA+ [25], among others. These tools have been used to generate new scientific discoveries (e.g. candidate genes for Parkinson's disease, a link between hypogonadism and diminished sleep quality); see [6] and [21] for recent reviews. However, confirmation of such discoveries via actual laboratory experiments remains rare.

Due to combination of these factors, LBD is not in wide use yet, despite its recognised potential for scientific research [15]. Although closer engagement with end-users, better consideration of end-users needs, and increased validation of findings in the context of laboratory experiments is needed, the fundamental bottleneck lies in the current LBD methodology which suffers from poor coverage as it is capable of identifying only some of the relevant information in literature.

2.2 Advanced Text Mining

LBD could be greatly improved via use of advanced TM. Combining methodology from Information Retrieval (IR), NLP and data mining, TM aims to automatically identify, extract and discover new information in written texts [20,19]. It can be used to organise vast amounts of unstructured textual data now generated through economic, academic and social activities into structured forms that are easily accessible and intuitive for users [15].

Given the rapid growth of scientific literature in biomedicine, biomedical TM has become increasingly popular over the past decade. Basic resources (e.g. lexicons, databases, annotated corpora, datasets) and NLP techniques such as part-of-speech (POS) tagging (i.e. classifying words) and parsing (i.e. analysing the syntactic structure of sentences) have been developed for biomedicine. IR (i.e. identification of relevant documents) and Information Extraction (IE) (i.e. identification of specific information in documents) is now developed, and relatively accurate techniques are now available for identification of named entities (e.g. concept name such as protein names, e.g. AntP), relations (e.g. specific interactions between AntP and BicD), and events (i.e. identifying facts about named entities, e.g. that the AntP protein represses BicD, repress(AntP,BicD)) in texts. Progress has also been made on increasingly complex tasks such as biological pathway or network extraction [10]. Not only direct evaluations against gold standard datasets but also evaluations in the context of practical tasks such as literature curation, literature review and semantic enrichment of networks have produced promising results, highlighting the great potential of deep TM in supporting biomedicine [20,19,9].

Much of recent TM research has focussed on enhancing TM further for demanding real-life tasks. In terms of accuracy, TM is challenged by the linguistic nature of biomedical texts. The biomedical language is characterized by heavy use of terminology and long sentences that have high informational and structural complexity (e.g. complex co-referential links and nested and/or inter-related relations). In addition, the mapping from the surface syntactic forms to basic semantic distinctions is not straightforward. For example, the same relation of interest may be expressed by nominalizations (e.g. phosphorylation of GAP by the PDGF receptor) and verbal predications (e.g. X inhibits/phosphorylates Y) which may not be easy to recognize and relate together.

NLP techniques such as statistical parsing and anaphora resolution which yields richer representations (e.g. internal structure of nominalisations, co-referential links in texts such as it, the protein, the AntP protein) are not challenged to the same extent as shallow techniques are [20,19]. Integration of lexical, semantic, and discourse analysis could help and improve accuracy further [4,12].

In terms of portability, TM has traditionally relied on expensive, manually developed resources (e.g. corpora consisting of thousands of sentences annotated for events by linguists) which are expensive to develop and therefore available for a handful of areas only (e.g. molecular biology, chemistry). Due to strong subdomain variation resources developed for one area are not directly applicable to others [13]. Researchers are now improving the adaptability of TM by reducing the need for manual annotations via minimally supervised machine learning [3] and use of declarative expert (e.g. task, domain) knowledge in guiding learning [4]. Because text mining components typically build on each other, traditional systems have a pipeline architecture where errors tend to propagate from one level to another. Leveraging mutual disambiguation among related tasks and avoiding error propagation, joint learning and inference of various TM tasks is also gaining popularity and has been shown to further improve accuracy [17].

2.3   Towards LBD Based on Advanced Text Mining

Based on much deeper analysis and understanding of texts, advanced TM could enable considerably more accurate, broader and dynamic LBD than current, largely dictionary-based methods. While this potential has been recognized, e.g. [15,6], very little work has been done on TM-based LBD, e.g. [25]. For long, application of TM to LBD has been challenged by the interdisciplinary nature of biomedical research - the fact that research in one area draws increasingly on that in many others, while TM has been typically optimised to perform well in a clearly defined area. However, given recent developments in the field aimed at optimising both accuracy and portability of TM (see the developments discussed in the section above), the approach is now ripe for application in real-life LBD. Whilst TM is challenging by nature and will not produce fully accurate output, errors can be reduced e.g. via statistical filtering to produce maximally accurate input to LBD. Filtering has proved effective in previous works which have demonstrated the usefulness of adaptive TM for practical tasks in biomedicine, e.g. [19,3]

The use of such enhanced, adaptive TM will enable targeting not only basic concepts (i.e. terms or named entities) like most previous LBD, but also complex concepts describing biomedical processes (i.e. events), and relations between concepts in diverse biomedical literature. The latter can be used to restrict search space by permitting direct connections only between concepts which are involved in specific relations [6]. All this information can be learned dynamically from relevant biomedical literature as science evolves, and LBD can be performed on the resulting complex network of concepts.

Open and closed LBD from such rich, TM-based data could also benefit from improved methodology for knowledge discovery. This methodology could be based on recent data mining techiques which enable considering all the intermediate concepts between target concepts. Just one example method is link prediction in complex networks [14] which has been applied successfully to to related problems in social network analysis [11] and web mining [2]. In comparison with most current LBD which is based on extensions of Swanson's ABC model [21] and considers only the most obvious intermediate concepts, such enhanced techniques could provide improved estimate of links between concepts. They could also generate data needed for calculating the likelihood of different concept pairs using statistical tests. This can be highly useful for scientists as it enables them to focus on highly confident hypotheses.

3   A Case Study in Cancer Biology

To illustrate the benefit of TM-based LBD we will describe how such an approach could be used to support the rapidly growing, literature-intensive area of cancer biology. Cancer biology is one of the "interdisciplinary" areas of biomedicine where knowledge discovery draws from advances made in a variety of sub-domains (rather than one well-defined sub-domain) of the field. This makes it particularly difficult for scientists to keep on top of all the relevant literature and highlights the need for automated LBD. From the perspective of TM-based LBD, an area such as cancer biology offers the research challenges needed for the development of adaptive TM technology as many sub-domains involved do not have annotated datasets that could be used for full supervision of systems.

The starting point is to gather relevant literature via PubMed – for example, all the MEDLINE abstracts and freely available full text articles from journals in relevant sub-areas of biomedicine (e.g. cell biology, toxicology, pharmacology, and medicine, among others). The resulting texts will be cleaned and processed using sophisticated NLP techniques such as part-of-speech tagging, parsing, semantic and discourse processing. Concepts of relevance to cancer research (e.g. cancer types, genes, proteins, drugs, physiological entities, symptoms, hallmarks of cancer) will then be extracted from NLP-processed texts, along with relations of interest (e.g. physical, spatial, functional, temporal) between the concepts.

While LBD that uses dictionary-based techniques can find mentions of simple concepts (e.g. gene names) and their known synonyms, TM can also find mentions of concepts "hidden" in anaphoric expressions, those appearing in complex linguistic constructions and those missing in resources such as UMLS, yielding more complete information for LBD. The concepts and relations would be extracted from rich NLP-annotated data using minimally supervised, adaptive TM-techniques. In the absence of relevant in-domain training data, TM can be guided by use of expert knowledge (e.g. constraints that capture task knowledge [4]) and joint inference of related tasks [17].

The network of concepts emerging from TM will be richer than that created by traditional methods. While it will also be noisier due to the challenging nature of advanced NLP and TM, previous work has demonstrated that the impact of noise on practical tasks, in particular after applying statistical noise filtering, will be minimal and unlikely to affect the usefulness of TM. Finally, sophisticated knowledge discovery techniques, e.g. [14], will be applied to the resulting complex network of concepts to conduct maximally accurate closed and/or open discovery.

Figure 1 illustrates how such a TM-based LBD tool could be used to support cancer biology. It shows an example that focuses on anti-carcinogenic effects of statins. Statins are known to have anti-carcinogenic properties but the underlying mechanism by which these drugs prevent cancer is not fully understood [18]. This problem can be studied by investigating whether specific proteins and hallmarks of cancer act as intermediate concepts between statins and different cancer cell types and if yes, whether such concepts could help to explain the mechanism. In the case study illustrated in Figure 1, cancer biologists use a TM-based

LBD tool for closed discovery to investigate the question *In which way do statins prevent prostate cancer?* The given concepts are

Concept A: Drug: Statin
Concept C: Cancer type: Prostate cancer

The tool will

1. gather literature: PubMed articles on "statin" and "prostate cancer",
2. identify Concepts B (Hallmarks, Proteins) in the resulting literature using TM,
3. build a concept map for Concepts A, B and C,
4. return B that link to both A and C.

The tool will also identify relevant relations between concepts:

Interacts with (Statin, Akt kinase)
Causes (Akt kinase, Prostate cancer)
Prevents (Statin, Sustaining proliferative signaling)
Causes (Sustaining proliferative signaling, Prostate cancer)
Exhibits (Akt kinase, Sustaining proliferative signaling)

The answer emerging from the tool is that *statins prevent prostate cancer by inhibiting cell proliferation via Akt kinase.*

To be useful, such TM-based technology should be integrated in a practical tool aimed at supporting cancer researchers in LBD. The tool should allow uploading articles of interest e.g. via PubMed, performing open and close discovery using a set of queries to define the scope of interest in terms of concepts and relations, visualising the results and the statistical trends in the data, and navigating

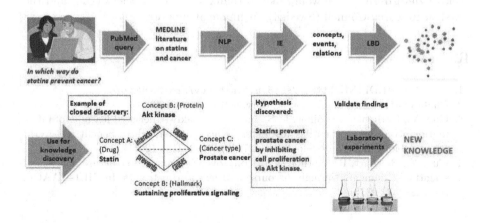

Fig. 1. TM-based LBD for cancer biology. The figure illustrates how LBD can discover the mechanism by which statins prevent prostate cancer.

through individual articles, highlighting the scientific evidence in its actual context. Such a tool should be developed in close collaboration with scientists to ensure optimal integration with existing research practices.

Finally, any new hypotheses or discoveries resulting from LBD should ideally be confirmed experimentally by scientists. In the case of cancer biology, one might validate the promising findings from LBD experimentally in vitro according to their nature. Such experimentation and subsequent publication in relevant journals can encourage the uptake of LBD by the research community, leading to further benefits.

4   Conclusion

In biomedicine a number of LBD tools have been developed to support the testing and discovery of research hypotheses in scientific literature. Although such tools could, in principle, greatly support scientists in their work, their uptake has remained limited. We have highlighted a number of issues that act as barriers to wider exploitation of LBD in scientific research, and have focused, in particular, on limitations related to the current LBD methodology. We have explained how use of advanced TM could enable the discovery of much richer information in scientific texts than is possible using current largely dictionary-based methods. This potential has been previously recognised, but TM has only recently reached the point where it can be realistically applied to diverse literature without costly creation of manually annotated in-domain training data. While the development of a fully optimal LBD approach based on TM will require considerable research effort, it is now realistic – and looking into the future, the approach could open the doors to much wider coverage LBD capable of better evolving with the development of biomedical science.

Acknowledgments. We would like to thank the Royal Society (UK) and the Swedish Research Council (Sweden) for financial support.

References

1. Semantic MEDLINE. http://skr3.nlm.nih.gov/Sem-MedDemo/
2. Chakrabarti, S.: Mining the web. Morgan Kaufmann (2002)
3. Guo, Y., Korhonen, A., Silins, I., Stenius, U.: Weakly supervised learning of information structure of scientific abstractsis it accurate enough to benefit real-world tasks in biomedicine? Bioinformatics 27(22), 3179–3185 (2011)
4. Guo, Y., Reichart, R., Korhonen, A.: Improved information structure analysis of scientific documents through discourse and lexical constraints. In: HLT-NAACL, pp. 928–937 (2013)

5. Hristovski, D., Peterlin, B., Mitchell, J.A., Humphrey, S.M.: Using literature-based discovery to identify disease candidate genes. International Journal of Medical Informatics 74(2), 289–298 (2005)
6. Hristovski, D., Rindflesch, T., Peterlin, B.: Using literature-based discovery to identify novel therapeutic approaches. Cardiovascular & Hematological Agents in Medicinal Chemistry (Formerly Current Medicinal Chemistry-Cardiovascular & Hematological Agents) 11(1), 14–24 (2013)
7. Humphreys, B.L., Lindberg, D.: The UMLS project: making the conceptual connection between users and the information they need. Bulletin of the Medical Library Association 81(2), 170 (1993)
8. Jin, W., Srihari, R.K., Ho, H.H., Wu, X.: Improving knowledge discovery in document collections through combining text retrieval and link analysis techniques. In: Seventh IEEE International Conference on Data Mining, ICDM 2007, pp. 193–202 (2007)
9. Kadekar, S., Silins, I., Korhonen, A., Dreij, K., Al-Anati, L., Högberg, J., Stenius, U.: Exocrine pancreatic carcinogenesis and autotaxin expression. PloS One 7(8), e43209 (2012)
10. Li, C., Liakata, M., Rebholz-Schuhmann, D.: Biological network extraction from scientific literature: state of the art and challenges. Briefings in Bioinformatics 15(5), 856–877 (2014)
11. Liben-Nowell, D., Kleinberg, J.: The link prediction problem for social networks. In: Proceedings of the Twelfth International Conference on Information and Knowledge Management, pp. 556–559 (2003)
12. Lippincott, T., Rimell, L., Verspoor, K., Korhonen, A.: Approaches to verb subcategorization for biomedicine. Journal of Biomedical Informatics 46(2), 212–227 (2013)
13. Lippincott, T., Séaghdha, D.Ó., Korhonen, A.: Exploring subdomain variation in biomedical language. BMC Bioinformatics 12(1), 212 (2011)
14. Lü, L., Zhou, T.: Link prediction in complex networks: A survey. Physica A: Statistical Mechanics and its Applications 390(6), 1150–1170 (2011)
15. McDonald, D., Kelly, U.: Value and benefits of text mining. JISC Publications (2012)
16. Özgür, A., Xiang, Z., Radev, D.R., He, Y.: Literature-based discovery of ifn-gamma and vaccine-mediated gene interaction networks. J. Biomed. Biotechnol. 2010, 426479 (2010)
17. Poon, H., Vanderwende, L.: Joint inference for knowledge extraction from biomedical literature. In: HLT-NAACL, pp. 813–821 (2010)
18. Roudier, E., Mistafa, O., Stenius, U.: Statins induce mammalian target of rapamycin (mtor)-mediated inhibition of akt signaling and sensitize p53-eficient cells to cytostatic drugs. Molecular Cancer Therapeutics 5(11), 2706–2715 (2006)
19. Shatkay, H., Craven, M.: Mining the Biomedical Literature. MIT Press (2012)
20. Simpson, M.S., Demner-Fushman, D.: Biomedical Text Mining. Springer US (2012)
21. Smalheiser, N.R.: Literature-based discovery: Beyond the abcs. Journal of the American Society for Information Science and Technology 63(2), 218–224 (2012)
22. Swanson, D.R.: Fish oil, raynaud's syndrome, and undiscovered public knowledge. Perspectives in biology and medicine 30(1), 7–18 (1986)

23. Swanson, D.R., Smalheiser, N.R.: An interactive system for finding complementary literatures: a stimulus to scientific discovery. Artificial intelligence 91(2), 183–203 (1997)
24. Tari, L., Anwar, S., Liang, S., Cai, J., Baral, C.: Discovering drug–drug interactions: a text-mining and reasoning approach based on properties of drug metabolism. Bioinformatics 26(18), 547–553 (2010)
25. Tsuruoka, Y., Miwa, M., Hamamoto, K., Tsujii, J., Ananiadou, S.: Discovering and visualizing indirect associations between biomedical concepts. Bioinformatics 119, 111–119 (2011)
26. Yetisgen-Yildiz, M., Pratt, W.: A new evaluation methodology for literature-based discovery systems. Journal of Biomedical Informatics 42(4), 633–643 (2009)

A New Feature Selection Methodology
for K-mers Representation of DNA Sequences

Giosuè Lo Bosco[1,2] and Luca Pinello[3,4]

[1] Dipartimento di Matematica e Informatica, Universitá degli studi di Palermo, Italy
[2] Dipartimento di Scienze per l'Innovazione e le Tecnologie Abilitanti,
Istituto Euro Mediterraneo di Scienza e Tecnologia, Palermo, Italy
[3] Department of Biostatistics, Harvard School of Public Health, Boston, MA, USA
[4] Department of Biostatistics and Computational Biology,
Dana-Farber Cancer Institute, Boston, MA, USA

Abstract. DNA sequence decomposition into k-mers and their frequency counting, defines a mapping of a sequence into a numerical space by a numerical feature vector of fixed length. This simple process allows to compare sequences in an alignment free way, using common similarities and distance functions on the numerical codomain of the mapping. The most common used decomposition uses all the substrings of a fixed length k making the codomain of exponential dimension. This obviously can affect the time complexity of the similarity computation, and in general of the machine learning algorithm used for the purpose of sequence analysis. Moreover, the presence of possible noisy features can also affect the classification accuracy. In this paper we propose a feature selection method able to select the most informative k-mers associated to a set of DNA sequences. Such selection is based on the *Motif Independent Measure (MIM)*, an unbiased quantitative measure for DNA sequence specificity that we have recently introduced in the literature. Results computed on public datasets show the effectiveness of the proposed feature selection method.

Keywords: k-mers, DNA sequence similarity, feature selection, DNA sequence classification.

1 Introduction

In biology, sequence similarity is traditionally established by using sequence alignment methods, such as BLAST [1] and FASTA [2]. This choice is motivated by two main assumptions: (1) the functional elements share common sequence features and (2) the relative order of the functional elements is conserved between different sequences. Although these assumptions are valid in a broad range of cases, they are not general. For example, in the case of cis-regulatory elements related sequences, there is little evidence suggesting that the order between different elements would have any significant effect in regulating gene expression. Anyway, despite recent efforts, the key issue that seriously limits the application of the alignment methods still remains their computational complexity. As such, the recently developed alignment-free methods [3] have emerged as a promising

The original version of this chapter was revised: The given name and family name of the authors has been corrected. The erratum to this chapter is available at DOI: 10.1007/978-3-319-24462-4_26

© Springer International Publishing Switzerland 2015
C. di Serio et al. (Eds.): CIBB 2014, LNCS 8623, pp. 99–108, 2015.
DOI: 10.1007/978-3-319-24462-4_9

approach to investigate the regulatory genome. One of the methods belonging to this latter class is based on substring counting of a sequence, and is generally named as *k-mers* or *L-tuple* representation. Informally, k-mers representation associates a sequence with a feature vector of fixed length, whose components count the frequency of each substrings belonging to a finite set of words. The main advantage is that the sequence can be represented into a numerical space where a particular distance function between the vectors can be adopted to reflect the observed similarities between corresponding sequences. K-mer representation have shown its effectiveness in several in-silico analysis applied to different genomics and epigenomics studies. In particular they have been used to characterize nucleosome positioning [4], to find enhancer functional regions [5], to characterize epigenetic variability [6], in sequence alignment and transcriptome assembly [7] and in gene prediction [8]. The interested reader can find the basic ideas of k-mer based methods to different biological problems in the following review [17]. Anyway, their use involve practical computational issues: they uses all the substrings of length k making the numerical biological data of exponential dimension [18]. Note that this represents a key problem, bioinformaticians frequently face the challenge of reducing the number of features of high dimensional biological data for improving the models involved in sequence analysis. To this purpose, *feature selection algorithms* can be successfully applied. The main goals of such algorithms are (1) speeding up the response of the model used for the analysis and (2) eliminate the presence of possible noisy features that could affect seriously the accuracy of the model. Commonly, feature selection methods belongs to the class of heuristics since they are based on the generation of a proper subset of features, and this latter problem has been shown to be computationally intractable [19]. There are mainly two classes of feature selection methods, i.e. the so called *wrapper* approaches which uses a predictive model to evaluate the feature subset, and *filter* approaches which score the subset just by looking at the intrinsic properties of data [20]. In this paper we propose a filter feature selection method able to select the most informative k-mers associated to a set of DNA sequences. Such selection is based on the weight assigned to each feature by a measure called *Motif Independent Measure (MIM)*. The effectiveness of the method has been tested on four public datasets with the purpose of sequence classification. In the next Section the methodology is formally described, recalling the definition of MIM and the idea of how using this in order to weight each feature. Finally, in Section 3 datasets, experiments and results are reported.

2   The Proposed Feature Selection Methodology

A generic DNA sequence s of length L can be represented as a string of symbols taken from a finite alphabet. We can think to a particular mapping function that project s into a vector x_s (the feature vector), allowing to represent s into a multi-dimensional space (the feature space) where a particular distance function between the vectors can be adopted to reflect the observed similarities between sequences. One of the most common ways of defining such mapping, is to consider a feature vector x_s that enumerates the frequency of occurrence of a finite set of pre-selected words $W = \{w_i, .., w_m\}$ in the string s. The simplest and most common definition of W is by using k-$mers$, i.e. a set containing any string of length k whose symbols are taken in the nucleotide alphabet $\Sigma = \{A, T, C, G\}$. In this case, each sequence s is mapped to a vector $x_s \in \mathbb{Z}^m$ with $m = 4^k$. The idea behind the proposed feature selection method is to assign a weight to each k-mer, and use this weights for their selection.

2.1   K-mers Weighting

In general, each numerical component x_s^j of the feature vector is set to the value f_s^j that represents the frequency of the $j - th$ k-mer w_k^j in s counted by a sliding window of length k that is run through the sequence s, from position 1 to $L - k + 1$. Another possible choice is to set x_s^j to the empirical probability $p_s^j = f_s^j/(L - k + 1)$.

Specifically, let $\mathbf{P}_S = (p_{s_i}^j)$ be the k-mer probability distributions corresponding to a set of n target sequences $S = \{s_i\}$ for a fixed length k, where $i = 1, .., n$, $j = 1, .., m$. Let us assume to have a process to generate a set of n $background$ sequences $B = \{b_i\}$ (analogously b_i represents a sequence in the set B). In the most simple case, the background set of sequences corresponds to random sequences that can be generated for example by randomly shuffling each one of the sequences belonging to the target set S.

Let $\mathbf{Q}_B = (q_{b_i}^j)$ be the k-mer probability distributions corresponding to B for a fixed length k. For each j, one can calculate the symmetrical Kullback-Leibler divergence between the empirical probabilities P_j and Q_j:

$$d_{kl}(P_j, Q_j) = \frac{\sum_i p_{s_i}^j log_2 \frac{p_{s_i}^j}{q_{b_i}^j} + \sum_i q_{b_i}^j log_2 \frac{q_{b_i}^j}{p_{s_i}^j}}{2} \tag{1}$$

We recall that this divergence is able to measure the difference between two probability distributions. The $Motif$ $Independent$ $Measure$ (MIM) value corresponding to a k-mer w_j is defined as the expected value $d_{kl}(P_j, Q_j)$, which is estimated by averaging over a finite set $N > n$ of background sequences, and is indicated as $MIM(w_j)$.

2.2   K-mers Selection

We can compute the MIM values for each k-mer w_j, obtaining a list $\mathcal{L}$ of $m = 4^k$ numerical values. Here we use their ranking as a guide to identify the most informative k-mers, in particular we sort $\mathcal{L}$ in ascending order, resulting in a ordered list of k-mers $w_{j_1}, .., w_{j_m}$. The criteria used to select the most informative k-mers among the possible 4^k is based on Z-$scores$ of the computed MIM values. Finally, let $A_\alpha = \{w_{j_i} | abs(Z(MIM(w_{j_i}))) > \alpha\}$ where Z indicate Z-score, the adopted selection criteria consists in selecting a number of k-mers equal to

$$r = max(|A_\alpha|, \beta * m) \tag{2}$$

with $\alpha, \beta < 1$. In particular, the α parameter represents the Z-score percentage, while β the percentage of number of features to select.

3   Experimental Results

To test the effectiveness of the proposed methods, we have considered two different domains of application: nucleosome identification and classification of bacteria. Let $S = \{s_1, .., s_n\}$ be the set of sequences and $T = \{t_1, .., t_n\}$ the corresponding preassigned classes, the numerical dataset is always indicated by the matrix D_S^k of size $n \times 4^k$ whose generic element $D_S^k(i,j)$ contains the empirical probabilities $p_{s_i}^j$ of the sequence s_i for a fixed k-mer length k. In all the experiments, we have used a *Support Vecotr Machine (SVM)* [11] with quadratic kernel in order to classify the data. The adopted kernel allows to project the data by a nonlinear function, and differently from other kernels, does not require to estimate any additional parameter. Finally, the classification results have been computed considering as input D_S^k, named the *FULL* dataset, and on three of its reductions acted by three different feature selection schema, named *random background selection (RB), negative background selection (NB)* and *random feature selection (WR)*. The RB selects the features by using the proposed methodology, assuming a k-mers background distribution obtained by first estimating the probability of each single nucleotide in the training set, and then calculating the probability of the j-th k-mer w_j as the product of the probability of its nucleotides. The NB uses the proposed methodology assuming as instead as background the set of negative sequences $B = \{s \in S \mid s \text{ is negative}\}$. Finally, the *(WR)* schema simply reduce the dataset by considering a random permutation of features. In the following, details about data-sets and related experiments will be given.

3.1   Nucleosome Identification

In this study we have considered three datasets of DNA sequences underlying nucleosomes from the following three species: (i) *Homo sapiens (HM)*; (ii) *Caenorhabditis elegans (CE)* and (iii) *Drosophila melanogaster (DM)*. The nucleosome is the primary repeating unit of chromatin, which consists of 147 bp of DNA wrapped 1.67 times around an octamer of core histone proteins [12]. Several studies have shown that nucleosome positioning plays an important role in gene regulation and that distinct DNA sequence features have been identified to be associated with nucleosome positioning [13]. Several specialized computational approach for the identification of nucleosome positioning have been recently proposed, thanks also to the development of genome-wide profiling technologies [14,15]. Details about all the step of data extraction and filtering of the three datasets can be found in the work by Guo et al [16] and in the references therein. Each of the three datasets is composed by two classes of samples: the nucleosome-forming sequence samples (positive data) and the linkers or nucleosome-inhibiting sequence samples (negative data). The HM dataset contains $2,273$ positives and $2,300$ negatives, the CE $2,567$ positives and $2,608$ negatives and the DM $2,900$ positives and $2,850$ negatives. The length of a generic sequence is 147 bp. We have computed the experiments for different k ranging from 5 to 7. Such range has been chosen due to the used classifier, since it has been noted that the SVM with the quadratic kernel does not lead the optimization to converge for $k < 5$. It is important to point out that the literature motivate the use of a k-mer length equal to 6 as a good choice to capture dependencies between adjacent nucleotides.

We have computed a total of 3 metrics to measure the performance of the classifier: *Sensitivity (Se), Specificity (Sp)* and *Accuracy (A)*. In the following, we recall their definitions:

$$Se = \frac{TP}{TP+FN}, \ Sp = \frac{TN}{FP+TN}, \ A = \frac{TP+TN}{TP+FN+FP+TN} \tag{3}$$

where the prefix T (true) indicates the number of correctly classified sequences, F (false) the uncorrect ones, P the positives class and N the negatives class.

In Table 1 we report the results of mean (μ) and standard deviation (σ) of the three metrics for all the three datasets, computed on 10 folds, following a 10 fold cross validation schema.

Results shows that for $k = 5$, in the case of CE and DM datasets the accuracy obtained by the RB and NB feature selection methods is comparable or superior to the $FULL$ and WR cases. Moreover, their sensitivity with respect to the $FULL$ case is improved (at least 4% for CE and at least 2% for DM). In the case of $k = 6$ it is observable a significant increase in sensitivity (at most 40%) and accuracy (at most 18%). Such improvements are not observable for the HM dataset, but both sensitivity and accuracy are comparable to the $FULL$ case. Finally, the use of $k = 7$ seems not the right choice for every considered dataset, but this is more visible for CE and DM. Finally we observed that the proposed feature selection method can decrease slightly the specificity.

3.2 Classification of Bacteria

Studies about bacteria species are based on the analysis of their 16S rRNA housekeeping gene [9], and the analysis of the related sequences has been carried out mainly by using alignment algorithms. In this experiment, the three most populous phyla belonging to the Bacteria domain have been considered: Actinobacteria, Firmicutes and Proteobacteria. The 16S rRNA sequences were downloaded from the RDP Ribosomal Database Project II (RDP-II) [10] and selected according to the type strain, uncultured and isolates source, average length of about $1200 - 1400$ bps, good RDP system quality and *class* taxonomy by NCBI. Finally, from the resulting sequences, 1000 sequences per phylum have been selected, so that we obtained a 16S sequences bacteria dataset of 3000 elements. The resulting number of classes is 3 and is established by the NCBI *phylum* taxonomy. The length of a generic sequence is around 1400 bp. We have computed the experiments for several k-mer lenghts k ranging from 3 to 7. Note that for this dataset, since the number of classes is 3 it is mandatory to use a

Table 1. In column, for each k in the range 5,..,7 the mean and standard deviation values of Specificity (Sp), Sensitivity (Se) and Accuracy (A) values computed on 10 folds in the cases of the Caenorhabditis elegans (CE), Drosophila melanogaster (DM) and Homo sapiens (HM) full (Full) and reduced (WR,RB,NB) datasets. The best values are in bold.

	A		Se		Sp	
K=5	μ	σ	μ	σ	μ	σ
CE-FULL	79,36	0,02	75,89	0,03	**82,78**	0,02
CE-WR	76,37	0,02	73,86	0,04	78,84	0,02
CE-RB	78,17	0,01	79,90	0,02	76,46	0,02
CE-NB	**80,08**	0,02	**80,76**	0,02	79,41	0,03
DM-FULL	**76,73**	0,01	70,59	0,03	**82,98**	0,02
DM-WR	73,72	0,01	69,24	0,02	78,28	0,02
DM-RB	75,51	0,02	**73,03**	0,03	78,04	0,03
DM-NB	76,09	0,01	72,55	0,03	79,68	0,02
HM-FULL	**84,38**	0,02	**91,72**	0,02	77,13	0,04
HM-WR	82,36	0,02	87,67	0,02	77,13	0,03
HM-RB	83,00	0,02	89,25	0,04	76,83	0,02
HM-NB	**84,38**	0,02	90,70	0,02	**78,13**	0,03
K=6	μ	σ	μ	σ	μ	σ
CE-FULL	62,94	0,02	31,01	0,03	**94,36**	0,02
CE-WR	69,97	0,02	49,04	0,03	90,57	0,01
CE-RB	74,82	0,02	63,14	0,04	86,31	0,02
CE-NB	**80,49**	0,02	**73,94**	0,02	86,93	0,02
DM-FULL	74,52	0,02	62,21	0,03	**87,05**	0,02
DM-WR	73,08	0,02	66,28	0,02	80,00	0,02
DM-RB	75,93	0,02	66,17	0,04	85,86	0,02
DM-NB	**78,16**	0,02	**72,86**	0,03	83,54	0,03
HM-FULL	84,81	0,02	91,63	0,02	**78,09**	0,03
HM-WR	83,61	0,01	88,55	0,01	78,74	0,03
HM-RB	84,11	0,01	92,60	0,01	75,74	0,03
HM-NB	**85,03**	0,01	**93,17**	0,01	77,00	0,03
K=7	μ	σ	μ	σ	μ	σ
CE-FULL	50,51	0	0,31	0	**99,92**	0
CE-WR	50,51	0	1,36	0,01	98,89	0,01
CE-RB	51,98	0,01	**5,10**	0,01	98,12	0,01
CE-NB	**50,67**	0,01	2,07	0,01	98,50	0,01
DM-FULL	52,30	0,01	14,31	0,02	90,95	0,01
DM-WR	53,53	0,01	14,41	0,02	93,33	0,03
DM-RB	52,96	0,01	9,41	0,02	**97,26**	0,01
DM-NB	**54,26**	0,02	**18,31**	0,03	90,84	0,02
HM-FULL	48,6	0	0,09	0	**96,48**	0,01
HM-WR	63,28	0,03	32,42	0,07	93,74	0,02
HM-RB	73,68	0,03	73,79	0,10	73,57	0,12
HM-NB	**73,98**	0,02	**93,96**	0,02	54,26	0,03

Table 2. In column, for each k in the range 3,..,7 the mean and standard deviation values of the Accuracy computed on 10 folds in the cases of the full (FULL) and reduced (WR,RB,NB) Bacteria dataset. The best values are in bold.

	K=3		K=4		K=5		K=6		K=7	
	μ	σ	μ	σ	μ	σ	μ	σ	μ	σ
BA-FULL	99,86	0,17	99,96	0,10	99,96	0,10	87,37	2,25	33,33	0
BA-WR	99.69	0,29	99,96	0,10	78,84	0,10	99,96	0,10	**50.33**	1,95
BA-RB	99.73	0,26	99,96	0,10	99,96	0,10	99,96	0,10	33,33	0
BA-NB	**99.86**	0,17	**99,96**	0,10	**100**	0	**100**	0	33,33	0

multi-class SVM paradigm. We have decided to use the so called *One versus all* paradigm, i.e. we have considered 3 binary SVM classifier each one trained on elements of class i as positives and elements of class $not - i$ as negatives. Finally, the test element is assigned to the class which has the maximum distance from the separation hyperplane.

In Table 2 we report the results of mean (μ) and standard deviation (σ) of the accuracy of the multiclass SVM classifier computed on 10 folds, following a 10 fold cross validation schema. Note that in the multiclass case, the accuracy is defined as the percentage of correct classified elements over the total number of elements.

Results shows that for $k = 3, 4, 5, 6$, the accuracy obtained by the NB is comparable or superior to the $FULL$, WR and RB paradigms. This improvement is not observable for $k = 7$, so that we conclude also for this dataset that this is not the right kmer length to choose. You can observe in this latter case that the WR paradigm has reached the best performance. In particular, we note that in this case the classifier process the dataset reduced by WR assigning correctly 50% of elements to Actinobacteria class and assigning all the rest to Firmicutes class, while RB and NB assign each element to Firmicutes class. This can be explained by the consideration that a big number of kmers m, cause their scores to assume similar values making the selection ineffective.

3.3   Empirical Complexity and Computation Time

In order to show the computational complexity advantage introduced by the proposed method, we have also computed the empirical computation time of the classifier for the test sequences in the case of $FULL$, RB, and NB for all the considered dataset. This is shown on Figure 1a,b,c,d. In the same figure, the number of features r used by the classifier for the three cases is also shown on top of each bar. Note that the used values of $\alpha = 0.7$ and $\beta = 0.5$ always reduce the number of features of RB and NB by a factor of 0.5, resulting in a significant reduction of computation time. Note that this experiment regard only the test phase, and we have not taken into account the computation time of the k-mer selection. Anyway, it is straightforward to note (see subsection 2.2) that the theoretical complexity of the selection in the case of RB and NB is linear on $n * 4^k$ where n represents the number of sequences in the set S.

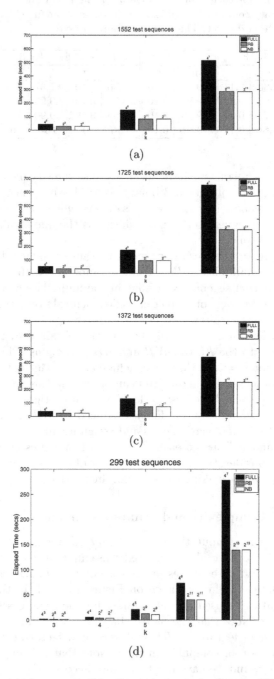

Fig. 1. The empirical computation times in seconds for the C.Elegans (a), D. Melanogaster (b), H. Sapiens (c) and bacteria (d) datasets. On top of each bar the number of features used by the SVM classifier.

4   Conclusion

In this paper we have presented a feature selection method for DNA sequences based on an unbiased quantitative measure for DNA sequence specificity called Motif Independent Measure (MIM). It uses the k-mers counting representation, and selects the most informative k-mers associated to a target set of DNA sequences by computing the *Kullback-Leibler* divergence between the k-mers distributions computed in the target set and in an opportunely generated background set. Results carried out on four datasets, have shown the advantage of the proposed feature selection method in terms of Sensitivity, Accuracy and computation time. In the future we plan to extend the experimental part on other dataset of sequences, also adopting other classification paradigms e.g. using other SVM kernels.

5   Funding

G. Lo Bosco was partially supported by Progetto di Ateneo dellUniversitá degli Studi di Palermo 2012-ATE-0298 *Metodi Formali e Algoritmici per la Bioinformatica su Scala Genomica.*

References

1. Altschul, S., Gish, W., Miller, W., et al.: Basic local alignment search tool. J. Mol. Biol. 25(3), 403–410 (1990)
2. Lipman, D., Pearson, W.: Rapid and sensitive protein similarity searches. Science 227(4693) (1985)
3. Vinga, S., Almeida, J.: Alignment-free sequence comparison–a review. Bioinformatics 19(4), 513–523 (2003)
4. Yuan, G.-C., Liu, J.S.: Genomic sequence is highly predictive of local nucleosome depletion. PLoS Comput. Biol. 4(1), e13 (2008)
5. Lee, D., Karchin, R., Beer, M.A.: Discriminative prediction of mammalian enhancers from DNA sequence. Genome Research 21(12), 2167–2180 (2011)
6. Pinello, L., Xu, J., Orkin, S.H., Yuan, G.-C.: Analysis of chromatin-state plasticity identifies cell-type specific regulators of H3K27me3 patterns. Proceedings of the National Academy of Sciences 111(3), 344–353 (2014)
7. Paszkiewicz, K., Studholme, D.J.: De novo assembly of short sequence reads. Briefings in Bioinformatics 11(5), 457–472 (2010)
8. Liu, Y., Guo, J., Hu, G.-Q., Zhu, H.: Gene prediction in metagenomic fragments based on the svm algorithm. BMC Bioinformatics 14(S-5), S12 (2013)
9. Drancourt, M., Berger, P., Raoult, D.: Systematic 16*S* rRNA Gene Sequencing of Atypical Clinical Isolates Identified 27 New Bacterial Species Associated with Humans. Journal of Clinical Microbiology 42(5), 2197–2202 (2004)
10. https://rdp.cme.msu.edu/
11. Cristianini, N., Shawe-Taylor, J.: An Introduction to Support Vector Machines and other kernel-based learning methods. Cambridge Univ. Press (2000)
12. Kornberg, R.D., Lorch, Y.: Twenty-five years of the nucleosome, fundamental particle of the eukaryote chromosome. Cell 98, 285–294 (1999)

13. Struhl, K., Segal, E.: Determinants of nucleosome positioning. Nat. Struct. Mol. Biol. 20(3), 267–273 (2013)
14. Yuan, G.-C., Liu, Y.-J., Dion, M.F., Slack, M.D., Wu, L.F., Altschuler, S.J., Rando, O.J.: Genome-scale identification of nucleosome positions in S. cerevisiae. Science 309(5734), 626–630 (2005)
15. Di Gesú, V., Lo Bosco, G., Pinello, L., Yuan, G.-C., Corona, D.V.F.: A multi-layer method to study genome-scale positions of nucleosomes. Genomics 93(2), 140–145 (2009)
16. Guo, S.-H., Deng, E.-Z., Xu, L.-Q., Ding, H., Lin, H., Chen, W., Chou, K.-C.: iNuc-PseKNC: a sequence-based predictor for predicting nucleosome positioning in genomes with pseudo k-tuple nucleotide composition. Bioinformatics 30(11), 1522–1529 (2014)
17. Pinello, L., Lo Bosco, G., Yuan, G.-C.: Applications of alignment-free methods in epigenomics. Briefings in Bioinformatics 15(3), 419–430 (2013)
18. Apostolico, A., Denas, O.: Fast algorithms for computing sequence distances by exhaustive substring composition. Algorithms for Molucular Biology 3(13), 1–9 (2008)
19. Kohavi, R., John, G.H.: Wrappers for feature subset selection. Artificial Intelligence 97(1–2), 273–324 (1997)
20. Saeys, Y., Inza, I., Larrañaga, P.: A Review of Feature Selection Techniques in Bioinformatics. Bioinformatics 23(19), 2507–2517 (2007)

Detecting Overlapping Protein Communities in Disease Networks

Hassan Mahmoud[1], Francesco Masulli[1,2], Stefano Rovetta[1],
and Giuseppe Russo[2]

[1] DIBRIS - Dip. di Informatica, Bioingegneria, Robotica e Ingegneria dei Sistemi
Università di Genova, Via Dodecaneso 35, 16146 Genova, Italy
{hassan.mahmoud,francesco.masulli,stefano.rovetta}@unige.it
[2] Sbarro Institute for Cancer Research and Molecular Medicine
College of Science and Technology, Temple University, Philadelphia, PA, USA
grusso@temple.edu

Abstract. In this work we propose a novel hybrid technique for overlapping community detection in biological networks able to exploit both the available quantitative and the semantic information, that we call *Semantically Enriched Fuzzy C-Means Spectral Modularity (SE-FSM) community detection* method. We applied *SE-FSM* in analyzing Protein-protein interactions (*PPIs*) networks of *HIV-1* infection and Leukemia in Homo sapiens. *SE-FSM* found significant overlapping biological communities. In particular, it found a strong relationship between *HIV-1* and Leukemia as their communities share several significant pathways, and biological functions.

Keywords: Community detection, clustering ensemble, semantic enrichment, fuzzy clustering, spectral clustering, modularity.

1 Introduction

Protein-protein interactions (*PPIs*) refer to physical contacts with molecular docking between proteins that occur in a cell or in a living organism. The interaction interface is intentional and evolved for a specific purpose distinct from totally generic functions such as protein production, and degradation [3].

In [11] we proposed the *Fuzzy c-means Spectral Modularity (FSM) community detection* method for network analysis. The FSM estimates the number of communities k^* using the maximization of modularity procedure depicted by Newman and Girvan in [14] and then performs data clustering in a subspace spanned by the first k eigenvectors of the graph Laplacian matrix [22]. The method is based on the spectral clustering approach described in [15], with the main difference consisting in using as a technique of clustering the Fuzzy C-Means [2] that makes it possible to identify significant overlapping protein communities. Moreover, in [21] we showed that using clustering ensemble boosts the quality of communities obtained.

This paper is organized as follows: A novel overlapping centrality measure termed *spreadability* is introduced in Sect. 2; The proposed Semantically Enriched

© Springer International Publishing Switzerland 2015
C. di Serio et al. (Eds.): CIBB 2014, LNCS 8623, pp. 109–120, 2015.
DOI: 10.1007/978-3-319-24462-4_10

Fuzzy c-means Spectral Modularity (SE-FSM) community detection method is illustrated in Sect. 3, while its application to the discovery of communities in the HIV-1 and Leumekia $PPIs$ networks of Homo sapiens is shown in Sect. 4; Sect. 5 contains the conclusions.

2   The Spreadability Measure

Spreadability ξ is a novel measure we propose here for estimating the node capability of spreading information among the different communities belonging to a network. A node s has a high ξ if it belongs to more than one community. Such nodes affect the network flow and information broadcasting in different communities. The spreadability measure depends on the dispersion in node memberships.

It is calculated using the following steps:

1. For each node s, having membership $U_{1..k}(s)$ in k communities and standard deviation σ, we measure the spreadability cut given by:

$$\varpi = \sigma(U_{1..k}(s)) - \sigma^2(U_{1..k}(s)). \tag{1}$$

2. Assign s to each community c_i having membership $> \varpi$, then estimate the number of belonging communities given by:

$$\lambda_s = |I_{/s}|, \ I_{/s} = |\{c_i|U_{ci}(s) > \varpi\}| \tag{2}$$

3. Nodes having $\lambda > 1$ are identified as fuzzy, and the more the $\lambda > 1$ the more the node is spreadable (a.k.a, has significant influence across the network communities), while nodes having $\lambda = 1$ are referred as crisp (a.k.a, located locally in their communities).

4. Spreadability for a fuzzy node s belongs to λ overlapping communities is given by:

$$\xi = \sum_{i=1}^{\lambda} U_i(s), \tag{3}$$

s.t, s is member in c_i, while for crisp node is given by:

$$\xi = 1 - max(U_{1..k}(s)). \tag{4}$$

It is worth noting that identifying fuzzy communities based ϖ is a robust and global measure and unlike other such as mean $\mu = \sum U_{1..k}(s)$, which is sensitive to noise and membership variation. Moreover, it does not have the limitation of other measures, e.g, method in [25], exponential entropy given by $\chi(s) = \Pi_{i=1}^{k} u_i(s)^{-u_i(s)}$ and bridgeness score in [13] given by $b(s) = 1 - \sqrt{k\sigma^2(U_{1..k}(s))}$, that requires an external parameter choice for tuning significant memberships [20].

3   The Semantically Enriched Fuzzy c-means Spectral Modularity (SE-FSM) Community Detection Method

The *SE-FSM* community detection method [10] infers the overlapping communities using the following steps:

1. The **Protein quantitative similarity estimation.** Given a set of l proteins, we can obtain their quantitative similarity using an ensemble of **quantitative information** on their interactions from $STRING^1$ [23] that is a public on-line repository incorporating different evidence sources for both physical and functional PPIs. *STRING* stores interaction evaluations for each pair of proteins m, n in different spaces (or features), including *homology, co-expression, experimental results, knowledge bases,* and *text mining.* In addition, *STRING* contains a *combined interaction score* between any pair of proteins calculated as:

$$i_{mn} = 1 - \prod_{f=1}^{h}(1 - a_{f,mn}). \tag{5}$$

This score is computed under the assumption of independence for the various sources, in a naive Bayesian fashion, and often has higher confidence than the individual sub-scores $a_{f,mn}$ [23].

For each feature f we build a *connectivity matrix* (or *similarity matrix*) $A_f = [a_{f,mn}]$ [8,18,12]. Then we combine the connectivity matrices obtaining a *consensus matrix* A with elements:

$$a_{mn} = \sqrt{\sum_{f=1}^{h}(a_{f,mn})^2} \quad f \in \{1, \ldots, h\}, \ m, n \in \{0, \ldots, l\}, \tag{6}$$

where h is the number of features. Finally from A we obtain the *quantitative ensemble similarity matrix* Q making use of a Gaussian kernel:

$$q_{mn} = 1 - e^{-a_{mn}} \quad m, n \in \{0, \ldots, l\}. \tag{7}$$

2. **Protein semantic similarity estimation.** We gather the **semantic information** from many web repositories containing annotated information about biological processes, molecular functions and cellular components including, *KEGG* pathway[2], *Reactome* pathway database[3], the pathway interaction database *(PID)* [4], and Gene Ontology *(GO)*[5]. Each Gene Ontology [1] term $t \in [1, p]$ has a set of annotated proteins related to it. The more semantically similar the gene function annotations between the interacting proteins,

---

[1] http://string-db.org
[2] http://www.genome.jp/kegg/
[3] http://www.reactome.org/
[4] http://pid.nci.nih.gov/
[5] http://www.geneontology.org

the more likely the interaction is physiologically relevant. For each protein we can build a binary valued indicator feature vector that refers to whether it contributes in any of the extracted biological terms or not, obtaining in this way a concurrence matrix $C = [c_{tm}]$ (see Fig. 1). Then we measure the *semantic distance* d_{mn} between each pair m, n of analyzed proteins given by:

$$d_{mn} = \sum_{t=1}^{T} |c_{tm} - c_{tn}| \qquad m, n \in \{0, \ldots, l\}, \tag{8}$$

where T refers to the number of semantically enriched Gene ontology terms or pathways used. Then we can obtain the *semantic similarity matrix S* whose elements are defined as:

$$s_{mn} = e^{\frac{-d_{mn}}{\nu}} \qquad m, n \in \{0, \ldots, l\}, \tag{9}$$

where ν is the dispersion parameter, it controls the width of the Gaussian "bell" and depends on the data distribution. There are many approaches to select the spread of the similarity function (ν). We select ν using histogram analysis of d_{mn} another possible choice is to tune the spread as done in [24].

GO terms, KEGG pathways,...

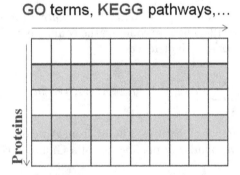

Fig. 1. Semantic co-association matrix construction.

3. **Quantitative and semantic similarities integration.** We **combine** the quantitative ensemble similarity matrix Q and the semantic similarity matrix S in a *hybrid similarity matrix* (or *semantically enriched similarity matrix*) H using the **_Evidence Accumulation Coding_ approach (_EAC_)** proposed by Fred and Jain in [6]. *EAC* is a clustering ensemble technique that builds the consensus partition by averaging the co-association matrices [6]. Co-association matrix based ensemble techniques are based on multiple data partitions, then apply a similarity-based clustering algorithm (e.g., single link and normalized cut) to the co-association matrix to obtain the final

Table 1. The FSM community detection method.

(a) Detect the number of cluster k using modularity measurement [14].
(b) Apply the spectral clustering (e.g., Ng et al. Normalized Spectral Clustering Algorithm [15]) and obtain the spectral space using top k eigen vectors.
(c) Cluster the resultant spectral space using Fuzzy C-Means [2].
(d) Assign vertices to clusters having members larger than the threshold ϖ (see Eq. 1) .

---

partition of the data. For each pair of data points, a co-association matrix indicates whether (or, if fuzzy, how much) they belong to the same cluster.

4. Finally, *SE-FSM* applies the Fuzzy C-Means Spectral Clustering Modularity (or *FSM*) community detection [11] (see Tab. 1 and Fig. 2) to infer the overlapping and semantically significant communities. *FSM* is derived by the Ng et al. [15] spectral clustering algorithm [22,5,4] with the main improvement consisting in the application of the Fuzzy c-means *(FCM)* algorithm [2] for clustering in the affine subspace spanned by the first k eigenvectors. In this work, we automate the parameter tuning in the original *FSM* [11] using the spreadability cut measure ϖ introduced in Sec.2 (Eq. 1): after *FCM* we remove nodes with membership to discovered communities below a threshold ϖ. This thresholding allows us to aggregate only proteins having strong memberships, and to remove the noise and possible outliers in communities. In extreme cases the spreadability thresholding allows us to eliminate insignificant communities including nodes with low membership only.

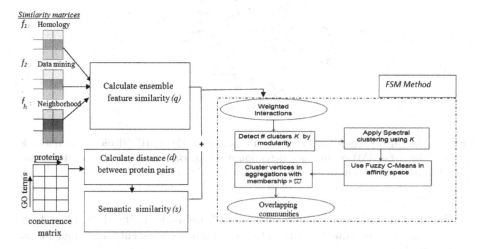

Fig. 2. The proposed Semantically Enriched Fuzzy c-means Spectral Modularity (SE-FSM) community detection method.

4   Experimental Validation

The experimental validation presented in this paper concerns the analysis of *PPI* in immunology networks.

The software was developed in Matlab R2013b[(C)] under Windows 7[(C)] 32 bit. The experiments were performed on a laptop with 2.00 GHz dual-core processor and 3.25 GB of RAM.

4.1   Application to HIV-1

In this study, we aim at detecting proteins annotated to the biological processes significantly related to Human immunodeficiency virus-1 *(HIV-1)* infection in *Homo sapiens* extracted from NCBI database[6] [19]. *HIV-1* is the etiologic agent of acquired immune deficiency syndrome (AIDS). The number of AIDS-related deaths was 2.1 million in 2007 alone [19]. In each experiment we applied the FSM community detection method on the *PPI* networks induced by the specified similarity matrices.

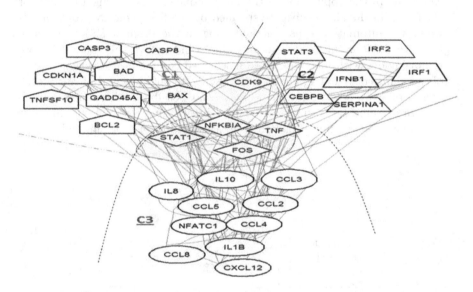

Fig. 3. Identified protein-protein interaction communities in HIV-1 biological network induced by (SE-FSM) community detection method. Bridge nodes significantly annotated to more than one community are framed by diamonds.

We started with the analysis of the *PPI* networks induced by Q and by $I = [i_{mn}]$. The obtained communities are highly consistent, and their Rand index [17] is .95.

---

[6] http://www.ncbi.nlm.nih.gov/RefSeq/HIVInteractions/

In order to identify the significant semantic terms annotated to *HIV-1* we considered *HIV-1* significant annotations based on biological processes (BP), molecular functions (MF), and cellular component (CC) of Gene Ontology.

Moreover, we selected the relevant protein pathways from *KEGG*, *Reactome*, and *PID*. Terms significance *(p-value)* is based on the hypergeometric test. For instance, the *GO* terms enrichment is obtained as $P(X = x) = \frac{\binom{g}{x}\binom{t-g}{r-x}}{\binom{t}{r}}$, where t denotes the total number of genes, g refers to the total number of genes belonging to the *GO* category of interest, r the number of differentially expressed *(DE)* genes, and x is the number of *DE* genes belonging to the *GO* category [16].

The analysis of the *PPI HIV-1* infection network inferred from *H* obtained three functional communities depicted in Fig. 3, showing proteins induced or activated by *HIV-1*. We validated our results with the cellular proteins induced by *HIV-1* infection reported by *QIAGEN*[7]. We highlight that proteins are assigned to different significant functional communities:

- proteins member of community *C*1 participate in many processes, including apoptosis, cell death, cell cycle and proliferation activities, cell cycle activities, protein dimerization activity *GO:0046983 (MF)*, amyotrophic lateral sclerosis (ALS), and prostate cancer KEGG pathways;
- proteins member of community *C*2 influence the transcription factors and regulators, STAT transcription factor, DNA-binding, T cell receptor signaling pathway, and Interferon alpha/beta signaling from reactome *RCTM38609*;
- proteins member of community *C*3 highly anticipate in viral activities such as response to virus and inflammatory response and homeostasis *GO:0055065 (BP)*, Chemokine signaling pathway, and cytokine activity *GO:0005125 (MF)*.

Proteins framed by diamonds in Fig. 3, we call *bridge nodes*, are proteins that our analysis annotated to more than one community. It is interesting to note that the information stored in *QIAGEN* reveal that the *bridge nodes* we found participate to the functions of the three communities: For instance, protein *STAT1* participates in protein dimerization activity *GO:0046983 (MF)* in community *C*1; STAT transcription factor and DNA-binding in community *C*2; and Chemokine signaling pathway in community *C*3.

In Tab. 2 we report the Rand indexes comparing the three communities obtained from the *PPI* networks induced by the similarity matrices *Q*, *S*, and *H*. We highlight that community *C*1 is identified by all approaches and its interactions are robust from the biological viewpoint.

It is worth to note that the semantic enrichment can enhance the *FSM* results, as proteins can be assigned to their significant functional communities as we consider their proteomics pathways as well. For instance, if we take into account the quantitative information only (stored in the similarity matrix *Q*), protein *CCL3* is assigned to community *C*2, but, when we consider the similarity matrix

---

[7] http://www.qiagen.com

Table 2. Rand indexes calculated from the results of the analysis performed on the *PPI* networks induced by Q, S, and H.

Rand	Q	S	H
Q	1	.84	.90
S	.84	1	.78
H	.90	.78	1

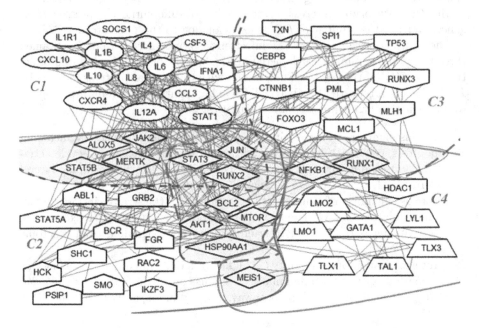

Fig. 4. Identified protein-protein interaction communities in leukemia biological network induced by (SE-FSM) community detection method. Bridge nodes significantly annotated to more than one community are framed by diamonds.

S including the information related to the semantically enriched terms such as *GO:0005125 (MF)*, the same protein *CCL3* is assigned to community *C3* where it participates in biological activities with higher relevance than those of *C2*.

4.2   Application to Leukemia

We applied the aforementioned procedure in analyzing Leukaemia *PPI*. Leukemia is a group of cancers that usually begins in the bone marrow and results in high numbers of abnormal white blood cells. These white blood cells are not fully developed and are called blasts or leukemia cells. There are four main common types of leukemia: acute lymphoblastic leukemia (ALL), acute myeloid leukemia (AML), chronic lymphocytic leukemia (CLL) and chronic myeloid leukemia (CML). In 2012 leukemia developed in 352,000 people globally and caused 265,000 deaths. It is the most common type of cancer in children [7] .

Semantic enrichment	HIV-1 (29 proteins)		Leukemia (58 proteins)	
	# proteins	p-value	# proteins	p-value
Biological Process				
GO:0002682 regulation of immune system process	12	1.89E-08	33	5.01E-29
GO:0045595 regulation of cell differentiation	10	3.60E-06	27	8.48E-21
GO:0006955 immune response	15	2.17E-11	26	2.04E-18
GO:2000026 regulation of multicellular organismal development	12	5.11E-08	25	7.22E-18
GO:0042129 regulation of T cell proliferation	4	3.71E-05	7	2.06E-08
GO:0042113 B cell activation	4	4.10E-05	5	1.33E-05
Molecular function				
GO: 0042379 chemokine receptor binding	5	7.69E-09	4	8.96E-07
GO:0005126 cytokine receptor binding	7	1.40E-08	8	1.70E-08
Cellular component				
GO:0000790 nuclear chromatin	3	2.69E-03	10	6.49E-11
GO:0031981 nuclear lumen	8	5.77E-03	23	2.27E-10
GO:0005740 mitochondrial envelope	4	6.09E-03	4	4.10E-02
KEGG pathways				
hsa05200 Pathways in cancer	11	1.71E-08	24	1.48E-29
hsa05220 Chronic myeloid leukemia	3	2.81E-03	11	1.97E-17
hsa04630 Jak-STAT signaling pathway	4	3.40E-03	13	9.47E-17

Fig. 5. Significant semantic annotations identified in communities of both HIV-1 and leukemia using SE-FSM.

We note that, the communities we discovered using *SE-FSM* emphasize the ptoteomics relation reported in many other experimental studies in addition to agree with the results reported in *QIAGEN*.

For instance, in *Goyama* et al. [9] discovered a dual role of RUNX1 in myeloid leukemogenesis using normal human cord blood cells and those expressing leukemogenic fusion proteins.

Goyama et al. [9] reported that, RUNX1 overexpression inhibited the growth of normal cord blood cells by inducing myeloid differentiation, whereas a certain level of RUNX1 activity was required for the growth of AML1-ETO and MLL-AF9 cells.

Moreover, using a mouse genetic model, they showed that the combined loss of Runx1/Cbfb inhibited leukemia development induced by MLL-AF9. RUNX2 could compensate for the loss of RUNX1. The survival effect of RUNX1 was mediated by BCL2 in MLL fusion leukemia.

In addition, *Goyama* et al. [9] study unveiled an unexpected prosurvival role for RUNX1 in myeloid leukemogenesis. Inhibiting RUNX1 activity rather than enhancing it could be a promising therapeutic strategy for AMLs with leukemogenic fusion proteins.

The application of the *SE-FSM* enabled us to infer the functional relationship between *HIV-1* and leukemia networks. We highlight that, the communities discovered using our approach, share several significant semantic annotations like those shown in Fig. 5.

5 Conclusions

The experimental validation performed in this paper demonstrates that the proposed Semantically Enriched Fuzzy c-means Spectral Modularity (SE-FSM) community detection method can characterize significant overlapping communities in *PPIs*.

We proposed a novel overlapping centrality measure called spreadability and used it in automating the parameter choice in *SE-FSM* and identifying nodes belong to communities with significant fuzzy memberships.

The proposed approach in analyzing *PPIs* of *HIV-1* Homo sapiens infection network, and Leukemia *PPIs* network allows us to infer the communities not only relying on the topological structure of interactome or the biological information, but also based on the semantic enrichment entailed in Gene Ontology and protein pathways.

Our proposed approach thus can boost the functional significance of the identified communities in overlapping protein interaction environments. Moreover, it allows us to find a significant semantic overlap between the detected communities of *HIV-1* and leukemia networks.

Acknowledgements. Work partially funded by a grant of the University of Genova and INDAM-GNCS. Hassan Mahmoud is a PhD student in Computer Science at DIBRIS, University of Genova.

References

1. Ashburner, M., Ball, C.A., Blake, J.A., et al.: Gene ontology: tool for the unification of biology. Nature Genetics 25(1), 25–29 (2000)
2. Bezdek, J.C.: Pattern recognition with fuzzy objective function algorithms. Kluwer Academic Publishers, Norwell (1981)
3. De Las Rivas, J., Fontanillo, C.: Protein-Protein Interactions Essentials: Key Concepts to Building and Analyzing Interactome Networks. PLoS Comput. Biol. 6(6), e1000807 (2010). doi:10.1371/journal.pcbi.1000807
4. Donath, W.E., Hoffman, A.J.: Lower bounds for the partitioning of graphs. IBM Journal of Research and Development 17(5964), 420–425 (1973)
5. Filippone, M., Camastra, F., Masulli, F., Rovetta, S.: A survey of kernel and spectral methods for clustering. Pattern Recognition 41, 176–190 (2008). ISSN: 0031-3203
6. Fred, A.L., Jain, A.K.: Combining multiple clusterings using evidence accumulation. IEEE Transactions on Pattern Analysis and Machine Intelligence 27(6), 835–850 (2005)
7. Hutter, J.J.: Childhood leukemia. Pediatrics in Review 31(6), 234–241 (2010)
8. Kuncheva, L.I.: Combining Pattern Classifiers. Methods and Algorithms. Wiley (2004)
9. Goyama, S., Schibler, J., Cunningham, L., Zhang, Y., Rao, Y., Nishimoto, N., Mulloy, J.C.: Transcription factor RUNX1 promotes survival of acute myeloid leukemia cells. The Journal of Clinical Investigation 123(9), 3876 (2013)
10. Mahmoud, H., Masulli, F., Rovetta, S., Russo, G.: Exploiting quantitative and semantic information in protein-protein interactions networks analysis. In: Computational Intelligence Methods for Bioinformatics and Biostatistics-11th International Meeting (CIBB 2014), Cambridge, UK, June 26-28, 2014
11. Mahmoud, H., Masulli, F., Rovetta, S., Russo, G.: Community detection in protein-protein interaction networks using spectral and graph approaches. In: Formenti, E., Tagliaferri, R., Wit, E. (eds.) CIBB 2013. LNCS, vol. 8452, pp. 62–75. Springer, Heidelberg (2014)
12. Monti, S., Tamayo, P., Mesirov, J., Golub, T.: Consensus clustering: A resampling based method for class discovery and visualization of gene expression microarray data. Machine Learning 52, 91–118 (2003)
13. Nepusz, T., Petrczi, A., Ngyessy, L., Bazs, F.: Fuzzy communities and the concept of bridgeness in complex networks. Physical Review E 77(1), 016107 (2008)
14. Newman, M.E., Girvan, M.: Finding and evaluating community structure in networks. Physical Review E 69(2), 026113 (2004)
15. Ng, A.Y., Jordan, M.I., Weiss, Y.: On spectral clustering: analysis and an algorithm. In: Proceedings of Neural Information Processing Systems, pp. 849–856 (2002)
16. Rivals, I., Personnaz, L., Taing, L., Potier, M.C.: Enrichment or depletion of a GO category within a class of genes: which test? Bioinformatics 23(4), 401–407 (2007)
17. Rand, W.M.: Objective criteria for the evaluation of clustering methods. Journal of the American Statistical Association 66, 846–850 (1971)

18. Strehl, A., Ghosh, J.: Cluster ensembles A knowledge reuse framework for combining multiple partitions. Journal of Machine Learning Research 3, 583–618 (2002)
19. Tastan, O., Qi, Y., Carbonell, J., Klein-Seetharaman, J.: Prediction of interactions between HIV-1 and human proteins by information integration. In: Pacific Symposium on Biocomputing, vol. 14, PubMed PMID: 19209727; PubMed Central PMCID: PMC3263379, pp. 516–527 (2009)
20. Havens, T.C., Bezdek, J.C., Leckie, C., Ramamohanarao, K., Palaniswami, M.: A soft modularity function for detecting fuzzy communities in social networks. IEEE Transactions Fuzzy Systems 21(6), 1170–1175 (2013)
21. Rosasco, R., Mahmoud, H., Rovetta, S., Masulli, F.: A quality-driven ensemble approach to automatic model selection in clustering. In: 23nd Italian Workshop on Neural Networks, WIRN 2013, Vietri, Italy (2013)
22. Von Luxburg, U.: A tutorial on spectral clustering. Statistics and Computing 17, 395–416 (2007)
23. Von Mering, C., Jensen, L.J., Snel, B., Hooper, S.D., Krupp, M., Foglierini, M., Jouffre, N., Huynen, M.A., Bork, P.: STRING: known and predicted proteinprotein associations, integrated and transferred across organisms. Nucleic Acids Res. 33(1), 433–437 (2005)
24. Zelnik-Manor, L., Perona, P.: Self-tuning spectral clustering. In: Advances in Neural Information Processing Systems, pp. 1601–1608 (2004)
25. Zhang, S., Wang, R.S., Zhang, X.S.: Identification of overlapping community structure in complex networks using fuzzy c-means clustering. Physica A: Statistical Mechanics and its Applications 374(1), 483–490 (2007)

Approximate Abelian Periods to Find Motifs in Biological Sequences

Juan Mendivelso[1], Camilo Pino[2], Luis F. Niño[2] and Yoan Pinzón[2]

[1] Fundación Universitaria Konrad Lorenz, Colombia
{juanc.mendivelsom}@konradlorenz.edu.co
[2] Universidad Nacional de Colombia, Colombia
{capinog,lfninov,ypinzon}@unal.edu.co

Abstract. A problem that has been gaining importance in recent years is that of computing the Abelian periods in a string. A string w has an Abelian period p if it is a sequence of permutations of a length–p string. In this paper, we define an approximate variant of Abelian periods which allows variations between adjacent elements of the sequence. Particularly, we compare two adjacent elements in the sequence using $\delta-$ and $\gamma-$ metrics. We develop an algorithm for computing all the $\delta\gamma-$ approximate Abelian periods in a string under two proposed definitions. We also show a preliminary application to the problem of identifying genes with periodic variations in their expression levels.

Keywords: Abelian periods, Parikh vectors, motifs, regularities, biological sequences.

1 Introduction

Finding patterns within long biological sequences is one of the foremost tasks in bioinformatics [22]. In particular, the classic string pattern matching problem is concerned with finding a *pattern string* within a *text string*. In order to support a wider range of application, the string matching problem has been extended into several variants. A well-known variant of the problem is $\delta\gamma-$ *Matching*, which is very effective in searching for all similar but not necessarily identical occurrences of a given pattern. In the $\delta\gamma-$matching problem, two integer strings of the same length match if the corresponding integers differ by at most a fixed bound δ and the total sum of these differences is restricted by γ. For example, strings $\langle 3, 8, 5, 4, 6 \rangle$ and $\langle 4, 7, 5, 4, 5 \rangle$ $\delta\gamma-$match, for $\delta = 1$ and $\gamma = 3$. This problem has been well-studied (*cf.* [7,10,12]) due to its applications in bioinformatics and music information retrieval [7]. Particularly, $\delta\gamma-$ matching has been used to determine if two biological sequences have similar composition [22].

On the other hand, another important problem in string processing, with relevant applications in bioinformatics, is finding periods in strings. Specifically, periodicity is a key component in the analysis of cell-cycle expression data taken in time series [24]; this task is carried out with the purpose of identifying genes with specific regulation behaviour. A string $u = u_{1...p}$ is a period of string

© Springer International Publishing Switzerland 2015
C. di Serio et al. (Eds.): CIBB 2014, LNCS 8623, pp. 121–130, 2015.
DOI: 10.1007/978-3-319-24462-4_11

$w = w_{1...n}$ if w is a prefix of u^k for a positive integer k. Then, w is conformed by the concatenation of $\lfloor n/k \rfloor$ complete instances of u and a prefix of u. In order to measure regularity of a string w, computing the length of its shortest period is an important problem. Such period can be calculated by using either Knuth, Morris and Pratt's algorithm (KMP) [18] or Breslauer and Galil's algorithm [6] for string pattern matching. Another problem of particular interest is computing all the periods of a string. Because it is important to analyse the abundant cyclic phenomena in the nature, this problem has important applications in different fields such as bioinformatics, astronomy, meteorology, biological systems and oceanography, to name some [4]. However, given that data in the real world may contain errors, several algorithms to find approximate periods under different models [1,2,3,20,21] have been developed.

Abelian periods constitute an interesting type of periods. Each occurrence of an Abelian period has the same multiplicity for each character in the alphabet disregarding the order of appearance. Because of this flexibility, Abelian periods are particularly useful in bioinformatics as they can represent consecutive fragments of a biological sequence with similar composition. The idea of Abelian periods in strings can be traced back to the problem defined by Erdös of finding the smallest possible alphabet for which an infinite string with no Abelian squares can be formed [14]. An example of such word on an alphabet of size 5 was given by Pleasants [23] and the proof for an alphabet of size 4 was presented by Keränen [17]. Later work defines the concept of Abelian periods and demonstrates some properties found on classical periods [9,5].

The first algorithm for computing Abelian periods was presented by Cummings and Smyth with $O(n^2)$ time complexity [13]. Later, Constantinescu and Ilie proved relevant combinatorial properties of Abelian periods using Fine and Wilf's periodicity theorem [9]. Recent work includes increasingly efficient algorithms for computing all the Abelian periods of a string [11,15,16,8,19]. Currently, the most efficient solution has $O(n \lg \lg n)$ time complexity [19]. However, none of the previous approaches has considered an approximate version of Abelian periods.

In this paper we define two approaches to approximate Abelian periods under $\delta\gamma$–distance (Section 2). Then, in Section 3, we present an algorithm that solves both variants of the problem. Furthermore, we show an application of this algorithm (Sections 2.3 and 3.2). Finally, we present the conclusions in Section 4.

2   Materials and Methods

In this section, we review the definition of Abelian periods. Then, we extend this definition by using δ– and γ– metrics to restrict the permitted error. In particular, we first present basic definitions (Section 2.1) and then the formulation of the new problems (Section 2.2). Finally, we describe the preprocessing of a gene expression's time series and its evaluation using the proposed approach to periodic patterns (Section 2.3).

2.1 Definitions

A *string* is a sequence of zero or more symbols from an alphabet Σ; the string with zero symbols is denoted by ϵ. Throughout the paper, the alphabet Σ is assumed to be an ordered set $\Sigma = \{a_1, \ldots, a_\sigma\}$; the cardinality of the alphabet is denoted as $\sigma = |\Sigma|$. A *text string* $w = w_{1\ldots n}$ is a string of length n defined over Σ. We use w_i to denote the *i-th element* of w; also, $w_{i\ldots j}$ is used as a notation for the *substring* $w_i w_{i+1} \cdots w_j$ of w, where $1 \le i \le j \le n$.

We also define Parikh vectors because the formal definition of Abelian periods is based on this concept. Let w be a string defined over alphabet Σ. The Parikh vector of w, denoted by $\mathcal{P}(w)$ (or simply $\mathcal{P}$ when w is clear from the context), is a σ–component vector where each $\mathcal{P}(w)[i]$, for $1 \le i \le \sigma$, expresses the number of occurrences of character a_i in w. The length of Parikh vector $\mathcal{P}$, denoted by $|\mathcal{P}|$, is the sum of its components. Equivalently, $|\mathcal{P}|$ is the length of a string whose Parikh vector is $\mathcal{P}$. Furthermore, given two strings x and y, we say that $\mathcal{P}(x) < \mathcal{P}(y)$ if we can obtain y from x by permuting and inserting certain characters, i.e. $\mathcal{P}(x)[i] \le \mathcal{P}(y)[i]$, for $1 \le i \le \sigma$ and $|\mathcal{P}(x)| < |\mathcal{P}(y)|$.

Each Abelian period in a string w is expressed as a pair (h, p). Specifically, a string $w = w_{1\ldots n}$ has an *Abelian period* (h, p) if it can be expressed as $u_0 u_1 \cdots u_{k-1} u_k$, where $|\mathcal{P}(u_0)| = h$ and $|\mathcal{P}(u_1)| = p$, such that $\mathcal{P}(u_0) < \mathcal{P}(u_1) = \cdots = \mathcal{P}(u_{k-1}) > \mathcal{P}(u_k)$. Notice that $u_1, \ldots, u_{k-1}$ are the full occurrences of the period. On the other hand, u_0 and u_k are strings that could constitute other occurrences of the period with the addition of certain characters; these strings are denoted as *head* and *tail*, respectively. For example, $aabbabab$ has an Abelian period $(1, 2)$ because it can be obtained from the concatenation of $u_0 = a$, $u_1 = ab$, $u_2 = u_3 = ba$ and $u_4 = b$. Note that, $\mathcal{P}(u_0) = (1, 0)$, $\mathcal{P}(u_4) = (0, 1)$, and $\mathcal{P}(u_1) = \mathcal{P}(u_2) = \mathcal{P}(u_3) = (1, 1)$ where the first component is the multiplicity of a and the second is the multiplicity of b.

Now, we define some notions to support $\delta\gamma-$ distances in Abelian periods. Let us consider strings x and y defined over alphabet $\Sigma = \{a_1, a_2, \ldots, a_\sigma\}$. We say that $\mathcal{P}(x) \overset{\delta}{=} \mathcal{P}(y)$ iff $\max_{i=1}^{\sigma} |\mathcal{P}(x)[i] - \mathcal{P}(y)[i]| \le \delta$ and we say that $\mathcal{P}(x) \overset{\gamma}{=} \mathcal{P}(y)$ iff $\sum_{i=1}^{\sigma} |\mathcal{P}(x)[i] - \mathcal{P}(y)[i]| \le \gamma$. Note that δ establishes the maximum difference for the multiplicity of each character while γ establishes the maximum sum permitted for the differences of the multiplicities of all characters. If both conditions are satisfied we say that $\mathcal{P}(x)$ and $\mathcal{P}(y)$ are $\delta\gamma$–*equal*, denoted as $\mathcal{P}(x) \overset{\delta\gamma}{=} \mathcal{P}(y)$. The transitive property does not hold for the $\delta\gamma$–equality: for three given strings x, y and z, the facts that $\mathcal{P}(x) \overset{\delta\gamma}{=} \mathcal{P}(y)$ and $\mathcal{P}(y) \overset{\delta\gamma}{=} \mathcal{P}(z)$ do not imply that $\mathcal{P}(x) \overset{\delta\gamma}{=} \mathcal{P}(z)$. However, it implies that $\mathcal{P}(x)$ and $\mathcal{P}(z)$ are $(2\delta, 2\gamma)$-equal.

Moreover, according to these δ and γ notions, we also say that x is $\delta\gamma-$ *extensible* to $\mathcal{P}(y)$, denoted as $\mathcal{P}(x) \overset{\delta\gamma}{\lesssim} \mathcal{P}(y)$, if $\mathcal{P}(x)$ may not be $\delta\gamma$–equal to $\mathcal{P}(y)$ but it is possible to obtain a string x', by inserting certain characters into x, such that $\mathcal{P}(x') \overset{\delta\gamma}{=} \mathcal{P}(y)$; furthermore, the length of x must be lower than $|\mathcal{P}(y)| + \gamma$. That is, x is $\delta\gamma$–extensible to $\mathcal{P}(y)$ if the following conditions are satisfied: (i) $\max_{i=1}^{\sigma}(\mathcal{P}(x)[i] - \mathcal{P}(y)[i]) \le \delta$; (ii) $\sum_{i=0}^{\sigma} \max(0, \mathcal{P}(x)[i] - \mathcal{P}(y)[i]) \le \gamma$; and (iii)

$|\mathcal{P}(x)| \leq |\mathcal{P}(y)| + \gamma$. Notice that if x is the empty string, then $\mathcal{P}(x) \overset{\delta\gamma}{\lesssim} \mathcal{P}(y)$ always holds.

2.2 New Problems

We propose two definitions of $\delta\gamma$-*approximate Abelian period* as the counterparts of the variants of regular approximate periods.

Definition 1. *A string* $w = w_{1...n}$ *is said to have a* $\delta\gamma$-*approximate Abelian period* (h, p) *if it can be expressed as* $w = u_0 u_1 \cdots u_{k-1} u_k$, *where* $|\mathcal{P}(u_0)| = h$ *and* $|\mathcal{P}(u_1)| = p$, *such that the following conditions are satisfied: (i)* $\mathcal{P}(u_0) \overset{\delta\gamma}{\lesssim} \mathcal{P}(u_1)$; *(ii)* $\mathcal{P}(u_1) \overset{\delta\gamma}{=} \mathcal{P}(u_i)$ *for* $1 < i < k$; *and (iii)* $\mathcal{P}(u_k) \overset{\delta\gamma}{\lesssim} \mathcal{P}(u_1)$.

In other words, the approximate Abelian period (h, p) refers to the Parikh vector of the length–p substring that starts at position $h + 1$ in w, where certain properties are satisfied. Specifically, w can be expressed as $w = u_0 u_1 \cdots u_{k-1} u_k$ such that: the head and the tail, *i.e.* u_0 and u_k, are $\delta\gamma$-extensible to $\mathcal{P}(u_1)$. Furthermore, $\mathcal{P}(u_1)$ is $\delta\gamma$-equal to u_2, ..., u_{k-1}. Then, the strings u_1, ..., u_{k-1} are called the $\delta\gamma$-*occurrences* of the period (h, p) in w, *i.e.* the $\delta\gamma$-occurrences of the Parikh vector of u_1 in w.

For example, given $\delta = 1$, $\gamma = 2$ and the alphabet $\Sigma = \{a, b, c\}$, the string $bccbacabbabcbcc$ has a $\delta\gamma$-approximate period $(1, 4)$ as it can be expressed as the concatenation $u_0 u_1 u_2 u_3 u_4$ where $u_0 = b$, $u_1 = ccba$, $u_2 = cabb$, $u_3 = abcbc$ and $u_4 = c$. We obtain that $\mathcal{P}(u_1) = (1, 1, 2)$, $\mathcal{P}(u_2) = (1, 2, 1)$ and $\mathcal{P}(u_3) = (1, 2, 2)$, where the first, second and third component correspond to the multiplicity of a, b and c, respectively. Notice that these multiplicity vectors are similar; this is because $\mathcal{P}(u_2)$ and $\mathcal{P}(u_3)$ are $\delta\gamma$-equal to $\mathcal{P}(u_1)$. Furthermore, u_0 and u_4 are $\delta\gamma$-extensible to $\mathcal{P}(u_1)$. It is important to remark that $\delta\gamma$– distance on Parikh vectors allows to find $\delta\gamma$-occurrences of different lengths; for instance the length of u_2 is 4 while the length of u_3 is 5.

Taking into account the aforementioned ideas, we can see that supporting δ– and γ– distances in Abelian periods allows to find interesting periodic properties in data from different areas, such as Computational Biology and Music Segmentation, that would not be found with the traditional definition of Abelian periods. In order to further extend flexibility of such properties, we next present an alternative definition of $\delta\gamma$-approximate Abelian period:

Definition 2. *A string* $w = w_{1...n}$ *is said to have a* $\delta\gamma$-*successively approximate Abelian period* (h, p) *if it can be expressed as* $w = u_0 u_1 \cdots u_{k-1} u_k$, *where* $|\mathcal{P}(u_0)| = h$ *and* $|\mathcal{P}(u_1)| = p$, *such that the following conditions are satisfied: (i)* $\mathcal{P}(u_0) \overset{\delta\gamma}{\lesssim} \mathcal{P}(u_1)$; *(ii)* $\mathcal{P}(u_i) \overset{\delta\gamma}{=} \mathcal{P}(u_{i-1})$ *for* $1 < i < k$; *and (iii)* $\mathcal{P}(u_k) \overset{\delta\gamma}{\lesssim} \mathcal{P}(u_{k-1})$.

This definition is similar to Definition 1 except for the fact that each substring is compared with its preceding substring rather than with u_1. For example, given $\delta = 1$, $\gamma = 2$ and the alphabet $\Sigma = \{a, b, c\}$, the string $bccbacabbabcbcc$ has a

$\delta\gamma$–successively approximate period $(0,3)$ given that it can be expressed as the concatenation of $u_0 = \epsilon$, $u_1 = bcc$, $u_2 = bac$, $u_3 = abb$, $u_4 = abc$, $u_5 = bcc$, and $u_6 = \epsilon$. Notice that $\mathcal{P}(u_4) = (1,1,1)$ is $\delta\gamma$-equal to $\mathcal{P}(u_3) = (1,2,0)$ where the first, second and third component correspond to the multiplicity of a, b and c, respectively. Likewise, $\mathcal{P}(u_1) \overset{\delta\gamma}{=} \mathcal{P}(u_2)$, $\mathcal{P}(u_2) \overset{\delta\gamma}{=} \mathcal{P}(u_3)$, $\mathcal{P}(u_4) \overset{\delta\gamma}{=} \mathcal{P}(u_5)$, $\mathcal{P}(u_0) \overset{\delta\gamma}{\lesssim} \mathcal{P}(u_1)$ and $\mathcal{P}(u_6) \overset{\delta\gamma}{\lesssim} \mathcal{P}(u_5)$. Note that both u_0 and u_6 are empty strings.

The main feature of Definition 2 is that, given that $\delta\gamma$–equality is not transitive, the last occurrences of the period may be greatly degenerated with respect to its first occurrence; it can be thought that evaluating an Abelian period with respect to its predecessor is equivalent to a Markov property. Therefore, it is appropriate to choose the definition of $\delta\gamma$–approximate Abelian period that is better for the requirements of the specific application problem. To the best of our knowledge, these are the first two approaches to approximate Abelian periods.

2.3   Identification of Periods in Expression Levels

With the purpose of testing our approach with the identification of approximate periods in real data, we used a dataset corresponding to expression levels of *M. cerevisiae* during the cell cycle [24]. The original microarray data was obtained at the author's website, already aggregated and normalized. There are 6178 measured genes in 6 different time-series, each corresponding to a different cell-cycle synchronization experimental method. We selected the CDC15 time series because its number of points: samples taken in intervals of 10 minutes during 300 minutes for a total of 31 data-points per gene with some missing measurements. More details can be found in the original publication.

In order to obtain a discrete representation suitable to be searched for approximate periods, we calculated a string for each gene of the target experiment, each position of such string correponding to an expression measurement. With the aim of defining an alphabet for each of those strings and the mapping between expression values and symbols, we clustered the data corresponding to each gene using k-means. Each cluster of expression values was assigned a symbol and the string was generated by taking the time-ordered list of expression values per gene and replacing each data point by the symbol of its cluster. This string can be seen as the transitions between expression states of each gene given the observed data. The number of clusters for each gene's time series was optimized using the silhouette coefficient during 100 runs. This preprocessing step was done using R and the FPC package.

The resulting strings were evaluated using the algorithm proposed in Algorithm 1. The value of the parameters was arbitrarily constrained in the following way: δ was assigned the value 1 in order to allow small changes in the presence of expression states between successive periods and γ was assigned 2 as the maximum number of overall changes between successive periods. Each string was evaluated using the following scores: (i) the number of ways that $\delta\gamma$–approximate Abelian periods can be found with the given parameters; (ii) its Shannon entropy score calculated as $S(x) = -\sum_{i=1}^{\sigma} \mathcal{P}(x)[i] log_2 \mathcal{P}(x)[i]$; (iii) the number of symbols in its alphabet (previously found by clustering); and (iv) the silhouette coefficient for such clustering.

3 Results

In this section, we first present a solution to solve both $\delta\gamma$–approximate Abelian periods variants (Section 3.1) and then present some experimental results (Section 3.2).

3.1 Algorithm

We first describe an algorithm to compute all the $\delta\gamma$–approximate Abelian periods in a string (Definition 1). Then, we show how to adapt this algorithm to find all the $\delta\gamma$–successively approximate Abelian periods of a string (Definition 2).

The pseudocode of the algorithm to compute all the Abelian periods of a string $w = w_{1...n}$ under Definition 1 is listed in Figure 1. We consider all the possible values for h, $i.e.$ the length of $\mathcal{P}(u_0)$ (line 2). These values are established by the condition (i) of Definition 1, which requires that $|\mathcal{P}(u_0)| \leq |\mathcal{P}(u_1)| + \gamma$. Then, in the upper limit, we have that $|\mathcal{P}(u_1)| = |\mathcal{P}(u_0)| - \gamma$. Replacing this expression in $|\mathcal{P}(u_0)| + |\mathcal{P}(u_1)| = n$, which constitutes the extreme case, we obtain that the upper limit for h is $\lfloor (n+\gamma)/2 \rfloor$. The lower limit corresponds to the empty string where $h = 0$. For p, the length of u_1, we also consider all the possible values according to the given h. Namely, the lower limit for p is $\max(1, |\mathcal{P}(u_0)| - \gamma)$, which is also obtained from condition (i) of Definition 1 (line 3). The upper limit for p is given by the fact that the sum of h and p must not exceed n (line 4). For each pair (h, p), the strings $w_{1...h}$ and $w_{h+1...h+p}$ correspond to u_0 and u_1, respectively. If $\mathcal{P}(u_0) \overset{\delta\gamma}{\lesssim} \mathcal{P}(u_1)$, then we evaluate if (h, p) is a $\delta\gamma$–approximate Abelian period by calling the method MATCHSUBSTRING() (line 5); in such case, we add it into the result set $\mathcal{R}$ (line 6).

The function MATCHSUBSTRING() is illustrated in Figure 2. The parameters are $w_{1...n}$ $w_{i...j}$, $w_{k...\ell}$, δ and γ. The goal of the function is determining whether $w_{i...n}$ can be expressed as the concatenation of substrings whose Parikh vectors $\delta\gamma$–match $w_{k...\ell}$; the first of such substrings is $w_{i...j}$. Let us consider the second substring. Its initial position is $j + 1$ while its final position m can vary from $j + |w_{k...\ell}| - \gamma$ to $j + |w_{k...\ell}| + \gamma$; this is because a string that $\delta\gamma$–matches $w_{k...\ell}$ has at least $|w_{k...\ell}| - \gamma$ characters and at most $|w_{k...\ell}| + \gamma$ characters. However, the former value must be at least $j + 1$ (the substring composed by one character) while the latter cannot exceed n (lines $5 - 6$). Then, for each of these possible substrings $w_{j+1...m}$, we evaluate two conditions: (i) if it is indeed $\delta\gamma$–equal to $w_{k...\ell}$; and (ii) if the string $w_{j+1...n}$ can also be expressed as the concatenation of substrings whose Parikh vectors $\delta\gamma$–match $w_{k...\ell}$ where the first of such substrings is $w_{j+1...m}$. Condition (ii) is evaluated by recursively calling MATCHSUBSTRING(). If both conditions are satisfied, for any of the substrings $w_{j+1...m}$, we conclude that $w_{i...n}$ can also be expressed as the concatenation of

---

Algorithm 1. DG-ABELIANPERIODS **Algorithm**

Input: $w = w_{1...n}$ defined over $\Sigma = \{a_1, \ldots a_\sigma\}$, and $\delta, \gamma \in \mathbb{N}$
Output: $\mathcal{R}$

---

1. $\mathcal{R} \leftarrow \{\}$
2.   **for** $h \leftarrow 0$ **to** $\lfloor (n+\gamma)/2 \rfloor$ **do**
3. $p \leftarrow \max(1, h - \gamma)$
4.   **while** $h + p \leq n$
5.   **if** $(\mathcal{P}(w_{1...h}) \overset{\delta\gamma}{\lesssim} \mathcal{P}(w_{h+1...h+p})$ **and**
                    $MatchSubstring(w_{1...n}, w_{h+1...h+p}, w_{h+1...h+p}, \delta, \gamma))$ **then**
6. $\mathcal{R}.Add(Pair(h, p))$
7. $p \leftarrow p + 1$
8.   **return** $\mathcal{R}$

Fig. 1. DG-ABELIANPERIODS Algorithm.

substrings whose Parikh vectors $\delta\gamma$–match $w_{k...\ell}$ where the starting substring is $w_{i...j}$ (lines 7-10).

The arguments for the first call to MATCHSUBSTRING() are $w_{1...n}$, u_1, u_1, δ, γ (Figure 1, line 5). Thus, this special case must be handled accordingly: the outcome is *true* if the final index is n. This is because it was already verified that $\mathcal{P}(u_0) \overset{\delta\gamma}{\lesssim} \mathcal{P}(u_1)$ and there is just a single $\delta\gamma$–occurrence of the period (namely, substring u_1). Notice that in later calls to MATCHSUBSTRING() (Figure 2, line 8), parameter i will take a greater value than parameter ℓ; thus, if $\mathcal{P}(w_{i...n}) \overset{\delta\gamma}{\lesssim} \mathcal{P}(w_{k...\ell})$ then $w_{i...n}$ is a valid u_k (Figure 2, line 3). Otherwise, and if $j = n$, we conclude that the corresponding (h, p) is not a period in $w_{1...n}$ (Figure 2, lines 3-4).

In order to compute all the $\delta\gamma$–successive approximate Abelian periods in string $w = w_{1...n}$ (see Definition 2), we can also use the pseudocode listed in Figures 1 and 2. The only difference is that we must compare each potential $\delta\gamma$–occurrence or tail of the period with respect to its preceding substring. This can be achieved by changing the arguments of the call to MATCHSUBSTRING() in line 8 of Figure 2. Specifically, the arguments must be $w_{1...n}$, $w_{j+1...m}$, $w_{i...j}$, δ and γ; hence each substring is compared with respect to the preceding occurrence of the period, *i.e.* $w_{i...j}$, rather than with u_1.

3.2   Discussion of the Experimental Results

Each expression string was scored by the number of ways that each possible $\delta\gamma$-approximate Abelian period could be found in it. The scores were filtered by

---

Algorithm 2. MATCHSUBSTRING **Function**

Input: $w_{1...n}$, $w_{i...j}$, $w_{k...\ell}$, δ, γ
Output: $true/false$

---

1. **if** ($\ell < i$ **and** $\mathcal{P}(w_{i...n}) \overset{\delta\gamma}{\lessgtr} \mathcal{P}(w_{k...\ell})$) **then return** $true$
2. **if** $j = n$ **then**
3.   **if** ($i = k$ **and** $j = \ell$) **then return** $true$
4.   **else return** $false$
5. $lower \leftarrow \max(j + |w_{k...\ell}| - \gamma, j + 1)$
6. $upper \leftarrow \min(j + |w_{k...\ell}| + \gamma, n)$
7. **for** $m = lower$ **to** $upper$
8.   **if** ($\mathcal{P}(w_{j+1...m}) \overset{\delta\gamma}{=} \mathcal{P}(w_{k...\ell})$ **and**
$$MatchSubstring(w_{1...n}, w_{j+1...m}, w_{k...\ell}, \delta, \gamma))$$
9.   **return** $true$
10. **return** $false$

Fig. 2. MATCHSUBSTRING Function.

silhouette coefficient using 0.9 as threshold with the purpose of eliminating uncertain mappings between symbols and expression levels. The distribution of the number of possible $\delta\gamma$–approximate Abelian periods does not discriminate cyclically expressed genes. Notwithstanding, the regularities found when the strings were sorted using this criterion, compared with respect to those present in cyclically expressed genes, suggest to confirm the same regularities in other expression data sets as future work.

4   Conclusions

We defined approximate Abelian periods in strings under δ– and γ– distances. The error between adjacent instances of the period is restricted by these metrics: δ is used to restrict the error on the multiplicities of each character and γ is used to restrict the sum of these errors. We solved the problem of finding all the $\delta\gamma$–approximate Abelian periods in a string under two variants of the problem. To the best of our knowledge, this problem has not been considered before. It is useful to find periodic properties of data in several areas, such as Bioinformatics, Music Segmentation and Complex Event Analysis. We also proposed an application of the presented algorithm to analyse gene expression's time series with the purpose of identifying periodic changes in expression levels in the cell-cycle of *M. cerevisiae*. We obtained encouraging results when compared with already identified genes.

References

1. Amir, A., Eisenberg, E., Levy, A.: Approximate periodicity. In: Cheong, O., Chwa, K.-Y., Park, K. (eds.) ISAAC 2010, Part I. LNCS, vol. 6506, pp. 25–36. Springer, Heidelberg (2010)
2. Amir, A., Eisenberg, E., Levy, A., Lewenstein, N.: Closest periodic vectors in L_p spaces. In: Asano, T., Nakano, S.-i., Okamoto, Y., Watanabe, O. (eds.) ISAAC 2011. LNCS, vol. 7074, pp. 714–723. Springer, Heidelberg (2011)
3. Amir, A., Eisenberg, E., Levy, A., Porat, E., Shapira, N.: Cycle detection and correction. ACM Transactions on Algorithms (TALG) 9(1), 13 (2012)
4. Amir, A., Levy, A.: Approximate period detection and correction. In: Calderón-Benavides, L., González-Caro, C., Chávez, E., Ziviani, N. (eds.) SPIRE 2012. LNCS, vol. 7608, pp. 1–15. Springer, Heidelberg (2012)
5. Blanchet-Sadri, F., Tebbe, A., Veprauskas, A.: Fine and wilf's theorem for abelian periods in partial words. In: Proceedings of the 13th Mons Theoretical Computer Science Days, vol. 6031 (2010)
6. Breslauer, D., Galil, Z.: An optimal o(\log\logn) time parallel string matching algorithm. SIAM Journal on Computing 19(6), 1051–1058 (1990)
7. Cambouropoulos, E., Crochemore, M., Iliopoulos, C., Mouchard, L., Pinzon, Y.J.: Algorithms for computing approximate repetitions in musical sequences. International Journal of Computer Mathematics 79(11), 1135–1148 (2002)
8. Christou, M., Crochemore, M., Iliopoulos, C.S.: Identifying all abelian periods of a string in quadratic time and relevant problems. International Journal of Foundations of Computer Science 23(06), 1371–1384 (2012)
9. Constantinescu, S., Ilie, L.: Fine and wilf's theorem for abelian periods. In: Bulletin of the EATCS, vol. 89, pp. 167–170 (2006)
10. Crochemore, M., Iliopoulos, C.S., Lecroq, T., Pinzon, Y.J., Plandowski, W., Rytter, W.: Occurrence and Substring Heuristics for d-Matching. Fundamenta Informaticae 56(1), 1–21 (2003)
11. Crochemore, M., Iliopoulos, C.S., Kociumaka, T., Kubica, M., Pachocki, J., Radoszewski, J., Rytter, W., Tyczyński, W., Waleń, T.: A note on efficient computation of all abelian periods in a string. Information Processing Letters 113(3), 74–77 (2013)
12. Crochemore, M., Iliopoulos, C.S., Navarro, G., Pinzon, Y.J., Salinger, A.: Bit-parallel (δ, γ)-matching and suffix automata. Journal of Discrete Algorithms 3(2), 198–214 (2005)
13. Cummings, L.J., Smyth, W.F.: Weak repetitions in strings. Journal of Combinatorial Mathematics and Combinatorial Computing 24, 33–48 (1997)
14. Erdös, P., et al.: Some unsolved problems. Magyar Tud. Akad. Mat. Kutató Int. Közl 6, 221–254 (1961)
15. Fici, G., Lecroq, T., Lefebvre, A., Prieur-Gaston, É.: Computing abelian periods in words. In: Proceedings of the Prague Stringology Conference 2011, pp. 184–196 (2011)
16. Fici, G., Lecroq, T., Lefebvre, A., Prieur-Gaston, É., Smyth, W.F.: Quasi-linear time computation of the abelian periods of a word. In: Proceedings of the Prague Stringology Conference 2012, pp. 103–110 (2012)
17. Keränen, V.: Abelian squares are avoidable on 4 letters. In: Kuich, W. (ed.) ICALP 1992. LNCS, vol. 623, pp. 41–52. Springer, Heidelberg (1992)
18. Knuth, D., Morris Jr., J., Pratt, V.: Fast pattern matching in strings. SIAM Journal on Computing 6, 323 (1977)

19. Kociumaka, T., Radoszewski, J., Rytter, W.: Fast algorithms for abelian periods in words and greatest common divisor queries. In: STACS, vol. 20, pp. 245–256 (2013)
20. Kolpakov, R., Bana, G., Kucherov, G.: Mreps: efficient and flexible detection of tandem repeats in dna. Nucleic Acids Research 31(13), 3672–3678 (2003)
21. Kolpakov, R., Kucherov, G.: Finding approximate repetitions under hamming distance. In: Meyer auf der Heide, F. (ed.) ESA 2001. LNCS, vol. 2161, pp. 170–181. Springer, Heidelberg (2001)
22. Mendivelso, J., Pinzon, Y.: A novel approach to approximate parikh matching for comparing composition in biological sequences. In: Proceedings of the 6th International Conference on Bioinformatics and Computational Biology (BICoB) (2014)
23. Pleasants, P.A.: Non-repetitive sequences. In: Proc. Cambridge Philos. Soc., vol. 68, pp. 267–274. Cambridge Univ. Press (1970)
24. Spellman, P.T., Sherlock, G., Zhang, M.Q., Iyer, V.R., Anders, K., Eisen, M.B., Brown, P.O., Botstein, D., Futcher, B.: Comprehensive identification of cell cycle–regulated genes of the yeast saccharomyces cerevisiae by microarray hybridization. Molecular Biology of the Cell 9(12), 3273–3297 (1998)

Sem Best Shortest Paths for the Characterization of Differentially Expressed Genes

Daniele Pepe[1], Fernando Palluzzi[2], and Mario Grassi[1]

[1]University of Pavia, Department of Brain and Behavioural Sciences,
via Bassi 21 – 27100 Pavia, Italy
[2] Politecnico di Milano, Department of Electronics, Information and Bioengineering,
via Ponzio 34/5 – 20133 Milano, Italy

Abstract. In the last years, systems and computational biology focused their efforts in uncovering the causal relationships among the observable perturbations
of gene regulatory networks and human diseases. This problem becomes even
more challenging when network models and algorithms have to take into account slightly significant effects, caused by often peripheral or unknown genes
that cooperatively cause the observed diseased phenotype. Many solutions,
from community and pathway analysis to information flow simulation, have
been proposed, with the aim of reproducing biological regulatory networks and
cascades, directly from empirical data as gene expression microarray data. In
this contribute, we propose a methodology to evaluate the most important
shortest paths between differentially expressed genes in biological interaction
networks, with absolutely no need of user-defined parameters or heuristic rules,
enabling a free-of-bias discovery and overcoming common issues affecting the
most recent network-based algorithms.

Keywords: SEM, disease module, shortest paths, gene expression, metabolic
pathways.

1 Introduction

Biological network analysis is particularly suited to study complex disorders, arising
from a combination of many different structural and functional perturbations. The key
problem is to identify causal and target genes having only a minor association with
the pathology under investigation. Although only slightly associated with the pathology, the combinatorial effect of those perturbed genes results in a variety of phenotypi-
cal variants, causing huge differences in prognosis, survival and drug response of affected individuals. Therefore, the only reasonable way of analyzing their
combinatorial effects is to identify perturbed paths or modules that are significantly
associated with the observed phenotype. The goal is to point out those paths connecting perturbed genes, involved in the observed phenotype, possibly highlighting previously unknown causal or target genes. Even though there are no straightforward
rules to extract significant sub-graphs from the original network, two major approaches have been successfully applied to solve the problem [1-3]: distance-based and

© Springer International Publishing Switzerland 2015
C. di Serio et al. (Eds.): CIBB 2014, LNCS 8623, pp. 131–141, 2015.
DOI: 10.1007/978-3-319-24462-4_12

information flow-based algorithms. The former is based on the idea that genes that are physically or functionally proximal tend to be involved in the same biological process, an assumption that is often called *guilt by association* [2]. The latter tries to simulate the information propagation in the biological network, by starting a random walk from a source node, passing through a number of intermediate nodes, and eventually ending in a sink node. The random walk propagation is controlled by the user through some defined parameter, such as walk length and edge weights [4, 6].

Classical distance-based methods have three limitations [3]: (i) they do not consider additional information, other than topology (e.g. gene expression), (ii) they consider only one shortest path per gene pair, (iii) the shortest path may not be the most informative one. In a recent work [4], the first two problems have been addressed using a single source k-shortest paths algorithm. On the other hand, the information flow model [5] has one important limitation: the weight normalization used to generate transition probabilities may cause loss of information, preventing the algorithm from detecting important paths, by flattening weights from different sub-graphs to artificially equal transition probabilities. This problem can be addressed by modifying transition probabilities using different damping factors for different nodes, thus reintroducing the diversity that has been erased by the normalization procedure. Indeed, assigning the proper importance to a given path is an issue also in the Shih and Parthasarathy's k-shortest path algorithm. The k-shortest paths detected are often very similar, having the majority of their connections in common. If one only takes into account the distance, this may introduce a strong bias towards highly similar shortest paths. In order to address this problem, the algorithm introduces a parameter, called diversity [4] that is proportional to the unique connections a path has. A path is said to be diverse if its diversity is higher than a user-defined threshold called λ. Again, similarly to the damping factor, introduced in the information flow method, the λ parameter must be defined a priori by the user. In this way, the user is forced to evaluate the impact of different parameter values only a posteriori, by repeating the same time consuming procedure to better tune the parameter itself. Furthermore, the method shows how the number of significantly enriched Gene Ontology (GO) [6] terms is not always clearly related to λ, especially for high parameter values.

In the method here illustrated, empirical data directly drive the selection of the most important nodes, while each path is evaluated using an annotation-driven model. In particular, expression data and metabolic pathways (using KEGG) [7] information are integrated to build the interactome. The analysis is similar to that proposed by Pepe & Grassi [8] with the relevant difference of the evaluation of equivalent shortest paths. Starting from differentially expressed genes (DEGs), extracted from a gene expression dataset, the proposed pipeline finds the biological context where they act and then tries to understand how they are connected by shortest paths. A shortest path between DEGs can contain other genes of the experiment not necessarily DEGs. This allows detecting genes that can be important as brokers. The point is that more equivalent shortest paths can connect two DEGs and using only topological information, it is not possible to detect the best one. The importance of each shortest path can be assessed using Structural Equation Models (SEM) [9]. The penalized goodness of fit scores returned by the SEM testing, provide a direct estimation of the importance of each shortest path. In this way, we overcome the three limitations of the classical distance based methods as: 1) data information are considered together with the topological information, given by "*a priori*" knowledge enclosed in biological pathways; 2) not only a shortest path is considered for each couple of genes as also the equivalent ones are included in the analysis; 3) for each couple of DEGs the best shortest path is found based on statistical criteria. We not only relieve the user from estimating *ad-hoc* thresholds, but we also provide an objective criterion to assess pathway importance on the base of its metabolic function. The fusion of the best shortest paths allows detecting the key area in the perturbed network that can explain the phenotype studied. Finally, the key network area is fitted with SEM to verify if the model proposed outlines well the data observed. The way to detect DEGs and perturbed pathways does not change the downstream analysis that we propose here. So every existing method could be used.

2   Material and Methods

The central part of our method is the selection of a disease module by SEM best shortest paths between DEGs. The procedure starts from DEGs and perturbed pathways identified, respectively, with a t-test (Benjamini-Hochberg adjustment) and through Significance Pathways Impact Analysis (SPIA) [10]. We used the KEGG pathway database for the pathway analysis. The significant pathways could be represented as directed graph $P = \{G, V\}$ where G are the set of genes and V are the edges that correspond to biological interactions between them. The selected pathways were merged to create a single graph with unique nodes, and edges are merged into a non-duplicate set. The merging procedure was performed with the function *merge-KEGGgraphs( )* of the R [11] package KEGGgraph [12]. The idea is to understand how DEGs are connected in the perturbed pathways to other microarray genes. A natural way to solve this problem is to select the shortest paths (geodesic distance) between DEGs. The geodesic distance between two DEGs, g_i and g_j, is defined as the minimum distance between these two genes. The function *get.all.shortest.paths( )* of the R package igraph [13] was used to compute all shortest paths between each couple of DEGs. The problem is that, in many cases, there is not a unique shortest path for couple of DEGs, but a set of equivalent shortest path. The goal here is to detect the equivalent shortest path that maximize the likelihood of the links between the considered genes. To select the best shortest path we employed SEM.

The first step is to specify a model describing the causal relationships among variables. In general, a SEM in which all the variables are observed can be represented as a system of linear equations:

$$Y_i = \sum_{j \in pa(i)} \beta_{ij} Y_j + U_i \quad i \in V$$

and a covariance structure:

$$cov(U_i; U_j) = \begin{cases} \psi_{ij} & \text{if } i = j \text{ or } j \in sib(i) \\ 0 & \text{otherwise} \end{cases}$$

The system of linear equations considers every i-th node (Y_i) as characterized by unidirected relationships with its "parents" set, $pa(i)$ quantified by path coefficients (β_{ij}).

The covariance structure describes the bi-directed relationships between the unmeasured component of the i-th node (U_i) and its "siblings" set, *sib(i)* quantified by their covariances (ψ_{ij}). In our specific case, every equivalent directed shortest path denotes a SEM, where there is an initial node represented by the source DEG connected to a destination DEG by intermediate DEGs or other microarray genes. Considering that the equivalent shortest paths are induced paths, every shortest path can be represented by m_k-1 recursive linear equations, where m_k is the number of nodes in the shortest path k:

$$Y_{2(k)} = \beta_{21(k)}Y_{1(k)} + U_{2(k)}; \quad Y_{3(k)} = \beta_{32(k)}Y_{2(k)} + U_{3(k)}; \text{ and so on}$$

For the estimation of the parameters $\theta=(\beta_{ij}; \psi_{ij})$ the Maximum Likelihood Estimation (MLE), assuming that all the measured expression genes have a multinormal distribution, are computed by using the R package `lavaan` [17]. The assessment of the best shortest path was obtained taking among the set of equivalent shortest path model, the one with the minimum Bayesian information criterion (BIC) [14]:

$$BIC_k = -2\log L(M_k) + t_k \log(n)$$

where $L(M_k)$ is the likelihood of the shortest path model M_k, t_k is the number of parameters of M_k, and n is the sample size. BIC is a penalized likelihood score that penalize short path models with many rather than few parameters. This is motivated by the idea that genes that functionally important for a given phenotype tend to be connected by few intermediate nodes.

After the selection of the best equivalent shortest paths between each couple of DEGs, the next step is to detect a disease module. At this aim, the unique nodes (genes) of the selected shortest paths were merged to create a unique graph that should represent the final "disease" module. The evaluation of the final module was obtained: 1) performing an enrichment analysis on the nodes present in the module, and 2) fitting the whole module with SEM. The enrichment analysis was obtained by the R package `clusterProfiler` [15]. While the SEM analysis was performed adding, in each structural equation of the whole module, the indicator variable of group condition (experimental/control), and testing which genes in the module are influenced by the different conditions. To note that the output of this last validation procedure is similar to the t-test with the important difference that the nodes are evaluated taking in consideration the model and not singularly. The badness of fit of the whole module with (p-1) genes is evaluated by the Standardized Root-Mean-square Residual (SRMR), a measure based on the differences between all pairwise sample (s) and model (σ) gene-gene and gene-group covariances

$$SRMR = \frac{\sum_{j=1}^{p}\sum_{k=j+1}^{p}(s_{jk} - \sigma_{jk})^2 / s_{jj}s_{kk}}{p(p+1)/2}$$

A value <0.10 is retained for adequate fitting approximation of the model to the data, whereas value <0.05 may be considered as a good fit [16]. If the number of genes in the gene set is greater than the number of samples, the shrinkage covariance proposed by Schafer and Strimmer [17] is applied to estimate the sample covariances given by R package parcor.

The proposed pipeline was performed using a gene expression experiment available on the Gene Expression Omnibus (GEO) database [18], with accession number GSE14580 [19]. Two groups were compared, patients affected by ulcerative colitis (UC) treated with infliximab (8 samples) against control (6 samples).

3   Results

Differential analysis using a t-test with a Benjamini-Hochberg correction, allowed us to individuate 1364 differentially expressed genes (DEGs). On these, a pathway analysis with the SPIA revealed five perturbed KEGG pathways: cytokine-cytokine receptor interaction, cell adhesion molecules, chemokine pathway, antigen processing and presentation, complement and coagulation cascades (Table 1).

Table 1. The KEGG perturbed pathways associated with the response of UC patients treated with infliximab.

Name	pSize	NDE	pNDE	pPERT	pGFdr	Status
Cytokine-cytokine receptor interaction	241	27	0,004	0,001	0,007	Activated
Cell adhesion molecules (CAMs)	88	16	0,000	0,266	0,019	Activated
Chemokine signaling pathway	178	18	0,041	0,001	0,019	Activated
Antigen processing and presentation	56	11	0,001	0,082	0,019	Activated
Complement and coagulation cascades	65	11	0,003	0,025	0,019	Inhibited

pSize is the number of genes in the pathway; NDE is the number of DEGs in the pathway; pNDE is the probability of obtaining a number of DE genes on the given pathway at least as large as the observed one; pPERT is the probability to observe the total accumulated perturbation of the pathway; pGFdr is the combination of pNDE and pPERT corrected for the false discovering rate.

The perturbed pathways are interesting considering that some of them play a fundamental role in the symptoms of UC, such as those relative to cytokines and chemokines [20].

All the perturbed pathways were merged in a unique graph. This allows to understand also how the gene belonging to different pathways interact between them, Starting from this graph, in which the DEGs act, the best shortest paths between each couple of DEGs were computed, as described in the methods. After merging the 5 pathways in a unique graph, only 22 DEGs over 83 were present in the perturbed network. The number of shortest equivalent paths for each couple of connected DEGs was in the range with a minimum value of 2 and a maximum value of 648. The best shortest paths were of 53 composed with a minimum of 3 nodes and a maximum of 11. The merging of all selected shortest paths gave a putative "disease" module composed by 41 genes and 47 edges, as illustrated in the Figure 1. The information about genes are reported in Table 2.

To identify which molecular functions are involved in the infliximab molecular action, GO enrichment analysis on the molecular function (MF) was performed. The significant (p-value <0.001) enriched MF terms are illustrated in the Figure 2. The main MFs are connected with chemokine and cytokine activity as expected (see Table 2). In fact the infliximab acts on the expression of the tumor necrosis factor α (TNF-α), a key proinflammatory cytokine resulted over-expressed in many diseases as Crohn's disease and UC [21]. In addition, the G-protein receptor binding function was associated with the intestinal inflammation [22].

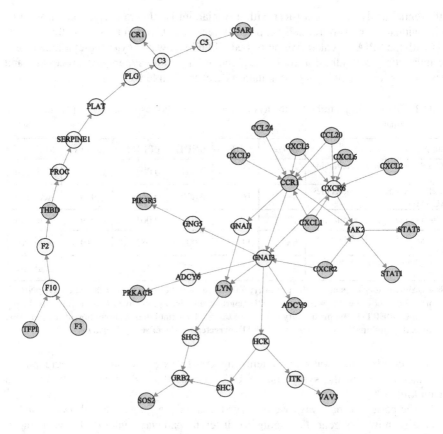

Fig. 1. The "disease" module obtained by merging of the best shortest paths between every pair of DEGs. The green nodes are DEGs, while the yellow nodes represent non-DEGs, present in the microarray.

Table 2. Description of genes that characterize the best equivalent shortest paths. Entrez id, gene symbols and gene name are reported.

ENTREZ ID	GENE SYMBOL	Gene Name
6655	SOS2	son of sevenless homolog 2 (Drosophila)
115	ADCY9	adenylate cyclase 9
2147	F2	coagulation factor II (thrombin)
112	Adcy6	adenylate cyclase 6
6774	Stat3	signal transducer and activator of transcription 3
2919	CXCL1	chemokine (C-X-C motif) ligand 1
6772	STAT1	signal transducer and activator of transcription 1, 91kDa
6464	SHC1	SHC transforming protein 1
2787	gng5	guanine nucleotide binding protein (G protein), gamma 5
1230	CCR1	chemokine (C-C motif) receptor 1
53358	SHC3	SHC transforming protein 3
3702	ITK	IL2-inducible T-cell kinase
6369	CCL24	chemokine (C-C motif) ligand 24
2770	Gnai1	guanine nucleotide binding protein (G protein)
728	C5ar1	complement component 5a receptor 1
727	C5	complement component 5
6372	Cxcl6	chemokine (C-X-C motif) ligand 6
5327	PLAT	plasminogen activator, tissue
3579	CXCR2	interleukin 8 receptor, beta
2152	F3	coagulation factor III (thromboplastin, tissue factor)
3055	HCK	hemopoietic cell kinase
2159	F10	coagulation factor X
7035	TFPI	tissue factor pathway inhibitor
8503	PIK3R3	phosphoinositide-3-kinase, regulatory subunit 3 (gamma)
2920	CXCL2	chemokine (C-X-C motif) ligand 2
5567	PRKACB	protein kinase, cAMP-dependent, catalytic, beta
2921	Cxcl3	chemokine (C-X-C motif) ligand 3
2773	GNAI3	guanine nucleotide binding protein (G protein)
5624	proC	protein C (inactivator of coagulation factors Va and VIIIa)
5054	SERPINE1	serpin peptidase inhibitor
5340	plg	plasminogen
3717	Jak2	Janus kinase 2
4283	Cxcl9	chemokine (C-X-C motif) ligand 9
6364	CCL20	chemokine (C-C motif) ligand 20
10663	CXCR6	chemokine (C-X-C motif) receptor 6
718	LOC653879	similar to Complement C3 precursor;
2885	GRB2	growth factor receptor-bound protein 2
1378	CR1	complement component (3b/4b) receptor 1
4067	LYN	v-yes-1 Yamaguchi sarcoma viral related oncogene
10451	VAV3	vav 3 guanine nucleotide exchange factor
7056	THBD	thrombomodulin

The module obtained was globally evaluated with SEM, adding the effect of the group conditions on each gene in the module, as showed in Table 3. The fitting of the model was adequate as the SRMR was of 0.073. The results confirms DEGs and the enrichment analysis. Many genes associated to the chemokine and cytokine activities as CXCL1, CXCL2, CXCL3, CXCL6, CXCR2 resulted have an effect ($P<0.05$, with Bonferroni multiple comparison correction) on the molecular mechanisms response to the infliximab treatment.

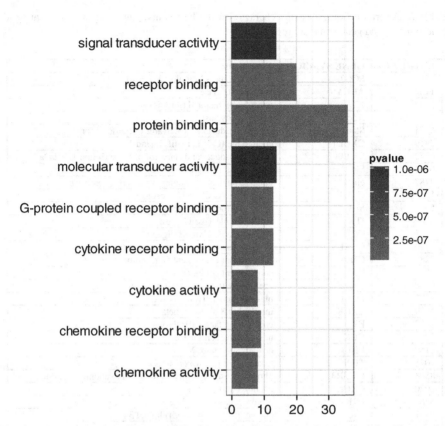

Fig. 2. GO molecular functions enriched by the genes present in the module. The x-axis represents the number of genes associated with the MF of each bar. The color is associated to the significance of the enrichment.

Table 3. Fitting of gene module with SEM, adding the indicator variable of the group condition. The analysis revealed which genes are influenced by the indicator variable representing the different biological conditions.

Y ← X	Estimate	SE	z-value	P*(>|z|)
CXCL2 ← group	2.355	0.271	8.687	0
CXCR6 ← group	0.107	0.403	0.265	1
Jak2 ← group	0.602	0.428	1.407	1
STAT1 ← group	0.913	0.23	3.968	0.002972
GNAI3 ← group	0.417	0.348	1.196	1
LYN ← group	1.306	0.278	4.703	0.000105
gng5 ← group	-0.259	0.056	-4.648	0.000137
PIK3R3 ← group	1.095	0.154	7.13	0
STAT3 ← group	0.909	0.219	4.157	0.001322
ADCY6 ← group	-0.495	0.053	-9.298	0
PRKACB ← group	-0.881	0.204	-4.309	0.000672
ADCY9 ← group	-0.657	0.068	-9.645	0
HCK ← group	1.051	0.152	6.923	0

Table 3. *(Continued)*

Y ← X	Estimate	SE	z-value	P*(>|z|)
SHC1 ← group	0.2	0.145	1.384	1
GRB2 ← group	0.51	0.099	5.172	0
SOS2 ← group	-0.842	0.211	-3.998	0.002619
ITK ← group	0.525	0.212	2.475	0.546268
VAV3 ← group	-1.31	0.23	-5.697	0
SHC3 ← group	0.097	0.186	0.524	1
THBD ← group	2.074	0.332	6.254	0
proC ← group	-0.765	0.648	-1.18	1
SERPINE1 ← group	0.519	0.401	1.294	1
PLAT ← group	0.372	0.118	3.147	0.067632
PLG ← group	-1.014	0.344	-2.95	0.130287
C3 ← group	1.64	0.54	3.035	0.098621
CR1 ← group	2.454	0.474	5.178	0
C5 ← group	-1.098	0.651	-1.686	1
C5AR1 ← group	1.685	0.272	6.184	0
CXCL9 ← group	1.896	0.211	8.976	0
CCR1 ← group	0.085	0.605	0.141	1
GNAI1 ← group	-0.224	0.219	-1.023	1
F3 ← group	2.301	0.219	10.501	0
F10 ← group	0.211	0.74	0.286	1
F2 ← group	0.839	0.19	4.417	0.000411
CXCL1 ← group	5.044	0.48	10.498	0
CCL20 ← group	2.811	0.284	9.897	0
CXCL6 ← group	4.083	0.44	9.286	0
CXCR2 ← group	3.247	0.361	8.986	0
CXCL3 ← group	4.04	0.389	10.386	0
TFPI ← group	1.669	0.181	9.23	0
CCL24 ← group	2.887	0.301	9.582	0

Estimate: mean gene differences adjusted by the gene topology of Figure1; SE,: standard error of the estimate; z-value: z.test=estimate/SE; P*(>|z|): pvalue (two-sided, adjusted with Bonferroni procedure)

4   Conclusions

Nowadays we know that rarely a disease is a consequence of the modification of the sequence or expression of a single gene, but rather it is the consequence of the perturbation of molecular interaction networks [1]. The reason is that proteins, nucleic acids, and small molecules form a dense network of molecular interactions in a cell. A critical level of the biological organization is the recognition of functional modules, networks composed by molecules which interactions and cooperation execute specific functions. In fact, it has been proposed that the biological networks are characterized by a modular architecture [23]. The identification of the perturbed modules is a critical part for the understanding of the biological phenomena. Efficient network reduction algorithms are proposed for this goal, such as the distance-based and information flow-based algorithms.

The first method is based on the principle that genes physically or functionally proximal cooperate for the same biological function. The second simulates the information propagation in the biological network, by starting a random walk from a source node, passing through transient nodes, and eventually ending in a sink node. The efficiency of these methods, however, depends on the choice of user-defined parameters, which are time-expensive to tune correctly. Moreover, it is often impossible to assess a clear correlation between the parameter and the corresponding outcome.

In this note, we proposed the identification of modules with the SEM best shortest path via the minimum BIC, a penalized likelihood score. Our approach is a mixed strategy based on hypothesis-driven of biological knowledge contained in KEGG database, and data-driven of the best shortest path between genes, among the set of equivalent shortest path. The procedure has been tested on a gene expression experiment, with the goal of clarifying the action of infliximab on patients affected by UC. The first step was to find DEGs and then performing pathway analysis to understand which pathways were involved in the drug response. The way to obtain DEGs and perturbed pathways can be changed on the basis of the preferences of the researcher. After identifying perturbed pathways, we considered the network obtained from merging significant pathways. We found all the best equivalent shortest paths involving pairs of DEGs, and then we generated the final module by merging the best shortest paths. The module obtained was composed by 41 genes and 47 edges. Our results were encouraging, considering that we found MF terms associated with chemokines and cytokines playing a key role in UC. The procedure tried to find a trade-off between complexity, an intrinsic characteristic of biological network, and the understanding of the role and importance of DEGs from a network point of view. The choice to consider one shortest path between two DEGs, with the mixed-approach proposed, outline this goal. However, information is surely lost, as there can be other ways that connect two DEGs. A possible extension could be to select the k-SEM best shortest paths also if in this way it is necessary to choose the k value, a choice that could be arbitrary. A better way to improve the procedure is to exploit the potentiality of SEM. For example adding in the shortest path equations the indicator variable of the group condition, thus minimum BIC score has the best mean differences between groups, or reducing the intermediate nodes in composite variable [24]. A further development would be to insert the procedure in the pipeline proposed by Pepe & Grassi [8], considering the problem of the equivalent shortest path in the generation of the pathway model.

References

1. Barabási, A.L., Gulbahce, N., Loscalzo, J.: Network medicine: a network-based approach to human disease. Nat. Rev. Genet. 12(1), 56–68 (2011). doi:10.1038/nrg2918.
2. Wang, X., Gulbahce, N., Yu, H.: Network-based methods for human disease gene prediction. Brief. Funct. Genomics 10(5), 280–293 (2011). doi:10.1093/bfgp
3. Cho, D.Y., Kim, Y.A., Przytycka, T.: Chapter 5: Network biology approach to complex diseases. PLoS Comput. Biol. 8(12), e1002820 (2012). doi:10.1371/journal.pcbi.1002820
4. Shih, Y.K., Parthasarathy, S.: A single source k-shortest paths algorithm to infer regulatory pathways in a gene network. Bioinformatics 28(12), i49–i58 (2012). doi:10.1093/bioinformatics
5. Stojmirović, A., Yu, Y. I.: Probe: analyzing information flow in protein networks. Bioinformatics 25(18), 2447–2449 (2009). doi:10.1093/bioinformatics

6. Gene Ontology Consortium, Gene Ontology annotations and resources. Nucleic Acids Research 41(D1), D530–D535 (2013)
7. Kanehisa, M., Goto, S., Kawashima, S., Okuno, Y., Hattori, M.: The KEGG resource for deciphering the genome. Nucleic Acids Res. 32(database issue), D277–D280 (2004)
8. Pepe, D., Grassi, M.: Investigating perturbed pathway modules from gene expression data via structural equation models. BMC Bioinformatics 15(1), 132 (2014)
9. Bollen, K.A.: Structural equations with latent variables. Wiley, New York (1989)
10. Tarca, A.L., Draghici, S., Khatri, P., Hassan, S., Mital, P., Kim, J., Kim, C., Kusanovic, J.P., Romero, R.: A novel signaling pathway impact analysis for microarray experiments. Bioinformatics 25, 75–82 (2009)
11. R Development Core Team R: A language and environment for statistical computing. R Foundation for Statistical Computing, Vienna, Austria. (2012). http://www.R-project.org/
12. Zhang, J.D., Wiemann, S.: KEGGgraph: a graph approach to KEGG PATHWAY in R and Bioconductor. Bioinformatics 25(11), 1470–1471 (2009)
13. Csardi, G., Nepusz, T.: The igraph software package for complex network research. Inter Journal, Complex Systems 1695 (2006). http://igraph.sf.net
14. Konishi, S., Kitagawa, G.: Bayesian information criteria. Information Criteria and Statistical Modeling, 211–237 (2008)
15. Yu, G., Wang, L.G., Han, Y., He, Q.Y.: clusterProfiler: an R package for comparing biological themes among gene clusters. Omics: A Journal of Integrative Biology 16(5), 284–287 (2012)
16. Hu, L.T., Bentler, P.M.: Cutoff criteria for fit indexes in covariance structure analysis: Conventional criteria versus new alternatives. Structural Equation Modeling: A Multidisciplinary Journal 6(1), 1–55 (1999)
17. Schäfer, J., Strimmer, K.: A shrinkage approach to large-scale covariance matrix estimation and implications for functional genomics. Statistical Applications in Genetics and Molecular Biology 4(1) (2005)
18. Barrett, T., Wilhite, S.E., Ledoux, P., Evangelista, C., Kim, I.F., Tomashevsky, M., Marshall, K.A., Phillippy, K.H., Sherman, P.M., Holko, M., Yefanov, A., Lee, H., Zhang, N., Robertson, C.L., Serova, N., Davis, S., Soboleva, A.: NCBI GEO: archive for functional genomics data sets–update. Nucleic Acids Res. 41(database issue), D991–D995 (2013)
19. Arijs, I., Li, K., Toedter, G., Quintens, R., et al.: Mucosal gene signatures to predict response to infliximab in patients with ulcerative colitis. Gut 58(12), 1612–1619 (2009)
20. Autschbach, F., Giese, T., Gassler, N., Sido, B., Heuschen, G., Heuschen, U., Meuer, S.C.: Cytokine/chemokine messenger-RNA expression profiles in ulcerative colitis and Crohn's disease. Virchows Archiv 441(5), 500–513 (2002)
21. Rutgeerts, P., Sandborn, W.J., Feagan, B.G., Reinisch, W., Olson, A., Johanns, J., Colombel, J.F.: Infliximab for induction and maintenance therapy for ulcerative colitis. New England Journal of Medicine 353(23), 2462–2476 (2005)
22. Sina, C., Gavrilova, O., Förster, M., Till, A., Derer, S., Hildebrand, F., Rosenstiel, P.: G protein-coupled receptor 43 is essential for neutrophil recruitment during intestinal inflammation. The Journal of Immunology 183(11), 7514–7522 (2009)
23. Hartwell, L.H., Hopfield, J.J., Leibler, S., Murray, A.W.: From molecular to modular cell biology. Nature 402, C47–C52 (1999)
24. Pepe, D., Grassi, M.: Pathway Composite Variables: A Useful Tool for the Interpretation of Biological Pathways in the Analysis of Gene Expression Data. Studies in theoretical and Applied statistics. Springer (2014)

The General Regression Neural Network to Classify Barcode and mini-barcode DNA

Riccardo Rizzo, Antonino Fiannaca, Massimo La Rosa, and Alfonso Urso

ICAR-CNR, National Research Council of Italy,
viale delle Scienze Ed.11, 90128 Palermo, Italy
{ricrizzo,fiannaca,larosa,urso}@pa.icar.cnr.it

Abstract. In the identification of living species through the analysis of their DNA sequences, the mitochondrial "cytochrome c oxidase subunit 1" (COI) gene has proved to be a good DNA barcode. Nevertheless, the quality of the full length barcode sequences often can not be guaranteed because of the DNA degradation in biological samples, so that only short sequences (mini-barcode) are available. In this paper, a prototype-based classification approach for the analysis of DNA barcode, exploiting a spectral representation of DNA sequences and a memory-based neural network, is proposed. The neural network is a modified version of General Regression Neural Network (GRNN) used as a classification tool. Furthermore, the relationship between the characteristics of different species and their spectral distribution is investigated. Namely, a subset of the whole spectrum of a DNA sequence, composed by very high frequency DNA k-mers, is considered providing a robust system for the classification of barcode sequences. The proposed approach is compared with standard classification algorithms, like Support Vector Machine (SVM), obtaining better results specially when applied to mini-barcode sequences.

Keywords: DNA barcode, Memory-based Neural Networks, GRNN, Classification.

1 Scientific Background

The identification of living species through the analysis of their DNA sequences is an open challenge. Because a massive comparison of a large collection of full genome sequences is not feasible, a bioinformatics approach to this problem is the analysis of some standard gene regions, containing enough information for the assignment to the proper taxa. The mitochondrial "cytochrome c oxidase subunit 1" (COI) gene is a comprehensive species-specific sequence library for all eukaryotes and it has proved to be a good marker for DNA sequences [8,13]; for this reason, it is considered as a DNA barcode for metazoan genomic.

Anyway, even though DNA barcode approach has proven to be useful for the identification and taxonomic rank assignment of very different species [6,12,11], its use can still be difficult if the biological samples are degraded. This is the case of archival specimen where biological samples can not guarantee the quality of

© Springer International Publishing Switzerland 2015
C. di Serio et al. (Eds.): CIBB 2014, LNCS 8623, pp. 142–155, 2015.
DOI: 10.1007/978-3-319-24462-4_13

the full length barcode sequence (650 bp) recovery. In fact, in many cases, only short sequences, also known as mini-barcode, are available (about 200 bp) [14].

In this paper, we propose a novel prototype-based classification approach based on the analysis of DNA barcode. Our method exploits a spectral representation of DNA sequences and a memory-based neural network for taxa estimation: spectral representation uses fixed-length DNA k-mers, whereas the neural network can store a set of prototypes (groups of k-mers) representing all the elements of the learning dataset.

In order to perform the barcode sequences classification, we introduce a modified version of General Regression Neural Network (GRNN) [19] that use, alternatively, a function derived from Jaccard distance and fractional distance (instead of the euclidean one) to compare learned prototypes against test sequences. The proposed approach implements these two kinds of distances and it is able to perform the classification task, even using only short fragments (200 bp) of the complete barcode sequence.

Finally, we compared our approach with the Support Vector Machine (SVM) [17] classification algorithm. Results show our method, implementing both Jaccard and fractional distances, is directly comparable with SVM in terms of classification metrics (accuracy, precision and recall) when considering full length sequences, whereas it overcomes SVM classifier when applied to short fragments of DNA barcode sequences.

2   Materials and Methods

The proposed method is based on two modified versions of the General Regression Neural Network. In the first subsection the basic principles of the GRNN are explained; the following two subsections present the proposed modifications, based on the Jaccard distance and the fractional distance; the last subsection describes the data sets used.

2.1   The General Regression Neural Network

The General Regression Neural Network [19] is a neural network created for regression i.e. the approximation of a dependent variable y given a set of sample $(\mathbf{x}, y)$, where $\mathbf{x}$ is the independent variable. In the following we will discuss the single output case, the extension to an output vector $\mathbf{y}$ is straightforward and can be found in [19]. In order to implement our classification tool for DNA sequences, we obtained the vector representation of the DNA sequences using a k-mer decomposition [10]. In this representation, sequences are coded by using fixed size vectors whose components are the number of occurrences of DNA snippets of k fixed-length, called k-mers. Considering $k = 5$, as proposed in [10], we have representing vectors $\mathbf{x} \in \Re^{1024}$.

The GRNN network has a one–pass training phase, it is just the memorization of all the training couples $(\mathbf{x_i}, y_i)$ each one in a neural unit i of the hidden layer. Fig. 1 shows a representation of the network: input layer has one neuron for each

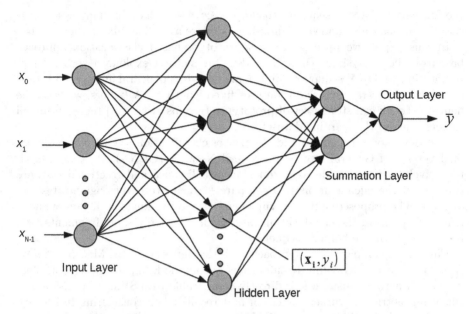

Fig. 1. The representation of the GRNN neural network.

component x_j in the input vector $\mathbf{x_i}$ and the hidden layer has one unit for each training sample.

During the test phase, when an unknown pattern $\mathbf{x}'$ is presented to the network, each hidden unit is excited according to the similarity of the pattern to the memorized input sample $\mathbf{x_i}$. The excitation level of the neural unit i is given by:

$$w_i = \exp\left\{-\frac{d(\mathbf{x}', \mathbf{x_i})}{2\sigma^2}\right\} \tag{1}$$

where $d(\mathbf{x}', \mathbf{x_i})$ is the distance between $\mathbf{x}'$ and $\mathbf{x_i}$, usually euclidean distance, and σ is the spread factor, representing the only parameter of the GRNN network. The hidden units have two outputs, $w_i * y_i$ and w_i, that are collected by the summation units. All the contributions of the hidden units are summed and normalized by the unit in the output layer, in order to obtain the network output y':

$$y' = \frac{\sum w_i * y_i}{\sum w_i}. \tag{2}$$

If we want to use this network as a classification tool we have to change our point of view: first of all the training couples are of the kind $(\mathbf{x_i}, c_h)$ where $c_h \in C$ is the class that is associated to $\mathbf{x_i}$ and C is the set of h classes. This means that all the classes must be coded in a real value vector $\mathbf{y_h}$ were each component y_h^i is given by:

$$y_h^i = \begin{cases} 0 \text{ if } i \neq h \\ 1 \text{ if } i = h \end{cases} \tag{3}$$

and requires a multiple output network. The σ value is the only parameter of the GRNN network. There are some studies on the optimal value of σ that can be a single value for the whole network or a specific value for each hidden unit. In [7] it is suggested a formula that depends on the maximum distance and number of patterns in the training set.

2.2   Jaccard Function

During experiments (see Section 3) we found that the euclidean distance used in the GRNN calculations was not enough "strong" for our purposes: in these kind of problems we have found that the presence or absence of a k-mer mean a lot and the euclidean distance does not emphasize this aspect [3].

Jaccard distance is defined among two sets A and B as

$$ J = \frac{\|A \cup B\| - \|A \cap B\|}{\|A \cup B\|} \tag{4} $$

where $\|A\|$ represents the number of elements of the set A; $J \in [0, 1]$ and if $J = 1$ the two sets have not elements in common.

In this work, we redefine the Jaccard distance as a new function between two vectors that considers we will define between two vectors is computed considering the number of components in common between them, so that A and B in Eq. 4 are the set of the indexes of non-zero components in the vectors. This distance, however, has still a low contrast for our purposes because the information related to the magnitude of each component in the vectors is discarded. The component magnitude can be taken into account again if we do not consider all the components but only the m biggest components in the vectors.

More formally the sets A and B are defined as the sets of indexes:

$$ s = \{s_0, s_1, ..., s_m\} \tag{5} $$

where s_j is the index of the x components that satisfies the ordering $x_{s_{j-1}} > x_{s_j} > x_{s_{j+1}}$ where $x_0 = max_{l=0,1,...N}\{x_l\}$

These sets can be compared using the Jaccard distance in Eq. 4.

Since Jaccard distance ranges from 0 to 1, it is necessary to map it in the interval $[0, \infty)$, using, for example, the following definition that we call J-function:

$$ Jf(\mathbf{x'}, \mathbf{x_i}) = \begin{cases} 0 & \text{if } J = 0 \\ \|\frac{1}{logJ}\| & \text{if } J \in (0, 1) \\ \infty & \text{if } J = 1 \end{cases} \tag{6} $$

This is necessary because w_i in eq. 1 should tend to zero if the distance is large. Moreover changing the distance method from euclidean to this normalised J-function distance stretches the original theory of the GRNN network; we leave a formal study of this problem to a future work.

2.3 Fractional Distances

High-dimensional spaces, such as the one defined by the sized vectors representing DNA sequences, are affected by the so called *curse of dimensionality*. In those spaces, in fact, the euclidean norm used to define the distance tend to *concentrate* [4]. That means all pairwise distances between high-dimensional objects appear to be very similar. In order to overcome this phenomenon, fractional norms can be used in place of euclidean norm [9,1]. Fractional norms are obtained from the Minkowski family norms defined as:

$$\|\mathbf{X}\|_p = \left(\sum_i |X_i|^p \right)^{\frac{1}{p}}. \tag{7}$$

With $p = 2$, the euclidean norm is obtained; whereas with $0 < p < 1$ Minkowski norms are called fractional norms, which induce fractional distances. In this work we adopted fractional norms, considering different values of p, in order to compute Eq. 1 and to limit the effects of the curse of dimensionality.

2.4 Barcode Dataset

We downloaded barcode sequences from the Barcode of Life Database (BOLD) [15]. In our study, we considered 10 barcode datasets belonging to different BOLD projects and living organisms. These datasets have been selected according to some criteria: we chose only *barcode compliant* dataset, i.e certified by BOLD as true barcode sequences, with sequence length not shorter than 500 bp and not longer than 800 bp. Following these criteria, we collected 2212 sequences. A full description of our barcode dataset can be retrieved in [3].

As discussed earlier, it is important to find a subset of the barcode gene in order to provide an effective identification mechanism for various animal or bacterial groups [6]. In fact, the recovery of full-length barcode sequence can be a problem in many cases: for example considering archival specimen, due to DNA degradation [5], or environmental samples [14]. There are studies, however, that tries to identify a specific location in the barcode gene, location that are called mini-barcode [14]. Our work is focused on the same idea but, instead of trying to identify a specific location in the gene, we explore the possibility of identifying species using small gene chunks. So that we fixed an amount of genetic material (200 bp) that could be enough to identify 95% of the species [14] and tried to understand what happens if this material comes not from a specific location of the gene, but it is scattered in two chunks of 100 bp (100x2), or in four subsequences of 50 bp (50x4). In both cases we do not check if these subsequences are overlapping or not, trying to reproduce laboratory conditions.

3 Results

Classification results obtained through our GRNN approach have been evaluated in terms of accuracy, precision and recall scores. We implemented both

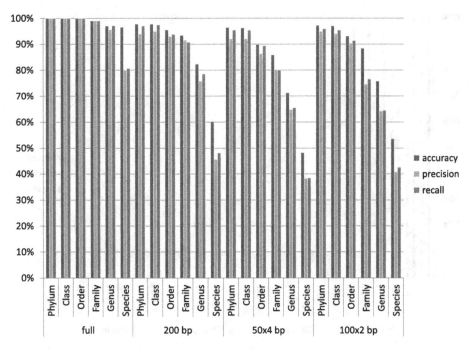

Fig. 2. Classification scores, in terms of accuracy, precision and recall, for the proposed approach based on the GRNN algorithm. The scores are arranged with regards to the taxonomic ranks and sequence sizes.

modified versions of GRNN algorithm using J-function and fractional distances, as explained in Sections 2.2 and 2.3, respectively. We performed two types of training/testing procedures. In the first scenario, we trained our classifiers with the whole full length dataset and then we tested it with all the sequence fragments. Our aim was in fact to assess if the GRNN classifier, trained with the full length sequences, is able to correctly classify the sequence fragments. It is important to underline that although in the training set (full length) and in the test set (fragments) there are the same number of sequences, their vector representations are completely different. In the second scenario, we adopted a ten fold cross validation scheme, considering as training set the full length sequences and as test set corresponding sequence fragments that did not belong to the training set. In this situation we wanted to assess if the GRNN classifier is able to classify sequence fragments even if it did not learned the corresponding full length patterns. Moreover we compared our method with another classifier used for nucleotide sequences classification [18]: the Support Vector Machine (SVM). We adopted the SVM implementation provided by the R package *e1071*, which is an interface to the *libsvm* library [2]. We used a Gaussian Radial Basis kernel, $k(\mathbf{x}, \mathbf{x'}) = exp(-\gamma \|\mathbf{x} - \mathbf{x'}\|^2)$. The parameter C and γ of the Gaussian kernel has been tuned through a grid search over a set of parameters values: γ ranging from 10^{-6} to 10^3; C ranging from 1 to 10^3, as suggested by the authors of *libsvm*.

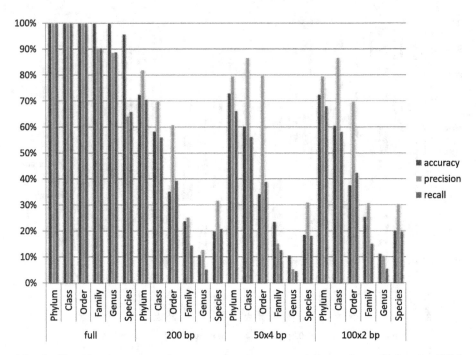

Fig. 3. Classification scores, in terms of accuracy, precision and recall, for the SVM classifier. The scores are arranged with regards to the taxonomic ranks and sequence sizes.

The best parameter values have been computed minimising the error measure using a 10-fold cross validation on the training set. The results obtained with the SVM classifiers have been carried out by means of the same two way training/testing procedure. In other experiments, not shown here, we also adopted Classification Tree, from the R package *rpart*, algorithm and we obtained very similar results to the ones obtained through SVM. For this reason, we do not present those results in this paper.

The GRNN outputs obtained using J-function have been obtained comparing the $m = 30$ biggest components between the vector prototypes and the test vectors. This value has been selected after a series of experiments, not presented here for lack of available space, using for comparison a number of components ranging from 20 to 50 and a σ value during the training phase ranging from 0.5 to 0.8. We reached a trade off between the best results and the smallest number of elements by using 30 components and $\sigma = 0.7$. A number of components $m > 30$ does not give a meaningful improvement in the results.

In the previous version of our work [16], we focused our attention on the comparison of classification results between the GRNN algorithm with J-function and the SVM classifier. Those results, arranged according to the test sequence sizes (full, 200 bp, 50x4 bp, 100x2 bp) and taxonomic ranks (from phylum down to species), are summarized in the charts shown in Fig. 2 and 3. We demonstrated

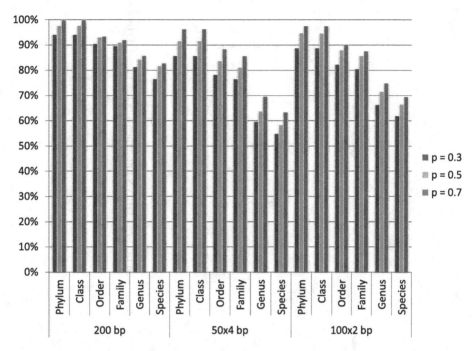

Fig. 4. Classification scores, in terms of accuracy, for the proposed approach based on the GRNN algorithm implementing fractional distances with different values of parameter p. The scores are arranged with regards to the taxonomic ranks and sequence sizes.

that our approach clearly outperforms SVM for the classification of sequence fragments; considering full length sequences, on the other hand, both GRNN and SVM classifiers reached very similar high scores, ranging between 100% and 95% of accuracy score.

In this work we analysed the performances of the GRNN algorithm implementing fractional distances in order to classify short barcode sequences of size 200 bp, 100x2 bp and 50x4 bp. Classification results, in terms of accuracy, precision and recall scores, have been compared with both the GRNN algorithm using J-function and the SVM classifier.

First of all, considering the first training/testing scenario, we studied how classification results change with regard to the parameter p of fractional distances (see Eq. 7). We carried out experiments with $p = 0.3$, $p = 0.5$ and $p = 0.7$ and the classification results, in terms of accuracy score are shown in Fig. 4. The most interesting result is that the best scores, ranging from 100% at phylum level to about 82% at species level, were obtained with $p = 0.7$ considering short fragments of 200 consecutive base pairs. With 50x4 bp and 100x2 bp sequence lengths, we obtained slightly lower scores. The analysis of precision and recall showed very similar scores to the accuracy one, not providing any further meaningful information: for this reason we did not report those results.

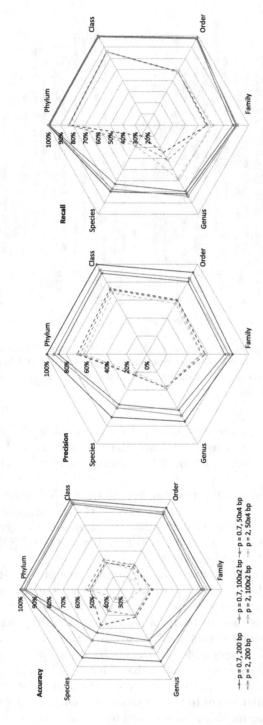

Fig. 5. Classification scores, in terms of accuracy, precision and recall, of the comparison between GRNN implementing fractional distance ($p = 0.7$) and euclidean distance ($p = 2$).

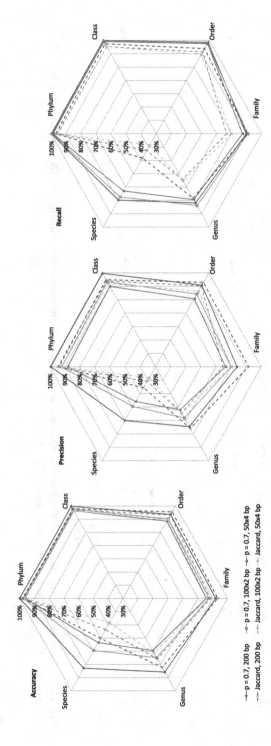

Fig. 6. Classification scores, in terms of accuracy, precision and recall, of the comparison between GRNN implementing fractional distance ($p = 0.7$) and Jaccard distance.

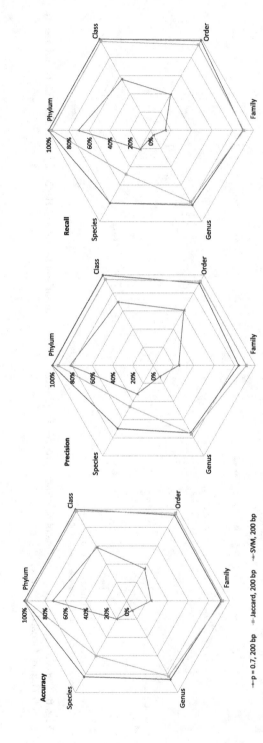

Fig. 7. Classification scores, in terms of accuracy, precision and recall, of the comparison among GRNN implementing both fractional distance ($p = 0.7$) and Jaccard distance, and SVM.

Table 1. Classification scores, in terms of accuracy, precision and recall, for the proposed approach based on the GRNN algorithm against the SVM classifier. The training/testing procedure refers to the second scenario, i.e. ten fold cross validation with full-length sequences as training set and test set composed of the sequence fragments whose corresponding complete sequences do not belong to the training set.

10-fold - 200 bp						
Distance	Phylum	Class	Order	Family	Genus	Species
ACCURACY						
J-function	85.8%	85.6%	78.5%	91.8%	66.7%	56.2%
SVM	75.9%	69.1%	36.8%	23.6%	11.3%	18.1%
p = 0.3	84.4%	84.0%	76.8%	89.7%	63.3%	57.1%
p = 0.5	85.1%	83.8%	79.9%	89.8%	67.4%	57.2%
p = 0.7	84.8%	86.7%	78.5%	91.3%	64.4%	57.6%
PRECISION						
J-function	79.4%	80.1%	78.1%	80.6%	58.4%	46.8%
SVM	45.5%	52.3%	48.7%	19.1%	4.4%	14.1%
p = 0.3	77.8%	80.2%	76.6%	87.0%	55.2%	46.7%
p = 0.5	77.2%	76.6%	77.8%	83.9%	60.6%	46.1%
p = 0.7	77.3%	77.2%	76.4%	80.8%	58.6%	49.9%
RECALL						
J-function	72.7%	75.2%	70.4%	75.0%	54.1%	45.0%
SVM	55.8%	53.7%	36.9%	14.6%	5.8%	14.5%
p = 0.3	67.8%	67.1%	65.3%	77.8%	48.4%	44.7%
p = 0.5	72.7%	72.1%	69.8%	78.3%	56.7%	45.1%
p = 0.7	75.6%	80.9%	72.6%	78.5%	56.1%	50.3%

Experiment trials with fractional distance with $p = 0.7$, therefore, is able to overcome the distance concentration phenomenon, as described in Section 2.3. In order to validate that thesis, we compared classification results obtained with GRNN using fractional distance ($p = 0.7$) and euclidean distance ($p = 2$). This comparison is presented by means of radar charts in Fig. 5. There it is clear how, at all taxonomic level and for each sequence length, euclidean distance is unable to provide acceptable results (accuracy score about 50%).

The comparison between classification results obtained using GRNN with fractional distance ($p = 0.7$) and Jaccard distance is summarized in the radar charts of Fig. 6. Once again the best results, at each taxonomic rank and for all sequence sizes, were reached by means of fractional distance. The most evident difference of the performances between those two approaches is at Species level, where with fractional distance we reached an accuracy score of about 80% against 60% with Jaccard distance.

The last comparison, showed in Fig. 7, was done considering GRNN with fractional distance ($p = 0.7$) and Jaccard distance against SVM. As discussed earlier in this Section, Fig. 3, both approaches implementing GRNN clearly outperforms the SVM classifier.

From these results it is evident that SVM is not able to deal with sequence fragments. In fact, the sequence fragments (mini-barcode) have a vector representation that is very different from the one computed for the original sequences, therefore SVM, from this point of view, can not correctly classify those fragments. Otherwise, our approach, considering both the Jaccard distance and the fractional distance, demonstrate the ability to to provide very reliable classification results when dealing with sequence fragments.

Finally, with regards to the second training/testing scenario, we obtained the classification results summarized in Table 1. We reported only the scores related to the 200 bp fragments because they are very similar to the ones obtained with 100x2 and 50x4 bp fragments. In this situation, although with lower scores, the GRNN algorithm, especially in the case of fractional distances, provides consistent classification performances and it outperforms the SVM classifier. Classification scores are lower with respect to the first scenario because the GRNN has never learned the full length sequences corresponding to the test fragments. In spite of that, our GRNN approach turned out to be robust enough to keep on providing acceptable classification scores.

4   Conclusion

In this work we presented a classification methodology for barcode DNA sequences based on the General Regression Neural Network algorithm. We introduced two modified versions of GRNN in order to overcome limitations of the standard euclidean approach: the first one implements the J-function derived from the Jaccard distance, the second one adopts fractional distances. We obtained very accurate and very robust classifiers with respect to sequence sizes. We tested our approaches, in fact, considering the so-called mini-barcode, that is a sequence fragment 200 bp long extracted from the original sequences. Classification results demonstrate that using fractional distances (with parameter $p = 0.7$) allows to reach the best scores in terms of accuracy, precision and recall. We compared also our methods with SVM classifier. Classification results at all taxonomic levels and for each sequence sizes clearly state that our classifiers outperforms SVM when applied to sequence fragments.

References

1. Aggarwal, C.C., Hinneburg, A., Keim, D.A.: On the surprising behavior of distance metrics in high dimensional space. In: Van den Bussche, J., Vianu, V. (eds.) ICDT 2001. LNCS, vol. 1973, pp. 420–434. Springer, Heidelberg (2000), doi:10.1007/3-540-44503-X
2. Chang, C.-C., Lin, C.-J.: Libsvm: A library for support vector machines. ACM Trans. Intell. Syst. Technol. 2(3), 1–27 (2011)
3. Fiannaca, A., La Rosa, M., Rizzo, R., Urso, A.: Analysis of DNA barcode sequences using neural gas and spectral representation. In: van Zee, G.A., van de Vorst, J.G.G. (eds.) EANN 2013. LNCS, vol. 384, pp. 215–224. Springer, Heidelberg (1989)

4. Francois, D., Wertz, V., Verleysen, M.: The Concentration of Fractional Distances. IEEE Transactions on Knowledge and Data Engineering 19(7), 873–886 (2007)
5. Hajibabaei, M., Smith, M.A., Janzen, D.H., Rodriguez, J.J., Whitfield, J.B., Hebert, P.D.N.: A minimalist barcode can identify a specimen whose DNA is degraded. Molecular Ecology Notes 6(4), 959–964 (2006)
6. Hajibabaei, M., Singer, G.A.C., Hebert, P.D.N., Hickey, D.A.: DNA barcoding: how it complements taxonomy, molecular phylogenetics and population genetics.. Trends in Genetics 23(4), 167–172 (2007)
7. Haykin, S.: Neural networks: a comprehensive foundation, 2nd edn. Prentice-Hall (1998)
8. Hebert, P.D.N., Ratnasingham, S., DeWaard, J.R.: Barcoding animal life: cytochrome c oxidase subunit 1 divergences among closely related species. Proceedings of the Royal Society. Series B, Biological Sciences 270(suppl.), S96–S99 (2003)
9. Hinnenburg, A., Aggarwal, C., Keim, D.: What is the nearest neighbor in high dimensional spaces?. In: Proceedings of the 26th International Conference on Very Large Data Bases, VLDB 2000, pp. 506–515. Morgan Kaufmann Publishers Inc. (2000)
10. Kuksa, P., Pavlovic, V.: Efficient alignment-free DNA barcode analytics. BMC Bioinformatics 10(suppl. 14), 9 (2009)
11. La Rosa, M., Fiannaca, A., Rizzo, R., Urso, A.: Alignment-free Analysis of Barcode Sequences by means of Compression-Based Methods. BMC Bioinformatics 14, S4 (2013)
12. La Rosa, M., Fiannaca, A., Rizzo, R., Urso, A.: A study of compression–based methods for the analysis of barcode sequences. In: Peterson, L.E., Masulli, F., Russo, G. (eds.) CIBB 2012. LNCS, vol. 7845, pp. 105–116. Springer, Heidelberg (2013)
13. Marshall, E.: Taxonomy. Will DNA bar codes breathe life into classification? Science 307(5712), 1037 (2005)
14. Meusnier, I., Singer, G.A.C., Landry, J.-F., Hickey, D.A., Hebert, P.D.N., Hajibabaei, M.: A universal DNA mini-barcode for biodiversity analysis.. BMC Genomics 9, 214 (2008)
15. Ratnasingham, S., Hebert, P.D.N.: bold: The Barcode of Life Data System (http://www.barcodinglife.org).. Molecular Ecology Notes 7(3), 355–364 (2007)
16. Rizzo, R., Fiannaca, A., La Rosa, M., Urso, A.: The General Regression Neural Network to Classify Barcode and mini-barcode DNA. In: Proceedings of CIBB (2014)
17. Scholkopf, B., Smola, A.: Learning with kernels. MIT Press, Cambridge (2002)
18. Seo, T.K.: Classification of nucleotide sequences using support vector machines. Journal of Molecular Evolution 71(4), 250–267 (2010)
19. Specht, D.F.: A general regression neural network. IEEE Transactions on Neural Networks 2(6), 568–576 (1991)

Transcriptator: Computational Pipeline to Annotate Transcripts and Assembled Reads from RNA-Seq Data

Kumar Parijat Tripathi*,**, Daniela Evangelista**, Raffaele Cassandra, and Mario R. Guarracino

Laboratory for Genomics, Transcriptomics and Proteomics (LAB-GTP),
High Performance Computing and Networking Institute (ICAR),
National Research Council of Italy (CNR), Via Pietro Castellino 111, Napoli, Italy
kumpar@na.icar.cnr.it

Abstract. RNA-Seq is a new tool, which utilizes high-throughput sequencing to measure RNA transcript counts at an extraordinary accuracy. It provides quantitative means to explore the transcriptome of an organism of interest. However, interpreting this extremely large data coming out from RNA-Seq into biological knowledge is a problem, and biologist-friendly tools to analyze them are lacking. In our lab, we develop a Transcriptator web application based on a computational Python pipeline with a user-friendly Java interface. This pipeline uses the web services available for BLAST (Basis Local Search Alignment Tool), QuickGO and DAVID (Database for Annotation, Visualization and Integrated Discovery) tools. It offers a report on statistical analysis of functional and gene ontology annotation enrichment. It enables a biologist to identify enriched biological themes, particularly Gene Ontology (GO) terms related to biological process, molecular functions and cellular locations. It clusters the transcripts based on functional annotation and generates a tabular report for functional and gene ontology annotation for every single transcript submitted to our web server. Implementation of QuickGo web-services in our pipeline enable users to carry out GO-Slim analysis. Finally, it generates easy to read tables and interactive charts for better understanding of the data. The pipeline is modular in nature, and provides an opportunity to add new plugins in the future. Web application is freely available at: www-labgtp.na.icar.cnr.it:8080/Transcriptator.

Keywords: RNA-Seq, QuickGO, DAVID, web-services, Python.

1   Scientific Background

The advent of new technologies in transcriptome studies such as RNA-Seq changes the face of traditional biological research approaches. Instead of studying one or

---

* Corresponding author.
** These authors contributed equally to this work.

© Springer International Publishing Switzerland 2015
C. di Serio et al. (Eds.): CIBB 2014, LNCS 8623, pp. 156–169, 2015.
DOI: 10.1007/978-3-319-24462-4_14

more genes at a time, researchers now simultaneously measure the genome wide changes and regulation of genes under a certain condition. RNA-Seq generates a large amount of biological data in the form of reads. There is a wide array of methodologies to computationally reconstruct the transcript structure and quantify it from raw reads [1]. However, interpreting this extremely large data into biological knowledge is still a challenging and daunting task. A large number of functional annotation pipelines and databases such as DAVID [2], QuickGO [4], ESTExplorer [8], FastAnnotator [5] and other methods [7], were independently developed to address the challenge of functional annotation of the large gene list coming out from RNA-Seq experiments. Both DAVID and QuickGO are very comprehensive databases and can provide putative functional and gene ontological term annotation for a transcript, based on sequence similarity to known genes. These are useful tools for understanding the biological inference of transcriptional response, as well as newly explored sequences. Despite their complex functionalities, both DAVID and QuickGO usually require many manual steps that are often not easy to implement for biologists who are unfamiliar with command line input. Previously, researchers also developed web tools such as FASTAnnotator [5] and ESTExplorer [8]. While ESTExplorer pipeline is specifically designed for EST analysis that includes the cleaning, assembly and clustering and functional annotation of ESTs, FASTAnnotator performs the GO term, enzyme and domain annotations on transcripts. These analyses are not comprehensive as they do not include annotations for pathways such as KEGG, Panther, BioCarta, and they also do not provide any information on protein-protein interactions and other functionalities. They also do not elaborate enrichment analysis for the functional annotation term for the given transcripts data set. The complex plethora of annotation tools and pipelines produces a confusing situation in front of the end users in deciding the most suitable enrichment tool for their analytic skills [3]. There is a need to develop a computational automated pipeline, with a user friendly interface which effectively translates assembled reads coming out from RNA-Seq experiments into biological interpretations such as functional annotation and GO enrichment analysis. We develop a computational pipeline to functional annotate an individual (differentially expressed) transcript, and carry out GO enrichment analysis of expression profiles, under the different treatment condition for organisms, which lacks the referenced genome. In this pipeline, we utilize the web services available for BLAST, QuickGO and DAVID tool for functional and gene ontology annotation and enrichment analysis. Our pipeline carries out automated BLAST run on the Refseq, Swiss-Prot and UniProt-TrEMBL databases to find the most similar genes/proteins for the assembled transcripts. Then, functional and gene ontology annotations are carried out by QuickGo [9] and DAVID web services [6]. The advantage of our pipeline is that it is very easy to use and informative in nature. It produces functional as well as gene ontological annotation for the given transcripts data set. It integrates the results from well established DAVID and QuickGO tools through web services. Our pipeline also provides a plethora of information about enriched pathways such as KEGG, Panther, and BioCarta. The pipeline offers a report on statistical analysis of GO enrichment. It enables a biologist to identify enriched biological themes, particularly GO terms related to biological process, molecular functions and cellular locations. It also provides information about the SMART, Panther, Prosite, Prodom, PFAM and InterPro domains along with protein interactions such as Mint, Bind for the annotated transcripts. Through our pipeline, we are providing an automated protocol to cluster differentially expressed transcripts based on functional annotation. It is modular in nature, so that it also provides a space for adding up new plugins in the future. All these utilities in our pipeline, deliver a platform for the biologist, to understand the humongous RNA-Seq data in a biological sense and in a straightforward way.

2   Materials and Methods

2.1   Web Interface

Trancriptator web application is designed using ZK framework (http://www.zkoss. org/download/zk) and J2EE (Java 2 Platform Enterprise Edition, www.oracle.com /technetwork/java/javaee) technologies. The modular and distributed J2EE platform is employed to integrate technologies for the exchange of information between different applications, such as XML and Web Services. The implementation of the graphical user interface (GUI) is obtained using ZK framework, Ajax web application open-source, with XUL/XHTML (XML User Interface Language/Extensible HyperText Markup Language) built-in based components. JFASTA library v. 2.1.2 (http://jfasta.sourceforge.net/) is implemented in order to handle FASTA format files (.fa). BIOJAVA3-ws module (http://www.biojava. org/ docs/api/org/ biojava3/ws) of BIOJAVA v. 3.0.7 API is used to provide analytical and statistical routines, sequences manipulation -such as BLAST alignment. Lastly, the Jython interpreter v. 2.5.3. (http://www.jython.org/) is used to integrate Pythons pipeline (Fig. 3) code on Javas platform.

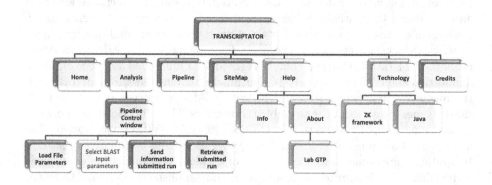

Fig. 1. Transcriptator block diagram

2.2   System Architecture of Transcriptator Pipeline

Trancriptator pipeline consist of three major components: (i) BLAST analysis, (ii) Gene ontology, (iii) Functional annotation, retrieval and statistical analysis of the data. It requires various levels of computational hardwares (Fig 2). This pipeline is embedded in web application written in Java and Python scripts. The front end user interface of Transcriptator is installed on LAB-GTP server. It helps the user's to submit their queries using our web application interface. The core engine of the pipeline is written in Python, it comprises of the blast analysis as well as different web services for functional annotation analysis from publicly available annotation databases such as DAVID and Quick GO. The core engine is locally installed on a interomics cluster which is connected to LAB-GTP server. For BLAST analysis, ncbi-blast.2.2.23 stand alone package is installed on the cluster. SwissProt and UniProt-trEMBL databases (http://www.uniprot.org/) are also installed for BLAST run. DAVID and Quick-GO webservices are installed on the cluster for the faster processing of results. The query of FASTA sequence datasets provided through web application on our web server is directly transferred to our interomics cluster. Local BLAST analysis is carried out on the local cluster implying BLAST X run on locally installed SwissProt and UniProt databases. BLAST results are analysed and top proteins hits id's are used as input for DAVID and QUICK-GO web-services to retrieve functional and gene ontological annotations. The retrieved data is processed and feeded to statistical analysis section of Transcriptator pipeline core engine. The results are provided in the form of graphs and tabular reports, and transferred to the LAB-GTP web server again. From the server, user can access to this information by using the job IDs provided by the server.

2.3   Pipeline Implementation

The Transcriptator pipeline is written in Python, bash and R scripts. It implements the web services available for DAVID and QuickGO tools. For DAVID web-services, it utilizes the available Python client source code. The Python client for DAVID web-services, which use light-weight soap client suds-0.4 module [https://pypi.python.org /pypi/suds] [6]. For QuickGO web-services, BioServices Python package is implemented in the pipeline [9]. It provides access to QuickGO and a framework to easily implement web service wrappers (based on WSDL/SOAP or REST protocols). In this pipeline, the annotation process comprises of four main parts: (i)finding the best hit in locally installed SwissProt and UniProt-Trembl database; (ii) assignment of functional annotation and gene ontology terms and their enrichment from DAVID; (iii) assignment of GO Slim terms and their analysis from QuickGO; (iv) integration and summarization of retrieved results from DAVID and QuickGO web services. Transcriptotator runs the first step of BLAST search on the local cluster. The second and third steps of pipeline simultaneously run to accelerate the annotation procedure. The last step retrieves the results, processes them and generates the statistical reports in the form of tables and charts.

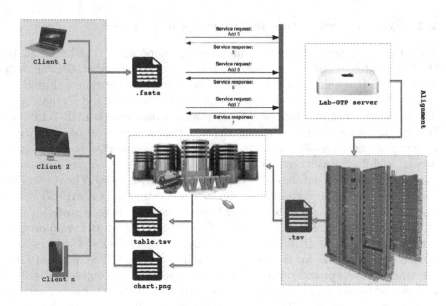

Fig. 2. System architecture of Transcriptator Pipeline

Identification of Best Hit. BLASTX program from locally installed ncbi-blast.2.2.23 stand alone package [10] is used (with threshold E-value 0.001) to identify the best hits for query sequences on locally installed SwissProt and UniProt-trEMBL databases (http://www.uniprot.org/). The main goal of the first step is to find the similar sequences within SwissProt and UniProt-trEMBL databases for the unannotated query from the user. The output of BLASTX run is an alignment file in a tsv format. The latter, using a bash script, is transformed into the protein list, which is the required input file for DAVID and QuickGO web services.

Assignment of Functional and Gene Ontology Annotation from DAVID. Python client source code for DAVID, retrieves the functional and gene ontology annotation for every single transcript in a query data set. These python scripts take the input protein list file from previous step and utilize DAVID database to obtain information in the form of ChartReport, ClusterReport, TableReport and Summary-ryReport. For a given query data set, Python source code implemented within the Transcriptator pipeline runs with default parameter for DAVID database search to obtain the enrichment statistics for each functional and GO term. ChartReport is an annotation-term-focussed view, which lists annotation terms and their associated genes under study. It also provides the Fischer exact statistics calculated for each annotation term and information about the statistically enriched annotation terms in the query data set. The ClusterReport displays the grouping of similar annotation terms along with their associated genes. The grouping algorithm is based on the hypothesis that similar annotations should have similar gene members.

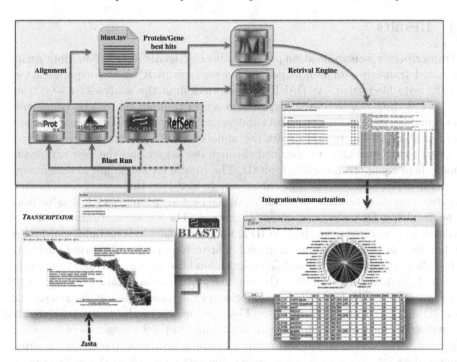

Fig. 3. Transcriptator pipeline: the lower panel boxes respectively show the input/output of the web interface, whereas the upper panel represents the steps of the Transcriptator engine.

Assignment and Analysis of GO Slim Terms from QuickGO. Transcriptator employs BioServices module from Python package, which provides access to many bioinformatics web services and a framework to easily implement web service wrappers (based on WSDL/SOAP or REST protocol). BioServices (bioservices.quickgo.QuickGO) is used to investigate the GO slims in the query data set. GO slim terms are the list of GO terms that have been selected from the full set of terms available from the gene ontology projects.

Processing of Retreived Annotation. Both DAVID and QuickGO web services can produce large amount of results for the given query data set. It is not possible for the users to understand and interpret this bulk amount of results in a simple way. For the integration and summarization of retrieved results from web-services, Python and R codes in Transcriptator are implemented to parse the results in simpler format. Transcriptator produces easy to read tables for enrichment analysis of GO and Functional terms, clustering analysis on transcripts and annotation assignment for every single transcript. R scripts are specifically implemented in the pipeline, to generate an interactive chart for the distribution of functional and GO terms such as biological process, molecular function and cellular components associated with the query data set of transcripts.

3   Results

Transcriptator web application provides a user a friendly interface to input unannotated transcripts or denovo assembled reads from RNA-Seq experiments in multi fasta file format. As DAVID web services limit the analyses to 3000 transcripts at time, our web server also allows a user to input up to 3000 transcripts for annotation. After a successful submission, a unique job ID is generated and provided as an identifier to start the annotation. All annotation results from DAVID and QuickGO are obtained through our server and the user can download them using the associated job ID. The results for single job id comprise of several tables and graphs. The tables are divided into three sections. The first section contains the table with the list of the best hit proteins with E-value from the databases for the corresponding transcript. The second section comprises of the tables generated from the DAVID annotation analysis. It includes four tables: ChartReport (for enrichment analysis); ClusterReport (for clustering analysis); TableReport for functional and GO annotation for every single transcripts in the dataset; SummaryReport with the summary of total annotation for the given query data set. The third section comprises of the table enlisting the assignment of GO slim terms on the transcripts. The pipeline also produces charts related to the distribution of GO terms specifically related to three categories of biological process, molecular function and cellular components, respectively, for the input query data set. For each job concluded, the related annotation results (Fig. 4) will be retained for one week on the Transcriptator server. User can access to this information by using the job IDs provided by the server.

3.1   *Case Study*

To demonstrate the utility of Transcriptator in biological studies, we have selected a sample dataset (five hunderd and forty four unannotated transcripts) of *Hydra vulgaris* transcriptome, downloaded from European Nucleotide Archieve (ENA) database (http://www.ebi.ac.uk/ena/). These transcripts are specifically differentially expressed in response to cadmium treatment (unpublished data of specific DE transcripts for cadmium treatment). Cadmium is a toxic element. It accumulates in the organisms body and produce pathogenic changes.To study the harmful effects of cadmium accumulation in the body, previously researchers studies the toxicity and chemical stress due to cadmium concentration in non model organism Hydra [11]. They have shown morphological, developmental and physical damage in Hydra due to the presence of high concentration of cadmium in the organism body. To undermine the molecular mechanism of cadmium poisoning in Hydra, we have investigated these cadmium specific differentially expressed transcripts through our pipeline Transcriptator. It annotates these transcripts for all the functional and gene ontology categories and produces results table and graphical charts for functional annotations as well as gene ontology enrichment analysis (Fig 5).

Our results from Transcriptator shows enrichment of Hedgehog signalling pathway and metal binding biological process functional terms with significant

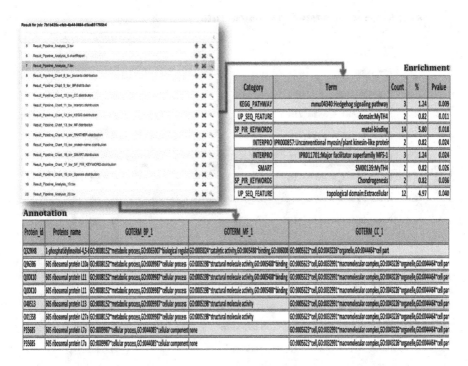

Fig. 4. Result files from the Transcriptator. It includes tables for BLAST results, functional and GO enrichment. Graphical chart for GO and functional annotation terms distribution in input data.

p-value of .009 and .018 respectively (Fig. 4: enrichment table).These results make biological sense as the Hedgehog (Hh) family of secreted signaling proteins plays a crucial role in development and morphogenesis of a variety of tissues and organs in *Hydra vulgaris*. Fig.6 shows biological process terms distribution chart generated by transcriptator for the sample data set. For the given data set. Transcriptator pipeline also provide distribution plots for both molecular functions as well as cellular components.

Molecular functions distribution plots suggests 42.6 percent of sample dataset of Hydra transcripts involved in binding function. It also shows transcription regulator activity (4.6 percent), transporter activity (6.6 percent), catalytic activity (16.7 percent) and molecular transducer activities (11.5 percent) are enriched in these transcripts dataset of Hydra in response to the cadmium toxicity (Fig. 7).

For the given query dataset, Transcriptator pipeline also provide cellular components distribution plot. Distribution of cellular componets associated to differentially expressed transcripts suggests the role of these transcripts in cellular composition. It also provide some information about other cellular components such as synapse, macromolecular complex region, organelles and membrane enclosed regions but unfortunately these cellular compents are not enriched in our query dataset (Fig 8).

Result for job: c41a5889-7534-45b0-bf37-ae7a8e39a6ff

ID	File Name			
1	Processed_fasta_file_1.fasta			
2	Result_BLAST_Analysis_2.tsv			
3	Result_BLAST_Analysis_3.tsv			
4	Result_BLAST_Analysis_4.tsv			
5	Result_BLAST_Analysis_5.tsv			
6	Result_Pipeline_Analysis_Chart_Report_6.chartReport			
7	Result_BLAST_Analysis_7.tsv			
8	Result_Pipeline_Chart_8_tsv_biocarta.distribution			
9	Result_Pipeline_Chart_9_tsv_BP.distribution			
10	Result_Pipeline_Chart_10_tsv_CC.distribution			
11	Result_Pipeline_Chart_11_tsv_Interpro.distribution			
12	Result_Pipeline_Chart_12_tsv_KEGG.distribution			
13	Result_Pipeline_Chart_13_tsv_MF.distribution			
14	Result_Pipeline_Chart_14_tsv_PANTHER.distribution			
15	Result_Pipeline_Chart_15_tsv_protein-name.distribution			
16	Result_Pipeline_Chart_16_tsv_SMART.distribution			
17	Result_Pipeline_Chart_17_tsv_SP_PIR_KEYWORD.distribution			
18	Result_Pipeline_Chart_18_tsv_Species.distribution			

Fig. 5. List of results obtained from Transcriptator: tabular results for functional and gene ontology terms enrichments. Graphical distributions plots for GO's terms and functional annotations for a given query data set.

Apart for Gene Ontological terms such as biological process, molecular functions and cellular components, Transcriptator pipeline also provides distribution plots for other different functional terms associated with the query dataset, such as Biocarta pathways, Panther pathways, KEGG pathways, proteins domains such as InterPro, PFAM, SMART. It also provide SP-PIR-keyword distributions. As Transcriptator uses DAVID web services, it can provide each and every relevant functional information related to our query dataset in the the form of distribution plots as well as summary and enrichments table. for example SP-PIR-keyword distribution plot (Fig. 9) shows a large number of keywords are associated to the query dataset, it includes terms like transcription,transducer,alternative splicing, differentiation, dna binding, developmental proteins, g-protein coupled receptor, nucleotide binding, signal and ion transport etc. All these terms associated to the dif-

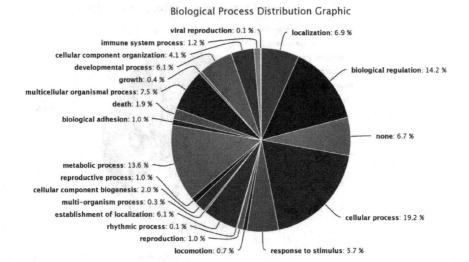

Fig. 6. Biological processes distribution in this case study dataset. It shows the significant biological activities, in which these transcripts (case study dataset) are involved. For example biological regulation, cellular process, stimulus response activities and developmental process are enriched within these transcripts.

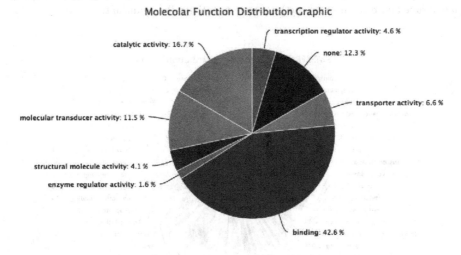

Fig. 7. Molecular function distribution in this case study dataset. It shows the significant molecular function activities, in which these transcripts (case study dataset) are involved. For example binding, molecular transducer activity, transcription regulator activity and catalytic activities are enriched within these transcripts.

ferentially expressed transcripts dataset suggest the most possible role of cadmium toxicity on differentiation, reproduction, developmental as well signal transduction processes in *Hydra vulgaris*. Transcriptator pipeline also provides the

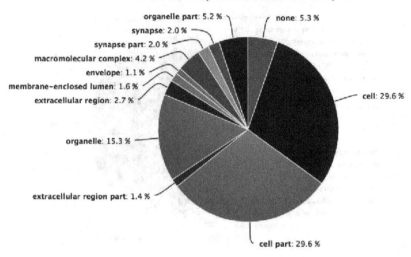

Cellular Component Distribution Graphic

organelle part: 5.2 %
synapse: 2.0 %
synapse part: 2.0 %
macromolecular complex: 4.2 %
envelope: 1.1 %
membrane-enclosed lumen: 1.6 %
extracellular region: 2.7 %
organelle: 15.3 %
extracellular region part: 1.4 %
none: 5.3 %
cell: 29.6 %
cell part: 29.6 %

Fig. 8. Cellular components distribution in this case study data set. It shows the significant cellular components, in which these transcripts (case study dataset) are involved. For example biological regulation,most of the differentially expressed transcripts from the query dataset are associated with cell organization. A small number of transcripts are also involved with structural composition of synapse, macromolecular complex, membrane and cellular organelle but are not statistically significant.

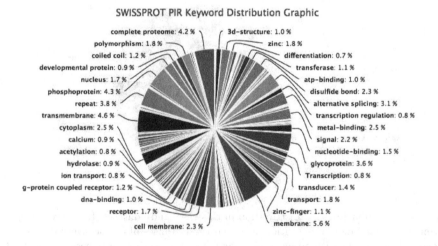

SWISSPROT PIR Keyword Distribution Graphic

complete proteome: 4.2 %
polymorphism: 1.8 %
coiled coil: 1.2 %
developmental protein: 0.9 %
nucleus: 1.7 %
phosphoprotein: 4.3 %
repeat: 3.8 %
transmembrane: 4.6 %
cytoplasm: 2.5 %
calcium: 0.9 %
acetylation: 0.8 %
hydrolase: 0.9 %
ion transport: 0.8 %
g-protein coupled receptor: 1.2 %
dna-binding: 1.0 %
receptor: 1.7 %
cell membrane: 2.3 %
3d-structure: 1.0 %
zinc: 1.8 %
differentiation: 0.7 %
transferase: 1.1 %
atp-binding: 1.0 %
disulfide bond: 2.3 %
alternative splicing: 3.1 %
transcription regulation: 0.8 %
metal-binding: 2.5 %
signal: 2.2 %
nucleotide-binding: 1.5 %
glycoprotein: 3.6 %
Transcription: 0.8 %
transducer: 1.4 %
transport: 1.8 %
zinc-finger: 1.1 %
membrane: 5.6 %

Fig. 9. SwissProt-PIR keywords distribution: Graphical representation of all the SP-PIR keywords.

Functional Terms	%	Counts
GENETIC_ASSOCIATION_DB_DISEASE	0,033	8
GENETIC_ASSOCIATION_DB_DISEASE_(	0,033	8
OMIM_DISEASE	0,05	12
COG_ONTOLOGY	0,071	17
PIR_SEQ_FEATURE	0,062	15
SP_COMMENT_TYPE	0,954	230
SP_PIR_KEYWORDS	0,979	236
UP_SEQ_FEATURE	0,975	235
ZFIN_ANATOMY	0,041	10
GOTERM_BP_1	0,809	195
GOTERM_BP_2	0,805	194
GOTERM_BP_3	0,788	190
GOTERM_BP_4	0,768	185
GOTERM_BP_5	0,718	173
GOTERM_BP_ALL	0,809	195
GOTERM_BP_FAT	0,784	189
GOTERM_CC_1	0,859	207
GOTERM_CC_2	0,838	202
GOTERM_CC_3	0,838	202
GOTERM_CC_4	0,822	198
GOTERM_CC_5	0,805	194
GOTERM_CC_ALL	0,859	207
GOTERM_CC_FAT	0,734	177
GOTERM_MF_1	0,809	195
GOTERM_MF_2	0,784	189
GOTERM_MF_3	0,73	176
GOTERM_MF_4	0,672	162
GOTERM_MF_5	0,556	134
GOTERM_MF_ALL	0,809	195
GOTERM_MF_FAT	0,734	177
PANTHER_BP_ALL	0,378	91
PANTHER_MF_ALL	0,378	91
CHROMOSOME	0,834	201
CYTOBAND	0,647	156
ENTREZ_GENE_SUMMARY	0,191	46
HOMOLOGOUS_GENE	0,772	186
OFFICIAL_GENE_SYMBOL	0,934	225
PIR_SUMMARY	0,548	132
SP_COMMENT	0,946	228
GENERIF_SUMMARY	0,402	97
HIV_INTERACTION_PUBMED_ID	0,004	1
PUBMED_ID	0,917	221
ENSEMBL_GENE_ID	0,726	175
ENTREZ_GENE_ID	0,954	230
BBID	0,008	2
BIOCARTA	0,029	7
EC_NUMBER	0,207	50
KEGG_PATHWAY	0,315	76
PANTHER_PATHWAY	0,237	57
REACTOME_PATHWAY	0,025	6
BLOCKS	0,602	145
COG_NAME	0,071	17
INTERPRO	0,971	234
PANTHER_FAMILY	0,705	170
PANTHER_SUBFAMILY	0,593	143
PFAM	0,963	232
PIR_SUPERFAMILY	0,51	123
PRINTS	0,423	102
PRODOM	0,232	56
PROFILE	0,589	142
PROSITE	0,78	188
SCOP_CLASS	0,033	8
SCOP_FAMILY	0,033	8
SCOP_FOLD	0,033	8
SCOP_SUPERFAMILY	0,033	8
SMART	0,515	124
SSF	0,315	76
TIGRFAMS	0,116	28
BIND	0,162	39
DIP	0,071	17
HIV_INTERACTION	0,004	1
HIV_INTERACTION_CATEGORY	0,004	1
MINT	0,158	38
NCICB_CAPATHWAY_INTERACTION	0,021	5
REACTOME_INTERACTION	0,017	4
UCSC_TFBS	0,191	46
CGAP_EST_QUARTILE	0,133	32
CGAP_SAGE_QUARTILE	0,129	31
GNF_U133A_QUARTILE	0,12	29
PIR_TISSUE_SPECIFICITY	0,241	58
UNIGENE_EST_QUARTILE	0,145	35
UP_TISSUE	0,722	174

Fig. 10. Summary table: it provides all the functional and GO terms associated with the given query dataset.

Category	Term	Count	%	Pvalue	List Total	Pop Hits	Pop Total	Fold Enrichment	Bonferroni	Benjamini	FDR
KEGG_PATHWAY	mmu04340:Hedgehog sig	3	1.244	0.0095	17	54	5738	18.75	0.326	0.326	8.5
UP_SEQ_FEATURE	domain:MyTH4	2	0.829	0.0114	47	4	16021	170.4	0.872	0.872	13.19
SP_PIR_KEYWORDS	metal-binding	14	5.809	0.0183	48	2682	17854	1.941	0.811	0.811	18.28
INTERPRO	IPR000857:Unconvention	2	0.829	0.0240	49	9	17763	80.55	0.961	0.961	24.81
INTERPRO	IPR011701:Major facilitat	3	1.244	0.0241	49	89	17763	12.21	0.962	0.805	24.86
SMART	SM00139:MyTH4	2	0.829	0.0263	28	9	9131	72.46	0.617	0.617	21.16
SP_PIR_KEYWORDS	Chondrogenesis	2	0.829	0.0362	48	14	17854	53.13	0.963	0.810	33.0
UP_SEQ_FEATURE	topological domain:Extrac	12	4.979	0.0404	47	2174	16021	1.881	0.999	0.975	39.84
INTERPRO	IPR008085:Thrombospon	2	0.829	0.0423	49	16	17763	45.31	0.996	0.855	39.77
UP_SEQ_FEATURE	topological domain:Cytop	14	5.809	0.0451	47	2780	16021	1.716	0.999	0.936	43.32
SP_PIR_KEYWORDS	zinc	10	4.149	0.0548	48	1886	17854	1.972	0.993	0.816	45.87
UP_SEQ_FEATURE	domain:TSP type-1 5	2	0.829	0.0567	47	20	16021	34.08	0.999	0.923	50.74
SP_PIR_KEYWORDS	osteogenesis	2	0.829	0.0563	48	22	17854	33.81	0.994	0.728	46.80
INTERPRO	IPR001111:Transforming	2	0.829	0.0578	49	22	17763	32.95	0.999	0.864	50.21
UP_SEQ_FEATURE	transmembrane region	18	7.468	0.0602	47	4113	16021	1.491	0.999	0.891	53.42
SP_PIR_KEYWORDS	alternative splicing	18	7.468	0.0603	48	4481	17854	1.494	0.996	0.674	49.22
UP_SEQ_FEATURE	domain:TSP type-1 4	2	0.829	0.0640	47	23	16021	29.64	0.999	0.861	55.70
INTERPRO	IPR001879:GPCR, family	2	0.829	0.0705	49	27	17763	26.85	0.999	0.859	57.5
UP_SEQ_FEATURE	domain:TSP type-1 3	2	0.829	0.0747	47	27	16021	25.24	0.999	0.862	61.56
SP_PIR_KEYWORDS	glycoprotein	15	6.224	0.0759	48	3600	17854	1.549	0.999	0.694	57.67
SP_PIR_KEYWORDS	cell membrane	9	3.734	0.0766	48	1713	17854	1.954	0.999	0.641	58.00
SMART	SM00008:HormR	2	0.829	0.0769	28	27	9131	24.15	0.944	0.763	51.04
INTERPRO	IPR000203:GPS	2	0.829	0.0830	49	32	17763	22.65	0.999	0.855	63.74
INTERPRO	IPR015615:Transforming	2	0.829	0.0830	49	32	17763	22.65	0.999	0.855	63.74
UP_SEQ_FEATURE	domain:GPS	2	0.829	0.0853	47	31	16021	21.99	0.999	0.864	66.64
UP_SEQ_FEATURE	domain:PH 2	2	0.829	0.0879	47	32	16021	21.30	0.999	0.839	67.80
SMART	SM00303:GPS	2	0.829	0.0905	28	32	9131	20.38	0.967	0.679	57.11
UP_SEQ_FEATURE	domain:PH 1	2	0.829	0.0906	47	33	16021	20.65	0.999	0.817	68.92
UP_SEQ_FEATURE	zinc finger region:RING-t	3	1.244	0.0907	47	176	16021	5.810	0.999	0.787	68.99
INTERPRO	IPR017948:Transforming	2	0.829	0.0953	49	37	17763	19.59	0.999	0.853	69.06
INTERPRO	IPR001839:Transforming	2	0.829	0.0953	49	37	17763	19.59	0.999	0.853	69.06
UP_SEQ_FEATURE	domain:TSP type-1 2	2	0.829	0.0984	47	36	16021	18.93	0.999	0.786	72.1
UP_SEQ_FEATURE	domain:TSP type-1 1	2	0.829	0.0984	47	36	16021	18.93	0.999	0.786	72.1

Fig. 11. Enrichment Chart: tabular results for functional and gene ontology terms enrichments. It provides the satistical P-values for the better interpretation of the results

summary information of all the functional and GO terms associated with the queried dataset along with the percentage and counts of transcripts. this information helps the user to undermine the associated functionalities for the given transcripts.

Through Transcriptator pipeline, for the given query dataset of Hydra transcripts, the complete tabular results for enriched functional and GO terms are obtained (Fig 11). It shows the different fucntional terms, such as pathways, protein domains, SP-PIR keywords which are enriched in the given query data set. It also provide the counts of transcripts and enriched P-values. Transcriptator pipeline also provide corrected P-values after Bonferroni and Benjamini correction. False discovery rate and fold enrichments information is also provided for the given fucntional terms associated with the transcripts dataset. It enables user to determine the statistical significance for the functionalities associated to their input transcripts dataset for the better biological interpretation of their transcriptomic data.

4   Conclusion

Transcriptator is a modular pipeline, which provides flexibility to user to carry out functional annotation of transcript's data. It allows users to choose two distinct types of web services for annotation purposes, as well as different BLAST databases for BLAST run. All these options help users to optimize their results, according to their needs. It provides the enrichment score for the functional terms and reports each and every annotation present in the given data set in the form of tables and interactive charts. In future, we will work on the addition of more modular functionalities and options in the pipeline, for both BLAST searches and annotation analysis.

Acknowledgements. We would like to thank the INTEROMICS flagship project, PON02-00612-3461281 and PON02-00619-3470457 for the funding support. Mario Guarracino work is conducted at National Research University Higher School of Economics and supported by RSF grant 14-41-00039.

References

1. Steijger, T., et al.: Assessment of transcript reconstruction methods for RNA-seq. Nat. Methods 10, 1177–1184 (2013)
2. Huang, D.W., et al.: Systematic and integrative analysis of large gene lists using DAVID bioinformatics resources. Nature Protocols 4, 44–57 (2009)
3. Huang, D.W., Sherman, B.T., Lempicki, R.A.: Bioinformatics enrichment tools: paths toward the comprehensive functional analysis of large gene lists. Nucleic Acids Res. 37(1), 1–13 (2009)
4. Binns, D., et al.: QuickGO: a web-based tool for Gene Ontology searching. Bioinformatics 15, 25(22), 3045–3046 (2009)
5. Chen, T.W., et al.: FastAnnotator- an efficient transcript annotation web tool. BMC Genomics 13(Suppl. 7), S9 (2012)
6. Jiao, X., et al.: DAVID-WS: a stateful web service to facilitate gene/protein list analysis. Bioinformatics 28(13), 1805–1806 (2012)
7. Wang, X., Cairns, M.J.: Gene set enrichment analysis of RNA-Seq data: integrating differential expression and splicing. BMC Bioinformatics 14(Suppl. 5), S16 (2013)
8. Nagaraj, S.H., et al.: ESTExplorer: an expressed sequence tag (EST) assembly and annotation platform. Nucleic Acids Res. 35(Web Server issue), 143–147 (2007)
9. Cokelaer, T., et al.: BioServices: a common Python package to access biological Web Services programmatically. Bioinformatics 29, 3241–3242 (2013)
10. Altschul, S.F., et al.: Basic local alignment search tool. J. Mol. Biol. 215, 403–410 (1990)
11. Karntanut, W., Pascoe, D.: The toxicity of copper, cadmium and zinc to four different Hydra (Cnidaria: Hydrozoa). Chemosphere 47, 1059–1064 (2002)

Application of a New Ridge Estimator of the Inverse Covariance Matrix to the Reconstruction of Gene-Gene Interaction Networks

Wessel N. van Wieringen[1,2,*] and Carel F.W. Peeters[1]

[1] Department of Epidemiology and Biostatistics, VU University medical center,
P.O. Box 7057, 1007 MB Amsterdam, The Netherlands
{w.vanwieringen,cf.peeters}@vumc.nl
[2] Deptartment of Mathematics, VU University Amsterdam,
1081 HV Amsterdam, The Netherlands

Abstract. A proper ridge estimator of the inverse covariance matrix is presented. We study the properties of this estimator in relation to other ridge-type estimators. In the context of Gaussian graphical modeling, we compare the proposed estimator to the graphical lasso. This work is a brief exposé of the technical developments in [1], focussing on applications in gene-gene interaction network reconstruction.

Keywords: Gaussian graphical model, Gene-gene interaction networks, Multivariate normal, Penalized estimation, Precision matrix.

1 Introduction

1.1 Scientific Background

Molecular biology aims to understand the molecular processes that occur in the cell. That is, which molecules present in the cell interact, and how are the interactions coordinated? For many cellular process, it is unknown which genes play what role.

A valuable source of information to uncover gene-gene interactions are (onco)genomics studies. Such studies comprise samples from n individuals with, e.g., cancer of the same tissue. Each sample is interrogated molecularly and the expression levels of many (p) genes are measured simultaneously. The resulting p-dimensional data vector is denoted $\mathbf{Y}_{i,*}$ for individual $i = 1, \ldots, n$.

From these data the gene-gene interaction network may be unraveled when the presence (absence) of a gene-gene interaction is operationalized as a conditional (in)dependency between the corresponding gene pair. Then, under the assumption of multivariate normality, $\mathbf{Y}_{i,*} \sim \mathcal{N}(\mathbf{0}_{p\times 1}, \mathbf{\Sigma})$, the absence of direct gene-gene interactions corresponds to zeros in the inverse covariance matrix $\mathbf{\Omega} \equiv \mathbf{\Sigma}^{-1}$ (also known as the precision matrix, whose elements are proportional to partial correlations). For instance, $(\mathbf{\Omega})_{1,2} = 0 \Leftrightarrow Y_1 \perp\!\!\!\perp Y_2 \,|\, Y_3, \ldots, Y_p$.

---

* Corresponding author.

© Springer International Publishing Switzerland 2015
C. di Serio et al. (Eds.): CIBB 2014, LNCS 8623, pp. 170–179, 2015.
DOI: 10.1007/978-3-319-24462-4_15

Hence, the gene-gene interaction network is found by inversion of the covariance matrix and (subsequent) determination of its support. When dealing with data, $\boldsymbol{\Sigma}$ is estimated by its sample counterpart: $\mathbf{S} = \frac{1}{n} \sum_{i=1}^{n} \mathbf{Y}_{i,*} \mathbf{Y}_{i,*}^{T}$.

In genomics the data are often high-dimensional, in the sense of $p > n$. In such situations the sample covariance matrix $\mathbf{S}$ is singular and the sample precision matrix is not defined. But even if $p < n$ and p approaches n, the sample precision matrix yields inflated partial correlations. Both situations require some form of regularization to obtain a well-behaved estimate of the precision matrix, and consequently of the gene-gene interaction network.

1.2   Ridge-Type Covariance Estimators

A penalized covariance estimator traditionally referred to as the 'ridge estimator' is:

$$\hat{\boldsymbol{\Sigma}}_{r_I}(\lambda_{r_I}) = \mathbf{S} + \lambda_{r_I} \mathbf{I}_{p \times p} \qquad \text{for } \lambda_{r_I} > 0.$$

It could be considered a ridge estimator in the sense that it is an ad-hoc fix of the singularity of $\mathbf{S}$, much like how ridge regression was originally introduced [2]. The inverse of $\hat{\boldsymbol{\Sigma}}_{r_I}(\lambda_{r_I})$ would then form the basis for inference on the gene-gene interaction network.

Alternatively, a 'ridge estimator' popularized by [3] in the field of genomics, is (cf. [4,5]):

$$\hat{\boldsymbol{\Sigma}}_{r_{II}}(\lambda_{r_{II}}) = (1 - \lambda_{r_{II}})\mathbf{S} + \lambda_{r_{II}}\boldsymbol{\Gamma} \qquad \text{for } \lambda_{r_{II}} \in (0, 1].$$

In this latter expression $\boldsymbol{\Gamma}$ is a $(p \times p)$-dimensional, symmetric positive definite (p.d.) target matrix. The target matrix is chosen prior to estimation. Its role is to serve as a 'null estimate' towards which the covariance estimate is shrunken as $\lambda_{r_{II}}$ tends to one. In the remainder we will mainly consider the following choice: $\boldsymbol{\Gamma}$ diagonal with $\text{diag}(\boldsymbol{\Gamma}) = \text{diag}(\mathbf{S})$. This represents a reasonable choice in the absence of any prior knowledge on the Gaussian process. Again, when determining the support of the precision matrix the inverse of this second 'ridge estimator' could be used.

Neither of the two ridge estimators above is a proper ridge estimator, in the sense that neither can be formulated as the result from the maximization of a loss function augmented with what is commonly perceived as the ridge penalty: the sum of the square of its elements.

1.3   Overview

In Section 2 an alternative ridge estimator for the inverse covariance matrix is presented. In Section 3 the proposed estimator is compared with the traditional ridge-type estimators and the graphical lasso. Section 4 illustrates, using oncoge-nomics data, practical usage of the proposed estimator in a graphical modeling setting. Section 5 carries some concluding remarks, while Section 6 closes with a small description of the accompanying software.

2   Materials and Methods

2.1   An Alternative Ridge Inverse Covariance Estimator

We consider estimation of the inverse covariance matrix with conventional ridge regularization. The alternative ridge estimator of the inverse covariance matrix maximizes the following penalized log-likelihood:

$$\mathcal{L}^{\text{pen}}(\boldsymbol{\Omega}; \mathbf{S}, \mathbf{T}, \lambda_a) = \ln |\boldsymbol{\Omega}| - \text{tr}(\mathbf{S}\boldsymbol{\Omega}) - f^{\text{pen}}(\boldsymbol{\Omega}, \mathbf{T}, \lambda_a), \qquad (1)$$

where λ_a is the penalty parameter, $\mathbf{T}$ denotes a symmetric p.d. target matrix, and $f^{\text{pen}}(\cdot, \cdot, \cdot)$ indicates the penalty function. The ridge penalty function amounts to:

$$f^{\text{pen}}(\boldsymbol{\Omega}, \mathbf{T}, \lambda_a) = \frac{1}{2}\lambda_a \text{tr}[(\boldsymbol{\Omega} - \mathbf{T})^{\text{T}}(\boldsymbol{\Omega} - \mathbf{T})]. \qquad (2)$$

In case $\mathbf{T} = \mathbf{0}_{p \times p}$, the penalty function reduces to $f^{\text{pen}}(\boldsymbol{\Omega}, \mathbf{T}, \lambda_a) = f^{\text{pen}}(\boldsymbol{\Omega}, \lambda_a) = \frac{1}{2}\lambda_a \sum_{j_1, j_2 = 1}^{p}[(\boldsymbol{\Omega})_{j_1, j_2}]^2$, which corresponds to the common perception of the ridge penalty. The penalty function (2) is thus a generalized ridge penalty.

We show (cf. [1]) that there is an explicit solution that maximizes the penalized log-likelihood (1) with the general ridge penalty (2):

$$\hat{\boldsymbol{\Omega}}^{\text{ridge}}(\lambda_a) = \left\{\left[\lambda_a \mathbf{I}_{p \times p} + \frac{1}{4}(\mathbf{S} - \lambda_a \mathbf{T})^2\right]^{1/2} + \frac{1}{2}(\mathbf{S} - \lambda_a \mathbf{T})\right\}^{-1}. \qquad (3)$$

This ridge precision estimator is p.d. when $\lambda_a \in (0, \infty)$ and can be viewed as a penalized maximum likelihood (ML) estimator. Moreover, in the low-dimensional case the ridge estimator (2) reduces to $\mathbf{S}^{-1}$ as $\lambda_a \downarrow 0$. When λ_a tends to infinity, $\hat{\boldsymbol{\Omega}}^{\text{ridge}}(\lambda_a)$ shrinks to $\mathbf{T}$, much like the covariance estimator of [3] shrinks to a user-specified target. Thus, when $\mathbf{T}$ is diagonal and $\text{diag}(\mathbf{T}) = 1/\text{diag}(\mathbf{S})$ the inverse of estimator (3) mimics the behaviour of the latter. Similarly, choosing $\mathbf{T} = \mathbf{0}_{p \times p}$ yields a ridge estimator of the precision matrix that shrinks to the null matrix as does the inverse of $\hat{\boldsymbol{\Sigma}}_{r_I}(\lambda_{r_I})$. The explicit form of our ridge estimator (3) allows us to calculate the moments of the estimator and prove its consistency [1].

2.2   Extracting an Interaction Network

When turning to the application of ridge estimation in Gaussian graphical modeling of gene-gene interaction networks, the proposed estimator (3) yields (after standardization) an estimate of the partial correlation matrix. In doing so, an informed choice of the penalty parameter needs to be made. Hereto we utilize an approximate leave-one-out cross-validation (LOOCV) procedure [6]. Finally, one needs to decide which elements of the partial correlation matrix are indistinguishable from zero, for which we employ the local false discovery rate (FDR) procedure of [3].

3   Results

3.1   Comparison with the Traditional Ridge Estimators

We compare the proposed ridge estimator (3) with the two other 'ridge estimators', $\hat{\mathbf{\Sigma}}_{r_I}(\lambda_{r_I})$ and $\hat{\mathbf{\Sigma}}_{r_{II}}(\lambda_{r_{II}})$. Analytically, we study the rate of shrinkage of the estimators. The proposed ridge precision estimator (3) with $\mathbf{T} = \mathbf{0}_{p \times p}$ displays slower shrinkage (with increasing penalty parameter) to the null target than $[\hat{\mathbf{\Sigma}}_{r_I}(\lambda_{r_I})]^{-1}$. As the target is degenerate, this behaviour is to be preferred. The opposite is seen when studying the shrinkage rate of estimator (3) with $\mathrm{diag}(\mathbf{T}) = 1/\mathrm{diag}(\mathbf{S})$ in relation to $[\hat{\mathbf{\Sigma}}_{r_{II}}(\lambda_{r_{II}})]^{-1}$ with $\mathbf{\Gamma} = \mathbf{T}^{-1}$. That is, the former shrinks faster to $\mathbf{T}$ than the latter. Whenever $\mathbf{T}$ is close to $\mathbf{\Omega}$, faster shrinkage is desirable. In a simulation study we turn to the comparison of the risk of the proposed ridge estimator and its contenders. For the scenario's studied, the former performs favourably.

3.2   Comparison with the Graphical Lasso

For the application to Gaussian graphical modelling, the inverse covariance matrix is often estimated by means of the graphical lasso [7,8], as it performs automated edge selection. The lasso precision estimator maximizes (1) under the alternative penalty $f'^{\mathrm{pen}}(\mathbf{\Omega}, \mathbf{T}, \lambda_l) = \lambda_l \|\mathbf{\Omega}\|_1 = \lambda_l \sum_{j_1, j_2 = 1}^{p} |(\mathbf{\Omega})_{j_1, j_2}|$. To accommodate the diagonal target matrix $\mathbf{T}$ (with $\mathrm{diag}(\mathbf{T}) = 1/\mathrm{diag}(\mathbf{S})$) this penalty function may be replaced by $f''^{\mathrm{pen}}(\mathbf{\Omega}, \mathbf{T}, \lambda_l) = \|\mathbf{\Lambda} \circ \mathbf{\Omega}\|_1$ in which $\circ$ denotes the Hadamard product and $(\mathbf{\Lambda})_{j_1, j_2} = \lambda_l$ when $j_1 \neq j_2$ and zero otherwise (as is implemented in the `glasso`-package [9] by the option `penalize.diagonal=FALSE`).

We compare the proposed ridge and lasso estimators of the standardized precision matrix, as this forms the basis for inference on the conditional independence graph (the standardized precision matrix equals the partial correlation matrix up to the sign of off-diagonal elements). This is done in a data-driven manner, to avoid bias towards any of the estimators. Five curated breast cancer studies with gene expression data generated by the same (or comparable) Affymetrix platform [10] are used for this purpose. The full data set is limited to sets of genes that map to a pathway (as defined by the KEGG repository [11]). High-dimensionality is then realized by drawing subsets of the pathway data at sample sizes $n = 5, 10, 25$. For each draw the covariance matrix is estimated by means of lasso and ridge procedures. For both the LOOCV is used to choose their penalty parameters. The ridge estimate is then subjected to the local FDR procedure to decide on the presence/absence of gene-gene interactions.

The sensitivity and specificity of the resulting ridge and lasso inferred conditional independencies are compared. Hereto we define a 'consensus truth' based on overlapping edges. The resulting sensitivity and specificity of edge retrieval is comparable between the proposed ridge and the lasso estimators. An alternative comparison focusses on the loss of the estimates of the standardized precision matrix. Then, the proposed ridge estimator clearly yields a lower loss. These observations are consistent over the sample sizes, pathways, and data sets considered. Figure 1 visualizes these observations for a particular pathway.

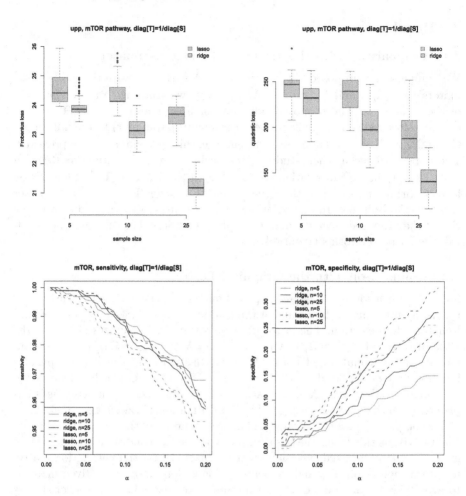

Fig. 1. The upper panels depict a loss comparison between the alternative ridge and the corresponding graphical lasso estimators for the mTOR-pathway on the UPP breast cancer data [10]. The loss is determined with a proxy of the standardized population precision matrix for the mTOR-pathway. The upper left-hand panel depicts Frobenius loss while the upper right-hand panel depicts quadratic loss. The lower panels depict a sensitivity and specificity comparison between the alternative ridge and graphical lasso estimators, again on the mTOR-pathway data. The evaluation of edge retrieval sensitivity and specificity requires knowledge of the true conditional dependencies. As such knowledge is absent we resort to defining a 'consensus truth', comprised of those conditional dependencies that appear in the top $100\alpha\%$ of at least 4 out of the 5 breast cancer data sets by both methods (graphical lasso and alternative ridge paired with local FDR edge selection). The parameter α ranges from .005 to .20, corresponding to what is believed to be biologically plausible (in terms of network density). Sensitivity (specificity) for a particular combination of n and α is then estimated as the median sensitivity (specificity) over the generated subsamples over all data sets. The lower left-hand panel gives sensitivity results while the lower right-hand panel gives specificity results.

4   Illustration

In this section we illustrate the reconstruction of a gene-gene interaction network from gene expression data using our R-implementation (see Section 6 below) of the proposed ML ridge estimator of the precision matrix. We employ breast cancer gene expression data by The Cancer Genome Atlas (TCGA) [12] of the mitogen-activated protein kinases (MAPK) pathway (as defined by KEGG).

For purposes of reproducibility we first provide the R-code that loads and 'processes' the data. It starts by activation of the necessary R-packages:

```
> library(biomaRt)
> library(cgdsr)
> library(KEGG.db)
> library(rags2ridges)
```

To get a list of all human genes and additional relevant information:

```
> ensembl = useMart("ensembl", dataset="hsapiens_gene_ensembl")
> geneList <- getBM(attributes=c("external_gene_name",
+                    "entrezgene"), mart = ensembl)
> geneList <- geneList[!is.na(geneList[,2]),]
```

Obtain the entrez IDs [13] of the genes that map to the MAPK pathway:

```
> kegg2entrez <- as.list(KEGGPATHID2EXTID)
> entrezIDs <- as.numeric(kegg2entrez[which(names(kegg2entrez)
+                    %in% "hsa04010")][[1]])
> entrez2name <- match(entrezIDs, geneList[,2])
> geneList <- geneList[entrez2name[!is.na(entrez2name)],]
```

Specify data set details (repository, TCGA study, samples, and profile):

```
> tcgaDB <- CGDS("http://www.cbioportal.org/public-portal/")
> cancerStudy <- "brca_tcga"
> caseList <- getCaseLists(tcgaDB, cancerStudy)[1,1]
> mrnaProf <- "brca_tcga_pub_mrna"
```

Extract the pathway expression data:

```
> Y <- getProfileData(tcgaDB, geneList[,1], mrnaProf, caseList)
> for (j in 1:ncol(Y)){
+     Y[,j] <- as.numeric(levels(Y[,j])[Y[,j]]) }
> Y <- data.matrix(Y)
```

Filter no-data samples and genes:

```
> sRemove <- which(rowSums(is.na(Y)) > ncol(Y)/10)
> Y <- Y[-sRemove,]
> gRemove <- which(colSums(is.na(Y)) > 0)
> Y <- Y[,-gRemove]
> Y <- sweep(Y, 2, colMeans(Y))
```

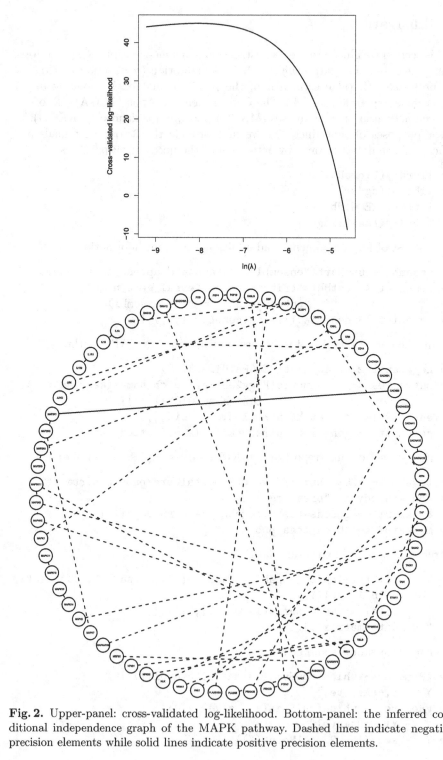

Fig. 2. Upper-panel: cross-validated log-likelihood. Bottom-panel: the inferred conditional independence graph of the MAPK pathway. Dashed lines indicate negative precision elements while solid lines indicate positive precision elements.

This concludes the executions in R required to obtain TCGA breast cancer data of the MAPK pathway as defined by KEGG. The gene expression data comprises $n = 496$ samples and $p = 259$ genes.

Finally, we turn to the reconstruction of the gene-gene interaction network of the MAPK pathway by means of the proposed ML ridge estimator of the precision matrix.[1] The target we use is $\mathbf{T} = \varphi \mathbf{I}_{p \times p}$, where φ denotes the average of the inverse (nonzero) eigenvalues of $\mathbf{S}$. Under this choice (3) is rotation equivariant, which is computationally advantageous (see Section 6). First, one needs to make an informed choice on the penalty parameter λ_a. This is done via the approximate LOOCV procedure (in which $\varphi \mathbf{I}_{p \times p}$ is the default target option):

```
> CVres <- optPenalty.aLOOCV(Y, 0.0001, 0.01, step=100)
```

The thus obtained cross-validated log-likelihood profile is plotted against the (logarithm of the) penalty parameter (see the upper-panel of Figure 2). The cross-validated log-likelihood achieves an optimum close to $\ln(\lambda_a) = -8.112$. This rather small value (little regularization) is due to the relative 'low-dimensionality' of the data.

With the optimal penalty parameter at hand the penalized ML ridge estimate of the precision matrix is obtained through:

```
> penPrec <- ridgeS(covML(Y), CVres$optLambda)
```

The object penPrec contains the desired estimate of the precision matrix that forms the basis for inferring the conditional independencies in the MAPK pathway data. Hereto the ML ridge estimate of the precision matrix is standardized to have a unit diagonal. The local FDR procedure of [3] is then applied to the off-diagonal elements of this standardized precision matrix. Edges corresponding to such elements with a posterior probability exceeding 0.95 are considered to be present in the gene-gene interaction network.

```
> P0 <- sparsify(penPrec, threshold="localFDR",
+                FDRcut=0.95)$sparsePrecision
```

The resulting sparsified precision matrix is visualized by its implied conditional independence graph:

```
> Ugraph(P0, type="fancy", prune=TRUE)
```

This gives an impression of the gene-gene interaction network underlying the MAPK pathway (see the bottom-panel of Figure 2).

5   Conclusion

We have presented a proper ML ridge estimator of the precision matrix, for which analytical properties can be proven. In a vis-á-vis comparison with other

---

[1] In the remainder of the illustration we use calls related to version 1.3 of our own R-package [15]. Please note that this package is in continual development so that certain calls may be depreciated or enhanced in future versions.

penalized inverse covariance estimators it was shown to yield a lower risk. More-over, its performance is on a par with the graphical lasso with respect to the sensitivity and specificity of selected conditional independencies. Hence, the pre-sented ridge estimator is a strong contender for inverse covariance estimation from high-dimensional data.

Currently, we are exploring the use of the target matrix $\mathbf{T}$. In this exposé we have limited ourselves to obvious choices. More sophisticated choices may be conceived. For instance, it may incorporate prior knowledge on the gene-gene interaction network as obtained from a pilot experiment or from repositories such as KEGG.

6 Software

The R [14] package `rags2ridges` [15] implements the proposed ridge precision estimator along with functions supporting subsequent graphical modeling. These additional functions enable, among others, (automated) penalty parameter se-lection, the evaluation of entropy and fit, support determination, (network) visu-alization, and network topology evaluation. The proposed estimator is analytic, making its computation friendly for a given penalty. When the chosen target implies rotation equivariance (i.e., the estimator leaves the eigenvectors of $\mathbf{S}$ unchanged), the search for an optimal penalty value and subsequent network extraction also become computationally efficient as the (relatively) expensive matrix square root can then be circumvented. In this situation only a single spectral decomposition and a single matrix inversion are required to obtain the complete solution path over any λ_a in the feasible domain. See the package doc-umentation [15] for more information. The package is freely available from the Comprehensive R Archive Network [16].

Acknowledgments. The research leading to these results has received fund-ing from the European Community's Seventh Framework Programme (FP7, 2007-2013), Research Infrastructures action, under the grant agreement No. FP7-269553 (EpiRadBio project).

References

1. van Wieringen, W.N., Peeters, C.F.W.: Ridge Estimation of Inverse Covariance Matrices From High-Dimensional Data. arXiv:1403.0904 [stat.ME] (2014)
2. Hoerl, A.E., Kennard, R.W.: Ridge Regression: Biased Estimation for Nonorthog-onal Problems. Technometrics 12, 55–67 (1970)
3. Schäfer, J., Strimmer, K.: A Shrinkage Approach to Large-Scale Covariance Matrix Estimation and Implications for Functional Genomics. Stat. Appl. Genet. Mo. B. 4, Article 32 (2005)
4. Ledoit, O., Wolf, M.: A Well-Conditioned Estimator for Large-Dimensional Co-variance Matrices. J. Multivariate Anal. 88, 365–411 (2004)

5. Fisher, T.J., Sun, X.: Improved Stein-Type Shrinkage Estimators for the High-Dimensional Multivariate Normal Covariance Matrix. Comput. Stat. Data An. 55, 1909–1918 (2011)
6. Vujačić, I., Abbruzzo, A., Wit, E.C.: A Computationally Fast Alternative to Cross-Validation in Penalized· Gaussian Graphical Models. arXiv: 1309.621v2 [stat.ME] (2014)
7. Banerjee, O., El Ghaoui, L., d'Aspremont, A.: Model Selection Through Sparse Maximum Likelihood Estimation for Multivariate Gaussian or Binary Data. J. Mach. Learn. Res. 9, 485–516 (2008)
8. Friedman, J., Hastie, T., Tibshirani, R.: Sparse Inverse Covariance Estimation with the Graphical Lasso. Biostatistics 9, 432–441 (2008)
9. Friedman, J., Hastie, T., Tibshirani, R.: **glasso**: Graphical Lasso-Estimation of Gaussian Graphical Models. R package, version 1.8 (2014), `http://cran.r-project.org/web/packages/glasso/index.html`
10. Schröder, M., Haibe-Kains, B., Culhane, A., Sotiriou, C., Bontempi, G., Quackenbush, J.: breastCancerMAINZ; breastCancerTRANSBIG; breastCancerUNT; breastCancerUPP; breastCancerVDX. R packages, versions 1.0.6 (2011), `http://compbio.dfci.harvard.edu/`
11. Kanehisa, M., Goto, S.: KEGG: Kyoto Encyclopedia of Genes and Genomes. Nucleic Acids Res. 28, 27–30 (2000)
12. The Cancer Genome Atlas, `http://cancergenome.nih.gov/`
13. The National Center for Biotechnology Information, `http://www.ncbi.nlm.nih.gov/gene`
14. R Development Core Team: R: A Language and Environment for Statistical Computing. R Foundation for Statistical Computing, Vienna, Austria (2011)
15. Peeters, C.F.W., van Wieringen, W.N.: **rags2ridges**: Ridge Estimation of Precision Matrices from High-Dimensional Data. R package, version 1.3 (2014), `http://cran.r-project.org/web/packages/rags2ridges/index.html`
16. Comprehensive R Archive Network, `http://www.R-project.org/`

Special Session: Computational Biostatistics for Data Integration in Systems Biomedicine

Estimation of a Piecewise Exponential Model by Bayesian P-splines Techniques for Prognostic Assessment and Prediction

Giuseppe Marano[1], Patrizia Boracchi[2], and Elia M. Biganzoli[1,2]

[1] Fondazione IRCSS Istituto Nazionale Tumori di Milano, Italy
[2] Dept. of Clinical Sciences and Community Health, University of Milan, Italy

Abstract. Methods for fitting survival regression models with a penalized smoothed hazard function have been recently discussed, even though they could be cumbersome. A simpler alternative which does not require specific software packages could be fitting a penalized piecewise exponential model. In this work the implementation of such strategy in Win-BUGS is illustrated, and preliminary results are reported concerning the application of Bayesian P-splines techniques. The technique is applied to a pre-specified model in which the number and positions of knots were fixed on the basis of clinical knowledge, thus defining a non-standard smoothing problem.

Keywords: Survival analysis, Hazard Smoothing, Bayesian P-splines, Piecewise Exponential Model.

1 Background

In bio-statistical applications, non-parametric and semi-parametric methods have been preferred over parametric ones for assessing the prognostic role of clinical/biological variables over time. The most widely adopted model is the Cox Model, in which no assumption of the functional form of the hazard function on time is made: however, such feature becomes a drawback if the interest lies on investigating the shape of the hazard function or in predictive modeling. To such ends, several methods for fitting survival models with a smoothed hazard function have been proposed (e.g.: Gray, R. J. 1992; Hastie, T. and Tibshirani, R. 1993; Kooperberg, C., Stone C. J. and Truong, Y. K. 1995; and Gamerman, D. 1991). A first approach was to apply a piecewise exponential (PE) model including time among predictors, by use of dummy variables. A further development was to obtain a smoothed estimate of the hazard function by including regression splines. This approach implies the definition of the number of basis and position of spline knots.

The use of penalized splines (P-splines) may be of advantage over other flexible polynomials, since it avoids an arbitrary definition of the number of spline bases (by using a high number of bases and eventually knots placed on a grid of equally spaced points). Furthermore, the degree of smoothing is controlled by regularized

© Springer International Publishing Switzerland 2015
C. di Serio et al. (Eds.): CIBB 2014, LNCS 8623, pp. 183–198, 2015.
DOI: 10.1007/978-3-319-24462-4_16

estimation, thus reducing the effective number of model degrees of freedom, and, therefore, giving a systematic method for protecting against overfitting.

Penalized methods for estimating survival models have been proposed in the frequentist framework by Cai, Hindman and Wand (Cai, J. T., Hyndman, R. J., and Wand, M. P. 2002) and Kauermann (Kauermann, G. 2005). The proposals are based on the relationship between P-splines and mixed model theory (cfr: Ruppert, D., Wand, M. and Carroll R. J. 2009). As a consequence, mixed model software routines may be used for estimation. However, the implementation of such a strategy could be cumbersome, because it requires the maximization of a likelihood function that includes a complex integral without a closed-form solution. An alternative is fitting a PE model by penalized GLM estimation routines, according to the link between the likelihood function of this model and the GLM family of distributions. It must be noted, however, that in this approach interval estimates are generally biased under standard large sample theory: thus, confidence intervals of model parameters and their functions are derived by a Bayesian approach (Wood, S. 2006).

The application of Bayesian P-splines techniques (Lang, S. and Brezger, A. 2004) in survival analysis has been discussed by Fahrmeir and Hennerfiend (Fahrmeir, L. and Hennerfiend, A. 2003); dedicated routines are available in the software package BayesX (Lang, S. and Brezger, A. 2000). In such a context, point and interval estimates of model parameters and survival functions may be obtained in a straightforward way from posterior density samples. To this aim Markov Chain Monte Carlo (MCMC) methods are needed, but they require efficient sampling algorithms in order to guarantee convergence of Markov chains, and may be computationally intensive. The estimation method in Fahrmeir and Hennerfiend's paper also deals with the approximation of a complex integral without closed form solution, and in order to solve this problem specific sampling schemes have been adopted for computation. Therefore, a dedicated software must be used to fit models in practice. An alternative could be fitting the PE model with general sampling algorithms for hierarchical (that is, mixed effects) GLM models: for example, see: Murray, T.A., Hobbs, B.P., Sargent, D.J., Carlin, B.P. 2014.

In this work we describe a procedure for estimating a PE model with a flexible specification of the hazard function and/or covariate effects. The particular assumptions on the shape of the hazard enable the use of general MCMC sampling routines without sacrificing flexibility. Thus, the described procedure could be helpful for developing robust parsimonious models for the purposes of exploratory assessment of prognostic factors, investigation of the hazard function and prediction.

The illustrated procedure has been applied to survival data from patients affected by soft tissue sarcoma (cfr: Ardoino, I. *et al* 2010) where the knots of the P-splines and the prognostic covariates were specified according to prior clinical knowledge. Therefore, we focused on flexible modeling of the hazard function only, since regularized estimation of covariate effects was not relevant in the present application. The computation was performed through WinBUGS

(Lunn, D. J. *et al* 2000), and, thus, the estimates were obtained by GIBBS sampling with the Adaptive Rejection Sampling (ARS) algorithm (Gilks, W. R. and Wild, P. 1992). Two relevant issues strictly concerning the implementation of the Bayesian method were evaluated: the sensitivity of estimates with respect to the hyper-prior of hazard parameters, and the use of modified penalties for accounting of unequal spacing of the observed follow-up times. More details about those points will be given in the methods section. Furthermore, for a comparison with frequentist methods, the same model was estimated through generalized additive model (GAM) techniques, via the *mgcv* package of R (Wood, S. 2006).

In the remainder part of the paper a formal definition of the penalized PE model is given. Then details are provided on the implementation of Bayesian estimation method. Results and discussion follow.

2 Methods

2.1 General Form of the Penalized PE Model

In the standard PE model the follow-up time is partitioned into a fixed set of H intervals: $T_h=[\tau_{h-1}, \tau_h)$; $h=1, \ldots, H$; and it is assumed that the time to failure has an exponential distribution, whose parameter depends on the time interval and the covariates:

$$\lambda(t \mid \mathbf{X_i}) = \left(\sum_{h=1}^{H} I(t \in T_h)\, \lambda_h \right) exp(\beta'\mathbf{X_i}) \quad ; i = 1, \ldots, n \tag{1}$$

where: $\lambda(t \mid \mathbf{X_i})$ is the parameter of the exponential distribution (hazard); $\mathbf{X_i}$ ($p \times 1$ vector) represents the set of covariates of the $i - th$ subject; $\lambda_1, \ldots, \lambda_H$ are the unknow values of the (piecewise constant) baseline hazard function; β is the $p \times 1$ vector of regression parameters; and $I(t \in T_h)$ is an indicator function being equal to 1 if $\tau_{h-1} \leq t < \tau_h$ and 0 otherwise.

Because of the relationship between the likelihood function of the PE model and the likelihood of a regression model with Poisson error (Aitkin, M., Laird, N., and Francis, B. 1983), the PE model may be estimated through GLM methods. To show the derivation of the GLM model, let t_i and δ_i be, respectively, the failure time of the $i - th$ subject ($i=1, \ldots,$ n) and the corresponding status at time t_i (0=censored, 1=failed). The response variable: Y_{ih}; is the status of subject i within the $h - th$ time interval (0=alive, 1=failed). Furthermore, to obtain estimates of the hazard through the fitted values of the model, the time spent in T_h by the subject must also be included as offset term. Expression (1) identifies a subclass of the family of proportional hazard (PH) models, for which: $\lambda(t \mid \mathbf{X_i}) = \lambda_0(t)exp(\beta'\mathbf{X_i})$. Thus, the likelihood function of the PE model can be derived from the general PH likelihood:

$$L(\boldsymbol{\lambda}, \boldsymbol{\beta} \mid \mathbf{t}, \boldsymbol{\delta}, \mathbf{X}) = \prod_{i=1}^{n} \left(\lambda(t_i \mid \mathbf{X_i})\right)^{\delta_i} exp\left(- \Lambda(t_i \mid \mathbf{X_i})\right) ;$$

where the $n \times 1$ vectors: $\mathbf{t}$ and $\boldsymbol{\delta}$, and the $n \times p$ matrix X represent the data; and $\Lambda(t \mid \mathbf{X_i}) = \int_0^t \lambda(s \mid \mathbf{X_i})ds$ is the cumulative hazard function.

By substituting the instantaneous and cumulative hazards with the corresponding expressions of the PE model, and by some algebraic simplifications one obtains:

$$L_{PE}(\boldsymbol{\lambda}, \boldsymbol{\beta} \mid \mathbf{t}, \boldsymbol{\delta}, \mathbf{X}) = \prod_{i=1}^{n} \prod_{h=1}^{H_i} \left(\lambda_h exp(\boldsymbol{\beta}'\mathbf{X_i}) \right)^{y_{ih}} exp\left(- \lambda_h exp(\boldsymbol{\beta}'\mathbf{X_i})\Delta_{ih} \right) ;$$

where H_i is the number of time intervals in which the $i-th$ subject is still alive or failed; y_{ih} represents the status of subject i within T_h (0=alive, 1=failed) ; and $\Delta_{ih} = min(t_i, \tau_h) - \tau_{h-1}$ is the time spent in T_h by the subject. Since y_{ih} can take only 0 or 1 values, the 'likelihood contribution' of subject i:

$$\prod_{h=1}^{H_i} \left(\lambda_h exp(\boldsymbol{\beta}'\mathbf{X_i}) \right)^{y_{ih}} exp\left(- \lambda_h exp(\boldsymbol{\beta}'\mathbf{X_i})\Delta_{ih} \right) ;$$

is proportional to the product of the likelihoods of H_i Poisson variables: Y_{ih}, with mean parameters: $\mu_{ih} = \lambda_h exp(\boldsymbol{\beta}'\mathbf{X_i})$; and offsets: Δ_{ih}. According to this property, inference on the parameters $\boldsymbol{\lambda}$ and $\boldsymbol{\beta}$ of the PE model is usually performed by estimation of the corresponding GLM model:

$$Y_{ih} \sim POISSON(\mu_{ih}) ; \quad log(\mu_{ih}) = \alpha_h + \boldsymbol{\beta}'\mathbf{X_i} + log(\Delta_{ih}) \qquad (2)$$

where: $\alpha_h = log(\lambda_h)$ is the the log-hazard in interval h. In order to estimate the parameters $\alpha_1, \ldots, \alpha_H$ in the GLM model (2), H dummy variables, corresponding to the H intervals T_h, must be included among the predictors. For more details about the relationships shown above, and about the estimation of the PE model through standard software routines, the reader may consult: Aitkin, M., Laird, N. and Francis, B. 1983; Lawless, J. F. 2011; and Congdon, P. 2007.

Because of the relationship between a P-spline estimation problem and mixed model theory (Lang, S. and Brezger, A. 2004) the Bayesian approach is based on a hierarchical model for expression (2), where penalty terms and parameters are included in the prior densities. A general expression of a model for regularized estimation of the hazard function is the following:

$$\begin{cases} Y_{ih} \sim POISSON(\mu_{ih}) ; \; log(\mu_{ih}) = \alpha_h + \boldsymbol{\beta}'\mathbf{X_i} + log(\Delta_{ih}) \\ \\ \boldsymbol{\beta} \sim \pi_\beta \\ \\ \boldsymbol{\alpha} \mid \tau^2 \propto exp\left(- \frac{\tau^2}{2} \boldsymbol{\alpha}'\mathbf{P}\boldsymbol{\alpha} \right) ; \; \tau^2 \sim \pi_\tau \end{cases} \qquad (3)$$

where $\mathbf{P}$ and τ^2 indicate, respectively, the penalty matrix (of dimension $H \times H$) and the 'tuning' parameter that governs the overall amount of smoothing. This model is equivalent to the general problem of penalized maximum likelihood estimation of the GLM model (2) (cfr: Lang, S., and Brezger, A. 2004) :

$$\arg\max_{\alpha,\beta} L_{GLM}(\alpha,\beta \mid \mathbf{t}, \boldsymbol{\delta}, \mathbf{X}) - p_{\tau^2}(\alpha) ;$$

where the penalty function $p_{\tau^2}(\alpha)$ has the form: $p_{\tau^2}(\alpha) = \frac{\tau^2}{2}\alpha'\mathbf{P}\alpha$. This highlights the correspondence between the choice of the penalty function and the specific form of P in the Bayesian model.

Model (3) is a particular case of the general model in: Fahrmeir, L. and Hennerfiend, A. 2003 (expressions 2 and 3); where a wider class of spline functions have been considered for modeling the baseline hazard. As in Fahrmeir and Hennerfiend's paper, the model may be readily extended for penalized estimation of covariate and time-dependent effects. Concerning the prior densities, several choices are available and the reader is referred to Ibrahim, J. G., Chen, M. H. and Sinha, D. 2005 and Congon, P. 2007 for a detailed discussion.

Estimation cannot be undertaken without specifying the penalty matrix P and the prior densities for β and τ^2, thus determining a particular model from the general expression (3). In our procedure a second-order difference penalty function and non-informative priors for the parameters have been adopted. A detailed discussion is provided in the following section.

2.2 Implementation of the Estimation Procedure on the Sarcoma Case Series

Data were collected from a case series of 192 subjects affected by soft tissue sarcoma in care at Istituto Nazionale dei Tumori di Milano who underwent surgical resection of primary localized disease (cfr: Ardoino, *et al.* 2010). The end-point of interest was time elapsed form surgery to death from any cause. Time was measured in months. The model was pre-specified according to prior clinical knowledge: 1) the time axis was partitioned in fifteen intervals according to the schedule of patients control visits; 2) predictors were five known prognostic factors: age, tumor size (continuous) histologic subtype, grading, surgical margins (categorical). A non linear effect of tumor size was modeled through a restricted cubic spline with three knots.

To determine a specific expression of the general model (3) non-informative (diffuse normal) priors for regression coefficients (n=10) and a non-informative prior for the tuning parameter have been adopted (further details about the latter will be given in subsequent discussion). The resulting model expression is:

$$\begin{cases} Y_{ih} \sim POISSON(\mu_{ih}) ; \; log(\mu_{ih}) = \alpha_h + \beta\mathbf{X_i} + log(\Delta_{ih}) \\[2mm] \beta \sim MVN_{10}(\mathbf{0}, 1000 \times \mathbf{I_{10}}) \\[2mm] \alpha \mid \tau^2 \sim \mathrm{MVN}_{15}\left(\mathbf{0}, \frac{1}{\tau^2}\mathbf{P}^{-1}\right) ; \; \tau^2 \sim \pi_{\tau^2} \end{cases} \qquad (4)$$

where MVN_k indicates a k-dimensional multivariate Gaussian density.

Concerning the expression of the matrix P, as a first choice we followed the approach of Eilers and Marx (Eilers, P. H. and Marx, B. D. 1996), adopting a second-order difference penalty function:

$$p_{\tau^2}(\boldsymbol{\alpha}) = \frac{\tau^2}{2} \sum_{h=1}^{H-2} \left(\Delta^2 \alpha_{h+2}\right)^2 = \frac{\tau^2}{2} \sum_{h=1}^{H-2} (\alpha_{h+2} - 2\alpha_{h+1} + \alpha_h)^2 ;$$

the term Δ^2 in the expression above denotes the second-order finite difference operator. Note that such penalty is coupled with a piecewise constant function (i.e. the baseline hazard of the PE model) that is equivalent to a B-spline of degree 0. The penalty function above can be expressed in matrix notation: $p_{\tau^2}(\boldsymbol{\alpha}) = \frac{\tau^2}{2}\boldsymbol{\alpha}'\mathbf{P}\boldsymbol{\alpha}$, by specifying:

$$\mathbf{P} = \begin{bmatrix} 1 & -2 & 1 & & & & & & \\ -2 & 5 & -4 & 1 & & & & & \\ 1 & -4 & 6 & -4 & 1 & & & & \\ & -1 & -4 & 6 & -4 & 1 & & & \\ & & \ddots & \ddots & \ddots & \ddots & \ddots & & \\ & & & -1 & -4 & 6 & -4 & 1 & \\ & & & & -1 & -4 & 6 & -4 & 1 \\ & & & & & 1 & -4 & 5 & 2 \\ & & & & & & 1 & 2 & -1 \end{bmatrix} \tag{5}$$

where the elements outside the second off-diagonals are equal to 0. Due to the particular structure of $\mathbf{P}$, the prior density for $\boldsymbol{\alpha}$ (expression (4) is equivalent to a random walk prior of order two (that is, the same order of the difference penalty). Such prior is frequently represented as a random walk process: $\alpha_{h+2} = 2\alpha_{h+1} - \alpha_h + u_h$; $h = 1, \ldots, 13$; where the u_h are i.i.d. Gaussian errors. Notably, the random walk prior is essentially an auto-regressive Gaussian density function with rank-deficient covariance matrix, and, thus, cannot be sampled directly. An efficient sampling method is based on a decomposition of $\boldsymbol{\alpha}$ in two vectors provided of a full-rank covariance matrix; for details see: Kneib, T., Fahrmeir, L. and Denuit, M. 2004.

It must be noted that the second-order difference penalty has been recommended for equally spaced knots (Eilers, P. H., and Marx, B. D. 1996) : however, this could be inadequate in our application. Therefore, a modified prior for $\boldsymbol{\alpha}$, discussed, amongst others, in Fahrmeir, L. and Lang, S. 2001, has been also adopted. This prior is based on a generalized expression for the random walk process that allows for unequal time gaps between adjacent variables. In particular, the expression for the second-order random walk is:

$$\alpha_{h+2} = \left(1 + \frac{D_{h+2}}{D_{h+1}}\right)\alpha_{h+1} - \frac{D_{h+2}}{D_{h+1}}\alpha_h + v_h ; \quad h = 1, \ldots, 13$$

where D_h denotes the time gap between α_h and α_{h-1}. It may be seen that by setting $D_h = 1$ for each h, the expression above reduces to the ordinary one. For including the modified prior into the model it is sufficient to insert the corresponding expression of $\mathbf{P}$ into the prior of $\boldsymbol{\alpha}$ in expression (4). For difference penalties of second-order (or higher) the expression is rather messy: therefore, for sake of simplicity, it has not been shown here. Finally, it must be noted that the prior of α is again a rank-deficient multivariate Gaussian density. Thus, the decomposition previously mentioned is required for sampling from the prior.

Concerning the non-informative priors for τ^2, it has been pointed out (see: Fahrmeir, L. and Kneib, T. 2009) that the choice of the density function may affect posterior estimates: thus, a sensitivity analysis is recommended for applications. To such aim two common densities in the context of Bayesian regularization were adopted in our work: a uniform(0,100) prior for $1/\tau$ and an inverse-gamma(1,1) prior for $1/\tau^2$. The latter one is less diffuse than the former, and, thus, may be considered as a more informative one.

Before concluding this section, some brief notes on software implememtation will be given. The code used for specifying the hierarchical model (4) in Win-BUGS is shown in Appendix A. As usual, the likelihood of the model is defined in the first part of the code, and the priors are defined in the subsequent part. In particular, the code for the likelihood is analogous to the code given by Congdon (cfr: Congdon, P. 2007) for the standard PE model, while some examples of P-spline estimation of GLM models are given in: Crainicenu, C., Ruppert, D., and Wand, M. P. 2005. Hobbs, Sargent and Carlin gave JAGS (Plummer, M. 2003) code for for P-spline estimation of PH models (Murray, T.A., Hobbs, B.P., Sargent, D.J., Carlin, B.P. 2014). The operations needed for calculating specific components of the model (such as the penalty matrix $\mathbf{P}$), and for improving the efficiency of the computation (such as centering continuous variables) were done in R (R Core Team 2013). One of the main advantages brought by this choice was the simplification of code writing in WinBUGS. Furthermore, by running WinBUGS under R via the *R2WinBUGS* package (Sturtz, S., Ligges, U. and Gelman, A. 2005), the posterior samples of model parameters and their functions are automatically imported in R, so that to avoid the cumbersome operations required for exporting *coda* files from WinBUGS.

2.3   Statistical Analysis

A total of four Bayesian models were estimated (2 penalties $\times$ 2 priors for the tuning parameter τ^2). The two penalties consist in a standard penalty and a modified version for accounting for the unequally spaced partition of follow-up time. The different priors were used for performing a sensitivity analysis, as discussed in the previous paragraph.

To avoid convergence problems the continuous covariates were centered, and the bases of the restricted spline (used for modeling the effect of tumor size) were orthogonalized. Results shown in the next section were obtained from one-chain samples with length 20,000, burn-in 2,000 and thinning by 5. Posterior means and 95% Credible Intervals were computed for regression coefficients, baseline hazard functions and survival probabilities at 120 months for some combination of predictors values. Moreover, the Deviance Information Criterion (DIC) was computed in order to compare the goodness of fit of the resulting models (Spiegelhalter, D., J. *et al* 2002).

The estimates from each model were compared to the estimates obtained by penalized estimation via GAM techniques. In this case the optimal amount of smoothing was determined through a prediction error criterion (Un-Biased Risk Estimator), which for additive models is essentially equivalent to Mallows Cp. The UBRE criterion is estimated for a trial set of smoothing parameters, and the model with optimal value of it is then selected (Wood, S. 2006). MCMC sampling was performed by executing WinBUGS under R via by the *R2WinBUGS* package; subsequent calculations were done by R with *coda* and *mgcv* packages added.

3   Results

Bayesian models were fitted without incurring in convergence problems. Standard diagnostic techniques showed appreciable mixing properties of the Markov Chains, both for model parameters and for their functions (baseline hazard and survival probabilities), and autocorrelations rapidly approaching zero. A slower convergence of autocorrelations was shown only for the tuning parameter τ^2, so that thinning was set to 5 for obtaining the final estimates. For each model the computation took about 45 minutes on a commercial laptop.

Overall, frequentist and Bayesian estimates of predictor effects (regression coefficients) were similar (not shown) ; this result was expected since in PH models such estimates are robust with respect to the functional form of the hazard function. Concerning the use of the modified penalty matrix, the estimates of regression coefficients, hazard parameters and survival probabilities were very close to the one obtained with the standard penalty. The estimates of hazard parameters were reported in Appendix B.

Concerning the hazard function, different behaviors were shown by frequentist and Bayesian estimates. In Fig. 1 the hazard function estimated by GAM techniques increases in the first half of follow-up period (0-60 months), after which it is constant. Bayesian estimates (posterior means) showed a two-peaked pattern with a rather strong decrease in the second half of follow-up (60-120 months). However, the ones obtained with the uniform prior seemed rather close to the frequentist ones, except for the second half of follow-up time (where few events were observed: n=10).

Bayesian estimates resulted to be sensitive to prior densities. Posterior means obtained with the inverse-gamma prior showed a greater spread than estimates with the uniform prior. This was due to the different degree of smoothing as determined by the posterior estimates of τ^2 (Fig. 2). The DIC criterion showed that a slight better fit was obtained with the uniform prior: DIC equal to 662.5 and 664.6 for uniform and inverse-gamma prior respectively.

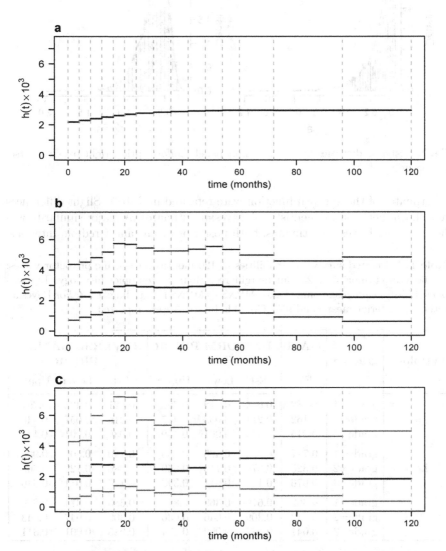

Fig. 1. Estimates of baseline hazard function. Panel a: GAM estimates; Panel b: posterior means for uniform prior on the deviance of hazard parameters, with respective 95% Credible Intervals; Panel c: posterior means, 95% CI for gamma-inverse prior.

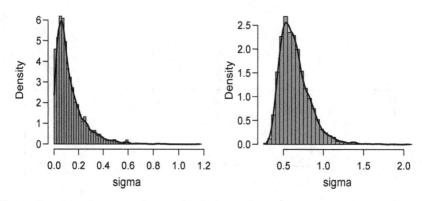

Fig. 2. Posterior densities of $\sigma = 1/\tau$. Left panel: uniform prior; right panel: inverse-gamma prior.

Estimates of the survival function were reported in Tab. 1. Slight differences were shown between frequentist and Bayesian estimates. A major similarity was shown among Bayesian estimates, both posterior means and Credible Intervals.

Table 1. Estimated survival probabilities at 120 months for specific predictor values: age: 60 year; tumor size: 20 cm; margin resection: microscopic; histology: Liposarcoma (Lipo), Leiomyosarcoma (Leio), MPNST (MPNST); grading: 1,2,3. For Bayesian methods, posterior means and 95% Credible Intervals were reported.

hystology	grading	GAM Est	UNIFORM PRIOR Est	Lower	Upper	INVERSE-GAMMA PRIOR Est	Lower	Upper
Lipo	grade=1	0.582	0.614	0.418	0.788	0.621	0.422	0.791
	grade=2	0.162	0.211	0.053	0.437	0.216	0.057	0.445
	grade=3	0.015	0.035	0.001	0.139	0.036	0.001	0.141
Leio	grade=1	0.717	0.734	0.495	0.895	0.740	0.500	0.900
	grade=2	0.325	0.379	0.119	0.649	0.386	0.129	0.669
	grade=3	0.076	0.120	0.011	0.342	0.126	0.011	0.346
MPNST	grade=1	0.662	0.679	0.369	0.891	0.680	0.374	0.890
	grade=2	0.248	0.309	0.046	0.636	0.309	0.047	0.643
	grade=3	0.041	0.087	0.002	0.324	0.085	0.001	0.311

4   Discussion

Penalized methods for smoothing the hazard function in proportional hazard models could provide useful tools for the assessment of the effect of prognostic variables and at the same time for assessing hazard and prediction. The piecewise exponential model may offer a simple alternative both in the frequentist and in the Bayesian framework. However, in the former one standard large sample results cannot guarantee the expected properties of estimates and it would be necessary to resort to non standard theory.

As a practical alternative we used a Bayesian method for estimating a PE model with a smoothed hazard function. Efficient computation was obtained with a standard sampling algorithm (ARS) and a popular software in a non-standard smoothing problem: in fact number and positions of knots were fixed according to relevant clinical knowledge. The reason of such efficiency is related to the particular features of the PE model: 1) the link with the widespread GLM family of distributions; 2) a simple expression of the hazard function which preserves flexibility without requiring restrictive distributional assumptions. Furthermore, results obtained with a modified version of the penalty matrix showed no sensible differences with respect to the standard procedure of estimation. Consequently, the standard procedure could be robust with respect to unequally spacing of the partition of the follow-up time.

Our results showed a different behavior between the hazard functions estimated by GAM and Bayesian P-splines methods, as a double peak was observed only in the latter case. It is worth of note that a double peaked hazard function over time is in agreement with previous findings with several frequentist approaches in the therapy follow-up of different solid tumor pathologies, so consistent interpretation of risk dynamics may be derived from them linked to the hypothesis of cancer dormancy. However, these first results deserve further exploration, since in the present work few hyper-priors for hazard parameters were assessed. Moreover, since fewer events were observed in the second half of follow-up, this could have limited the interpretation of the results in later follow-up time. A more extensive evaluation could be properly carried out in a simulation setting. For future assessments, measures of model fit would be needed in order to compare the various alternative hyper-priors. In our case a preliminary comparison according to the DIC criterion showed that a slightly better fit was obtained by assuming a uniform density with respect to a rather informative inverse-gamma density.

As expected by cumulative function estimation, a major degree of robustness to prior specification was shown by the estimated survival function. The differences between frequentist and Bayesian point estimates may be explained by the fact that posterior density of survival probabilities were rather asymmetric. Therefore, the comparison should be extended, e.g. to maximum posterior density estimates. Overall, the proposed method can extend in a practical way the tools available for the flexible modeling of survival functions allowing the study of the disease dynamics in complex frameworks like cancer follow-up studies.

Acknowledgements. This work was funded by the Italian Association for Cancer Research (AIRC). IG 2012 rif: 13420, Statistical Tools for Prognosis and Prediction in Cancer: Assessments and Application to a Sarcoma Case Series. Elia Biganzoli was Principal Investigator. Giuseppe Marano was a fellow of AIRC. The authors thank Dr. Alessandro Gronchi for his support to the project.

References

Aitkin, M., Laird, N., Francis, B.: A reanalysis of the Stanford heart transplant data. Journal of the American Statistical Association 78(382), 264–274 (1983)

Ardoino, I., Miceli, R., Berselli, M., Mariani, L., Biganzoli, E.M., Fiore, M., Colini, P., Stacchiotti, S., Casali, P.G., Gronchi, A.: Histology-specific nomogram for primary retroperitoneal soft tissue sarcoma. Cancer 116(10), 2429–2436 (2010)

Cai, J.T., Hyndman, R.J., Wand, M.P.: Mixed model-based hazard estimation. Journal of Computational and Graphical Statistics 11(4), 784–798 (2002)

Congdon, P.: Bayesian statistical modeling. John Wiley & Sons (2007)

Crainiceanu, C., Ruppert, D., Wand, M.P.: Bayesian analysis for penalized spline regression using WinBUGS (2005)

Eilers, P.H., Marx, B.D.: Flexible smoothing with B-splines and penalties. Statistical Science, 89–102 (1996)

Fahrmeir, L., Lang, S.: Bayesian inference for generalized additive mixed models based on Markov random field priors. Journal of the Royal Statistical Society: Series C (Applied Statistics) 50(2), 201–220 (2001)

Fahrmeir, L., Hennerfiend, A.: Nonparametric Bayesian hazard rate models based on penalized splines, Discussion paper//Sonderforschungsbereich 386 der Ludwig-Maximilians-Universität München (2003)

Fahrmeir, L., Kneib, T.: Propriety of posteriors in structured additive regression models: Theory and empirical evidence. Journal of Statistical Planning and Inference 139(3), 843–859 (2009)

Gamerman, D.: Dynamic Bayesian models for survival data. Applied Statistics, 63–79 (1991)

Gilks, W.R., Wild, P.: Adaptive rejection sampling for Gibbs sampling. Applied Statistics, 337–348 (1992)

Gray, R.J.: Flexible methods for analyzing survival data using splines, with applications to breast cancer prognosis. Journal of the American Statistical Association 87(420), 942–951 (1992)

Hastie, T., Tibshirani, R.: Varying-coefficient models. Journal of the Royal Statistical Society. Series B (Methodological), 757–796 (1993)

Murray, T.A., Hobbs, B.P., Sargent, D.J., Carlin, B.P.: Flexible Bayesian Survival Modeling with Semiparametric Time-Dependent and Shape-Restricted Covariate Effects. Bayesian Analysis (2015)

Ibrahim, J.G., Chen, M.H., Sinha, D.: Bayesian survival analysis. John Wiley & Sons, Ltd. (2005)

Kauermann, G.: Penalized spline smoothing in multivariable survival models with varying coefficients. Computational Statistics & Data Analysis 49(1), 169–186 (2005)

Kneib, T., Fahrmeir, L., Denuit, M.: A mixed model approach for structured hazard regression Discussion paper//Sonderforschungsbereich 386 der Ludwig-Maximilians-Universität München (2004)

Kooperberg, C., Stone, C.J., Truong, Y.K.: Hazard regression. Journal of the American Statistical Association 90(429), 78–94 (1995)

Lang, S., Brezger, A.: BayesX-Software for Bayesian Inference based on Markov Chain Monte Carlo simulation techniques (2000)

Lang, S., Brezger, A.: Bayesian P-splines. Journal of Computational and Graphical Statistics 13(1), 183–212 (2004)

Lawless, J.F.: Statistical models and methods for lifetime data, vol. 362. John Wiley & Sons (2011)

Lunn, D.J., Thomas, A., Best, N., Spiegelhalter, D.: WinBUGS-a Bayesian modelling framework: concepts, structure, and extensibility. Statistics and Computing 10(4), 325–337 (2000)

Plummer, M.: JAGS: A program for analysis of Bayesian graphical models using Gibbs sampling. In: Proceedings of the 3rd International Workshop on Distributed Statistical Computing, Vienna, vol. 124, p. 125 (2003)

R Core Team: R: A language and environment for statistical computing. R Foundation for Statistical Computing, Vienna, Austria (2013). http://www.R-project.org/ ISBN 3-900051-07-0

Ruppert, D., Wand, M., Carroll, R.J.: Semiparametric regression during 2003-2007. Electronic Journal of Statistics 3, 1193–1256 (2009)

Spiegelhalter, D.J., Best, N.G., Carlin, B.P., Van Der Linde, A.: Bayesian measures of model complexity and fit. Journal of the Royal Statistical Society: Series B (Statistical Methodology) 64(4), 583–639 (2002)

Sturtz, S., Ligges, U., Gelman, A.: R2WinBUGS: A Package for Running WinBUGS from R. Journal of Statistical Software 12(3), 1–16 (2005)

Wood, S.: Generalized additive models: an introduction with R. CRC Press (2006)

A   WinBUGS Code for Estimation of the Penalized PE Model

Note: the log-hazard parameters (alpha) in model (4) have been reparameterized in order to obtain samples from the random walk prior (see the section: Implementation of the estimation procedure on the sarcoma case series). Thus, two new sets of parameters appear in the code below. The first set includes two elements, namely alpha.unp[1] and alpha.unp[2], that must not be subjected to penalized estimation. The latter set include 13 penalized parameters: alpha.pen[1:13]. For more details, see Kneib, T., Fahrmeir, L. and Denuit, M. 2004.

```
# CONSTANTS:
# N : overall number of time intervals in which the n subjects are
#       alive or failed
# T : number of time intervals (15)
#
# PARAMETERS:
# alpha.unp[1:2]          : unpenalized log-hazard parameters
# alpha.pen[1:13]         : penalized log-hazard parameters
# beta[1:10]              : regression coefficients
# tau.square              : tuning parameter
#
# VARIABLES:
# age             ([1:N]) : patient age
# t.size1, t.size2 ([1:N]) : spline bases for tumor size
# Dhist2  Dhist9  ([1:N]) : dummy variables for histology
# Dgrad2, Dgrad3  ([1:N]) : dummy variables for grading
# Dmarg2          ([1:N]) : dummy variable for surgical margins
# timespent[1:N]  ([1:N]) : offset term (delta_ih)
#
# psi.unp[1:N]            : fixed coefficients for alpha.unp[2]
# psi.pen[1:N,1:(T-2)]    : fixed coefficients for alpha.pen

model {

### likelihood ###
for (i in 1:N) {

  y[i]~ dpois( theta[i] ) ;

  theta[i] <- timespent[i] *
          exp( random[i] +
              beta[1]*age[i]     + beta[2]*t.size1[i] +
              beta[3]*t.size2[i] + beta[4]*Dhist2[i]  +
              beta[5]*Dhist3[i]  + beta[6]*Dhist4[i]  +
```

```
                beta[7]*Dhist9[i]  + beta[8]*Dgrad2[i]  +
                beta[9]*Dgrad3[i]  + beta[10]*Dmarg[i] ) ;


  random[i] <- alpha.unpen[1]              + alpha.unpen[2]*psi.unp[i] +
            alpha.pen[1]*psi.pen[i,1]      + alpha.pen[2]*psi.pen[i,2] +
            alpha.pen[3]*psi.pen[i,3]      + alpha.pen[4]*psi.pen[i,4] +
            alpha.pen[5]*psi.pen[i,5]      + alpha.pen[6]*psi.pen[i,6] +
            alpha.pen[7]*psi.pen[i,7]      + alpha.pen[8]*psi.pen[i,8] +
            alpha.pen[9]*psi.pen[i,9]      + alpha.pen[10]*psi.pen[i,10] +
            alpha.pen[11]*psi.pen[i,11]    + alpha.pen[12]*psi.pen[i,12] +
            alpha.pen[13]*psi.pen[i,13] ) ;
}


### priors ###
for (i in 1:10) {
  beta[i] ~ dnorm(0,0.0001) ;
}


alpha.unpen[1] ~ dflat() ;
alpha.unpen[2] ~ dflat() ;


# multivariate gaussian prior for the penalized parameters
for (i in 1:(T-2)) {
  mu.vec[i] <- 0 ;
  for (j in 1:(T-2)) {
    prec.mat[i,j] <- equals(i,j)*tau.square ;
  }
}
alpha.pen[1:(T-2)] ~ dmnorm(mu.vec[],prec.mat[,]) ;


# hyperprior
tau.square <- pow(sigma,-2) ;
sigma ~ dunif(0,100) ;
}
```

B   Bayesian Estimates of Hazard Parametersl

Table 2. Posterior means and respective standard deviations of the parameters of the baseline hazard function ($exp(\alpha_h)$; h=1, ..., 15) are reported in columns 2-5. The estimates were obtained by adopting two hyper-priors for the the tuning paramater (uniform and inverse-gamma) and two penalties accounting for equal and unequal spacing of the partition of follow-up time. The corresponding 15 time intervals are shown in column 1; interval length (in months) is reported in parenthesis.

	UNIFORM PRIOR		INVERSE-GAMMA PRIOR	
Interval	equal Est (SD)	unequal Est (SD)	equal Est (SD)	unequal Est (SD)
0-4 (4)	0.0019 (0.0009)	0.0020 (0.0009)	0.0017 (0.0009)	0.0017 (0.0009)
4-8 (4)	0.0021 (0.0009)	0.0022 (0.0009)	0.0019 (0.0009)	0.0019 (0.0009)
8-12 (4)	0.0024 (0.0010)	0.0025 (0.0010)	0.0026 (0.0012)	0.0027 (0.0012)
12-16 (4)	0.0026 (0.0010)	0.0027 (0.0010)	0.0026 (0.0012)	0.0027 (0.0012)
16-20 (4)	0.0029 (0.0012)	0.0029 (0.0011)	0.0034 (0.0015)	0.0034 (0.0015)
20-24 (4)	0.0030 (0.0012)	0.0029 (0.0011)	0.0035 (0.0015)	0.0034 (0.0015)
24-30 (6)	0.0029 (0.0011)	0.0029 (0.0011)	0.0027 (0.0012)	0.0028 (0.0012)
30-36 (6)	0.0029 (0.0011)	0.0029 (0.0010)	0.0025 (0.0012)	0.0025 (0.0011)
36-42 (6)	0.0029 (0.0011)	0.0029 (0.0011)	0.0024 (0.0011)	0.0024 (0.0011)
42-48 (6)	0.0029 (0.0011)	0.0030 (0.0011)	0.0027 (0.0013)	0.0027 (0.0012)
48-54 (6)	0.0031 (0.0012)	0.0031 (0.0012)	0.0036 (0.0018)	0.0037 (0.0017)
54-60 (6)	0.0031 (0.0012)	0.0030 (0.0012)	0.0038 (0.0018)	0.0038 (0.0018)
60-72 (12)	0.0029 (0.0012)	0.0029 (0.0011)	0.0035 (0.0016)	0.0035 (0.0015)
72-96 (24)	0.0026 (0.0011)	0.0025 (0.0010)	0.0022 (0.0011)	0.0023 (0.0011)
96-120 (24)	0.0025 (0.0013)	0.0022 (0.0013)	0.0018 (0.0012)	0.0017 (0.0012)

Use of q-values to Improve a Genetic Algorithm to Identify Robust Gene Signatures

Daniel Urda[1], Simon Chambers[2], Ian Jarman[2], Paulo Lisboa[2],
Leonardo Franco[1], and Jose M. Jerez[1]

[1] University of Málaga, Spain
durda@lcc.uma.es
[2] Liverpool John Moores University, United Kingdom
s.j.chambers@ljmu.ac.uk

Abstract. Several approaches have been proposed for the analysis of DNA microarray datasets, focusing on the performance and robustness of the final feature subsets. The novelty of this paper arises in the use of q-values to pre-filter the features of a DNA microarray dataset identifying the most significant ones and including this information into a genetic algorithm for further feature selection. This method is applied to a lung cancer microarray dataset resulting in similar performance rates and greater robustness in terms of selected features (on average a 36.21% of robustness improvement) when compared to results of the standard algorithm.

Keywords: DNA microarrays, Evolutionary algorithms, t-test, q-values, Feature selection.

1 Scientific Background

DNA microarray technology has been widely used for gene expression profiling and prediction of cancer. Analysis of such data involves facing a problem commonly referred to as the curse of dimensionality [9] where each sample is described by thousands of features (genes) with few samples - often fewer than a hundred - available. Several approaches have been proposed to identify relevant genes with good performance in classifying the disorder under investigation. However, these approaches lack a desirable feature when identifying gene expression profiles - robustness. A common feature of such methods is instability of results with high variability of identified features when repeated executions of the algorithm are made. To tackle this problem, recent works have proposed different methodologies that try to achieve robust feature subset selections with good performance rates in test data [7,10].

Use of statistical tests with multiple features against some null hypothesis is common practice with the expectation that a proportion of such features would be incorrectly considered significant [8]. In such circumstances it is important to use some form of false discovery rate technique to either adjust the p-values [1] or use a different measure which takes into account false positives such as

© Springer International Publishing Switzerland 2015
C. di Serio et al. (Eds.): CIBB 2014, LNCS 8623, pp. 199–206, 2015.
DOI: 10.1007/978-3-319-24462-4_17

the q-value [8]. Use of such a measure allows focus to be placed on features which can be considered to satisfy a null hypothesis in further analysis. In the original paper [8] this methodology reduced the number of features identified in the Hedenfalk dataset from 605 to 162 within a total feature set of 3170.

In this paper a modified t-test and q-values [8] are incorporated into a feature selection procedure similar to the genetic algorithm (GA) described in [7] with the purpose of identifying genes that are significant in differentiating lung cancer microarray expressions. In their approach, biological information from KEGG [5,6] database was included into the GA resulting in more robust feature subsets with good performance rates. The expectation of introducing a subset of genes, selected using q-values, into the GA would be for better and more robust solutions than the original results from the GA.

The rest of the paper is structured as follows: Section 2 describes the dataset used in this study as well as the methodology applied; Section 3 shows the results obtained in this work and a comparison to previous results of one similar approach; and Section 4 provides some conclusions.

2   Materials and Methods

A freely available[1] high dimensional biomedical dataset has been used throughout this work, comprising 181 tissue samples of two types of lung cancer, malignant pleural Mesothelioma (MPM) and Adenocarcinoma (ADCA) [4]. Samples are unbalanced with 31 corresponding to MPM and 150 ADCA, described in each case by 12533 genes. The Affymetrix ID for the lung cancer DNA microarray dataset is hgu95a and the R package "hgu95a.db" [2] was used to manage and pre-process biological information related to this microarray. For the analysis, the dataset was separated into training and test sets, comprising 80 samples and 101 samples respectively with care taken to keep the same proportion of both MPM and ADCA classes.

The novelty of this approach is the introduction of a more robust statistical method with the expectation of an improvement in the robustness of the final obtained subset of features with direct biological relevance, evidenced by the maintaining of good generalisation in the validation results.

2.1   Significance Testing

A permutation based modified t-test [8] was used to evaluate the null hypothesis that there is no difference in expression between the two different groups (MDM and ADCA) accounting for the different variance within each group. The two sample t-statistic for a given gene is expressed as in (1), and the p-values estimated as per (2). In this case $\bar{x}_1$ and $\bar{x}_2$ represent the means of group 1 and group 2, with s_1^2 and s_2^2 being the respective variances. B is the number of re-samples taken for the modified t-test (a value of B=100 was used), n the

---

[1] http://cilab.ujn.edu.cn/datasets.htm

number of features and t the value for a given t-statistic (t_1^{0b} to t_n^{0b} are the set of null statistics calculated using the resampling procedure).

$$t = \frac{\bar{x}_1 - \bar{x}_2}{\sqrt{\frac{s_1^2}{n_1} + \frac{s_2^2}{n_2}}} \qquad (1)$$

$$p_i = \sum_{b=1}^{B} \frac{\#\left\{j : \left|t_j^{0b}\right| \geq |t_i|, j = 1, \ldots, n\right\}}{n \times B} \qquad (2)$$

This approach has as the null hypothesis that there is no difference in expression between the two genes and that the t-statistic holds to the same distribution across both [8].

Using the p-values from this study, an FDR-based significance measure, q-values [8], was used to select only those genes significant at the 5% level for inclusion in the model estimation stage. These q-values are an important tool in determining significance of features, and particularly so in genome studies, as they explicitly and systematically account for multiple testing, and allow for a more accurate determination of the expected false-positives from the inclusion of a particular feature.

2.2   Model Estimation

In this paper, the strategy proposed in [7] is used with some slight changes. The strategy consists of two separate stages:

– The 228 pathways identified as related to lung cancer disease are analysed to produce a ranking which allows selection of the most promising pathways to be further analyzed. In contrast to the first stage in [7] only the training dataset of 80 samples is analysed by applying a 10-fold cross-validation strategy. The purpose at this stage is to obtain an accuracy measure and identify the number of keywords for each pathway using a text-mining procedure, following the same process as [7] but refined using genes identified as significant in Section 2.1.

$$fitness(\mathbf{x}) = (1 - \lambda - \beta)(1 - ACC(\mathbf{x})) + \lambda\frac{k}{100} + \beta score(\mathbf{x}), \qquad (3)$$

$$score(\mathbf{x}) = \frac{\left(1 - \frac{i}{M}\right) + \left(1 - \frac{j}{N-M}\right)}{2}, \qquad (4)$$

– Using pathways identified from the first stage as being of importance, a genetic algorithm is applied using the fitness function from (3) where $\lambda, \beta \in [0, 1)$ and $\lambda + \beta < 1$, k is the number of selected features, 100 is a normalization factor due to the limited number of active features in a chromosome, and function $score(\mathbf{x})$ which estimates the biological relevance of the selected features. This function has been modified to (4) where M and N are normalization factors representing the number of significant genes on the pathway and total number of significant genes on all pathways respectively. i is the number of selected significant genes included in the pathway being analysed and j the number of selected significant genes not in the pathway such that $i + j = k$. To obtain the accuracy rate for (3), Linear Discriminant Analysis (LDA) [3] is used by applying 10-k-fold cross validation to each chromosome analysed within the GA execution.

- The final step validates the performance of the selected model by performing LDA on the training data and applying the results to the larger test dataset to obtain the prediction accuracy. LDA was chosen in order to make a fair comparison to results previously published in [7] as well as for its simplicity.

3   Results

The predictive capability and the number of relevant keywords were calculated for each of the 228 pathways. Table 1 details the best pathways according to Accuracy (Acc) values using the genes identified as being significant at $q < 0.05$ as the first sorting criterion, and the number of keywords found during the text mining of the pathway descriptions in the KEGG database as the second criterion. Bold rows correspond to pathways which ranked in the top 10 found in [7] during the first stage. Those in bold-italic are the six best pathways selected to be analysed on the second stage of the methodology previously. Of note is that the top 10 pathways from the original work are ranked in the overall top 27 pathways ($< 12\%$ of total), and pathway "04610" being the only one exhibiting minimal decline in ranking.

Instead of selecting the best pathways using this ranking as previously done, for comparative purposes the six best pathways from the previous work were selected [7] for analysis using the modified GA presented in Section 2.2 using the test/train datasets for model estimation and validation. This stage of the analysis was repeated 100 times for each of the six pathways to obtain estimates of the model accuracy.

Table 2 shows on average a perfomance comparison using the GA approach published in [7] and our proposal in this paper. In terms of prediction accuracy, both approaches obtain similar performance (approximately 95% depending on the analysed pathway). However, the main advantage of the present approach arises while analyzing the robustness of the subset of features selected. This robustness measure is obtained by averaging each gene frequency of appearance over the 100 GA executions, discarding those genes that do not appear more than 5% of the time. In this sense, it should be highlighted that in two out of six pathways analysed, a 78.33% and 59.93% of improvement is reached in terms of robustness (pathways "04010" and "04514" respectively). Pathway "04530" is the one with lowest improvement (just 11.86%), while for the remaining pathways the robustness was approximately increased by a 20%.

The top eleven final selected features for each of the pathways as shown in Table 4 can be directly compared to those obtained in [7]. Consistency is apparent in these as at least four genes are present in the previous work

Table 1. Pathways ranked by prediction ability using significant genes and number of keywords found. Bold rows correspond to pathways which ranked in the top 10 according to [7] (See the text for more details).

Rank	Code	Pathway	#Genes	Acc	#Genes 0.05	Acc 0.05	#Keywords
1	04020	Calcium signaling pathway	246	0.933	28	0.9975	0/1116
2	*04144*	*Endocytosis*	*244*	*0.99*	*32*	*0.99*	*0/506*
3	04650	Natural killer cell mediated cytotoxicity	172	0.915	13	0.9875	3/871
4	*04010*	*MAPK signaling pathway*	*423*	*0.935*	*37*	*0.9875*	*2/609*
5	**04062**	**Chemokine signaling pathway**	**254**	**0.945**	**29**	**0.9875**	**1/901**
6	04141	Protein processing in endoplasmic reticulum	181	0.908	14	0.9875	1/458
7	**01100**	**Metabolic pathways**	**970**	**0.975**	**146**	**0.9875**	**0/116**
8	00230	Purine metabolism	148	0.93	19	0.9875	0/271
9	04270	Vascular smooth muscle contraction	149	0.973	28	0.9875	0/891
10	**00240**	**Pyrimidine metabolism**	**83**	**0.975**	**15**	**0.9875**	**0/150**
11	04510	Focal adhesion	320	0.955	42	0.985	1/824
12	*05200*	*Pathways in cancer*	*557*	*0.96*	*63*	*0.9825*	*11/4504*
20	*04530*	*Tight junction*	*158*	*0.965*	*27*	*0.9775*	*1/545*
25	04360	Axon guidance	166	0.975	22	0.975	0/427
27	*04514*	*Cell adhesion molecules (CAMs)*	*154*	*0.95*	*25*	*0.975*	*0/921*
45	*04610*	*Complement and coagulation cascades*	*73*	*0.978*	*14*	*0.965*	*3/660*

Table 2. Comparison of original approach and proposed approach. Columns 2-4 show the mean of each of No. of Genes, genes in pathway and significant genes in pathway with standard deviations. Additionally, column 5 shows the robustness of results and the last column the accuracy.

	Pathway	#Genes	#Genes in pathway	#Genes significant in pathway	Robustness	Accuracy
Original GA	04144	4.43±1.00	2.87±1.22	1.75±1.19	0.1225	0.9568±0.0229
	04530	4.22±1.05	2.79±1.11	2.04±1.04	0.14125	0.9630±0.0248
	04514	3.96±1.07	2.32±1.29	1.49±0.95	0.135	0.9463±0.025
	04610	3.85±1.02	2.94±1.24	1.63±1.00	0.24455	0.9398±0.0197
	04010	4.04±1.29	1.71±1.09	0.88±0.83	0.086667	0.9445±0.0275
	05200	3.86±1.52	1.51±1.20	0.68±0.79	0.079	0.9450±0.0249
Our modified GA	04144	4.82±1.31	3.96±1.54	3.66±1.63	0.148	0.9590±0.0201
	04530	4.63±1.54	3.94±1.64	3.86±1.57	0.158	0.9732±0.0192
	04514	5.59±1.55	4.94±1.79	4.89±1.76	0.21591	0.9458±0.0261
	04610	4.29±1.04	3.94±1.23	3.12±1.21	0.29846	0.9439±0.02
	04010	3.54±1.14	2.58±1.33	2.28±1.32	0.15455	0.9431±0.0302
	05200	3.08±1.24	1.63±1.28	1.34±1.20	0.098182	0.932±0.0261

Table 3. Frequency of selection for the most frequently picked features in each pathway previously identified [7] as important. The notes column highlights whether the gene is significant in pathway (*), not significant but in pathway (**) or out of pathway(***).

ID	Symbol	Probe Set ID	Freq.(%)	Note
2520	**CLTB**	32522_f_at	13	**
1035	KDR	1954_at	14	*
633	**ERBB3**	1585_at	15	*
1182	**ERBB3**	2089_s_at	16	*
11052	PARD3	40973_at	16	*
2521	**CLTB**	32523_at	22	*
12020	DAB2	479_at	22	*
9758	**SH3GLB1**	39691_at	27	*
967	NTRK1	1892_s_at	28	*
3893	RAB11FIP5	33882_at	41	*
9863	**AP2M1**	39795_at	43	*

(a) Lung Pathway 04144

ID	Symbol	Probe Set ID	Freq.(%)	Note
6335	PRKCZ	362_at	13	*
7453	MYH11	37407_s_at	14	*
12312	MYH11	773_at	14	*
3916	CLDN3	33904_at	15	*
4174	**ACTG1**	34160_at	19	*
8537	**CLDN7**	38482_at	20	*
11052	**PARD3**	40973_at	23	*
8393	**RRAS**	38338_at	26	*
2039	**PRKCD**	32046_at	32	*
3844	**SPTAN1**	33833_at	33	*
5301	**CLDN4**	35276_at	57	*

(b) Lung Pathway 04530

ID	Symbol	Probe Set ID	Freq.(%)	Note
1248	PECAM1	268_at	24	*
1718	HLA-DOA	31728_at	25	*
3173	**NEO1**	33169_at	25	*
3916	CLDN3	33904_at	25	*
8509	ICAM2	38454_g_at	25	*
10372	ICAM3	402_s_at	27	*
8537	**CLDN7**	38482_at	29	*
1143	**CDH2**	2053_at	30	*
8508	**ICAM2**	38453_at	34	*
4217	PVRL3	34202_at	38	*
5301	**CLDN4**	35276_at	66	*

(c) Lung Pathway 04514

ID	Symbol	Probe Set ID	Freq.(%)	Note
6581	**F3**	36543_at	11	*
5783	PROS1	35752_s_at	13.00	*
6821	**SERPINA1**	36781_at	14	*
8496	**CD46**	38441_s_at	14	*
12146	VWF	607_s_at	22	*
12211	SERPINE1	672_at	23	**
9474	C1R	39409_at	36	*
5727	CFI	35698_at	45	*
8178	**SERPINE1**	38125_at	59	**
5853	**CFB**	35822_at	67	*
9843	**SERPING1**	39775_at	68	*

(d) Lung Pathway 04610

ID	Symbol	Probe Set ID	Freq.(%)	Note
6700	CD14	36661_s_at	6	*
620	PDGFB	1573_at	7	*
957	MECOM	1882_g_at	7	*
5104	FGF9	35081_at	9	*
3250	**MAPK13**	33245_at	11	*
5997	HSPA6	35965_at	14	*
909	RRAS2	1838_g_at	16	*
8393	**RRAS**	38338_at	16	*
5356	**FLNC**	35330_at	17	*
967	NTRK1	1892_s_at	23	*
667	**FGF9**	1616_at	44	*

(e) Lung Pathway 04010

ID	Symbol	Probe Set ID	Freq.(%)	Note
8813	FADD	38755_at	6	*
12200	GAS1	661_at	6	***
6616	BIRC2	36578_at	7	*
411	RARB	1381_at	8	*
7849	SEMA3C	377_g_at	8	***
8370	**ALDH1A2**	38315_at	9	***
5104	**FGF9**	35081_at	10	*
967	NTRK1	1892_s_at	11	*
1136	**JUP**	2047_s_at	13	*
10532	STAT5A	40458_at	14	*
667	**FGF9**	1616_at	16	*

(f) Lung Pathway 05200

(those highlighted in bold). Furthermore, because of the use of q-values to limit features to those which exhibit significant difference in expression, this approach yields results that contains a larger number of significant selected genes belonging to the top 11 features of a given pathway. These significant genes are also picked by the GA with greater frequency than those shown in the previous work, and thus the robustness of the present method should be higher, as indeed is as seen in Table 2.

Table 4. Frequency of selection for the most frequently picked features in each pathway previously identified [7] as important. The notes column highlights whether the gene is significant in pathway (*), not significant but in pathway (**) or out of pathway(***).

ID	Symbol	Probe Set ID	Freq. (%)	Note
2520	CLTB	32522_f_at	13	**
1035	KDR	1954_at	14	*
633	ERBB3	1585_at	15	*
1182	ERBB3	2089_s_at	16	*
11052	PARD3	40973_at	16	*
2521	CLTB	32523_at	22	*
12020	DAB2	479_at	22	*
9758	SH3GLB1	39691_at	27	*
967	NTRK1	1892_s_at	28	*
3893	RAB11FIP5	33882_at	41	*
9863	AP2M1	39795_at	43	*

(a) Lung Pathway 04144

ID	Symbol	Probe Set ID	Freq. (%)	Note
6335	PRKCZ	362_at	13	*
7453	MYH11	37407_s_at	14	*
12312	MYH11	773_at	14	*
3916	CLDN3	33904_at	15	*
4174	ACTG1	34160_at	19	*
8537	CLDN7	38482_at	20	*
11052	PARD3	40973_at	23	*
8393	RRAS	38338_at	26	*
2039	PRKCD	32046_at	32	*
3844	SPTAN1	33833_at	33	*
5301	CLDN4	35276_at	57	*

(b) Lung Pathway 04530

ID	Symbol	Probe Set ID	Freq. (%)	Note
1248	PECAM1	268_at	24	*
1718	HLA-DOA	31728_at	25	*
3173	NEO1	33169_at	25	*
3916	CLDN3	33904_at	25	*
8509	ICAM2	38454_g_at	25	*
10372	ICAM3	402_s_at	27	*
8537	CLDN7	38482_at	29	*
1143	CDH2	2053_at	30	*
8508	ICAM2	38453_at	34	*
4217	PVRL3	34202_at	38	*
5301	CLDN4	35276_at	66	*

(c) Lung Pathway 04514

ID	Symbol	Probe Set ID	Freq. (%)	Note
6581	F3	36543_at	11	*
5783	PROS1	35752_s_at	13.00	*
6821	SERPINA1	36781_at	14	*
8496	CD46	38441_s_at	14	*
12146	VWF	607_s_at	22	*
12211	SERPINE1	672_at	23	**
9474	C1R	39409_at	36	*
5727	CFI	35698_at	45	*
8178	SERPINE1	38125_at	59	**
5853	CFB	35822_at	67	*
9843	SERPING1	39775_at	68	*

(d) Lung Pathway 04610

ID	Symbol	Probe Set ID	Freq. (%)	Note
6700	CD14	36661_s_at	6	*
620	PDGFB	1573_at	7	*
957	MECOM	1882_g_at	7	*
5104	FGF9	35081_at	9	*
3250	MAPK13	33245_at	11	*
5997	HSPA6	35965_at	14	*
909	RRAS2	1838_g_at	16	*
8393	RRAS	38338_at	16	*
5356	FLNC	35330_at	17	*
967	NTRK1	1892_s_at	23	*
667	FGF9	1616_at	44	*

(e) Lung Pathway 04010

ID	Symbol	Probe Set ID	Freq. (%)	Note
8813	FADD	38755_at	6	*
12200	GAS1	661_at	6	***
6616	BIRC2	36578_at	7	*
411	RARB	1381_at	8	*
7849	SEMA3C	377_g_at	8	***
8370	ALDH1A2	38315_at	9	***
5104	FGF9	35081_at	10	*
967	NTRK1	1892_s_at	11	*
1136	JUP	2047_s_at	13	*
10532	STAT5A	40458_at	14	*
667	FGF9	1616_at	16	*

(f) Lung Pathway 05200

4   Conclusion

In this work, a lung cancer disease microarray dataset has been analysed in order to obtain a subset of genes with good predictive performance by using a previously published genetic algorithm modified to include a significance test based on the use of q-values. It has been shown that the inclusion of this information into the GA, to identify those genes having a significant difference in expression, has yielded results that are similar in performance to the original method but exhibiting improved robustness in terms of the selected features with an improvement between 11.86%-78.33% (average 36.21%). This higher robustness observed is achieved as the search in the GA is now guided to genes previously identified as significant without discarding the potential utility of other genes. Moreover, these results are consistent with the original ones since in the top 11 most selected genes, 4 to 6 genes were also included within the results in the original work. Further work should consider a deeper biological analysis of these results and also further investigation of the predictive abilities of pathways either alone or in combination with those genes identified as significant in this study.

Acknowledgments. The authors acknowledge support through Grants TIN2010-16556 from MICINN-SPAIN and P08-TIC-04026 (Junta de Andalucía), all of which include FEDER funds.

References

1. Benjamini, Y., Hochberg, Y.: Controlling the False Discovery Rate: A Practical and Powerful Approach to Multiple Testing. J. R. Stat. Soc. Ser. B 57(1), 289–300 (1995)
2. Carlson, M.: hgu95a.db: Affymetrix Human Genome U95 Set annotation data (chip hgu95a) (2011)
3. Duda, R.O., Hart, P.E., Stork, D.G.: Pattern Classification, 2nd edn. Wiley (2000)
4. Gordon, G.J., Jensen, R.V., Hsiao, L.-L., Gullans, S.R., Blumenstock, J.E., Ramaswamy, S., Richards, W.G., Sugarbaker, D.J., Bueno, R.: Translation of microarray data into clinically relevant cancer diagnostic tests using gene expression ratios in lung cancer and mesothelioma. Cancer Res. 62(17), 4963–4967 (2002)
5. Kanehisa, M., Goto, S.: KEGG: kyoto encyclopedia of genes and genomes. Nucleic Acids Res. 28(1), 27–30 (2000)
6. Kanehisa, M., Goto, S., Sato, Y., Kawashima, M., Furumichi, M., Tanabe, M.: Data, information, knowledge and principle: back to metabolism in KEGG. Nucleic Acids Res. 42(Database issue), D199–D205 (2014)
7. Luque-Baena, R.M., Urda, D., Gonzalo Claros, M., Franco, L., Jerez, J.M.: Robust gene signatures from microarray data using genetic algorithms enriched with biological pathway keywords. J. Biomed. Inform., January 2014
8. Storey, J.D., Tibshirani, R.: Statistical significance for genomewide studies. Proc. Natl. Acad. Sci. U.S.A. 100(16), 9440–9445 (2003)
9. West, M.: Bayesian factor regression models in the large p, small n paradigm. In: Bernardo, J.M., Dawid, A.P., Berger, J.O., West, M., Heckerman, D., Bayarri, M., Smith, A.F. (eds.) Bayesian Stat. 7 - Proc. Seventh Val. Int. Meet., pp. 723–732. Oxford University Press (2003)
10. Xu, J.-Z., Wong, C.-W.: Hunting for robust gene signature from cancer profiling data: sources of variability, different interpretations, and recent methodological developments. Cancer Lett. 296(1), 9–16 (2010)

Special Session: Computational Intelligence Methods for Drug Design

Drug Repurposing by Optimizing Mining of Genes Target Association

Aicha Boutorh[1], Naruemon Pratanwanich[2], Ahmed Guessoum[1], and Pietro Liò[2]

[1] University of Science and Technology Houari Boumediene
Laboratory for Research in Artificial Intelligence (LRIA)
Algiers, Algeria
{aboutorh,aguessoum}@usthb.dz
[2] University of Cambridge
Artificial Intelligence Group of the Computer Laboratory
Cambridge,United Kingdom
{np394,Pietro.Lio}@cam.ac.uk

Abstract. A major alternative strategy for the pharmacology industry is to find new uses for approved drugs. A number of studies have shown that target binding of a drug often affects not only the intended disease-related genes, leading to unexpected outcomes. Thus, if the perturbed genes are related to other diseases this permits the repositioning of an existing drug. Our aim is to find hidden relations between drug targets and disease-related genes so as to find new hypotheses of new drug-disease pairs. Association Rule Mining (ARM) is a well-known data mining technique which is widely used for the discovery of interesting relations in large data sets. In this study we apply a new computational intelligence approach to 288 drugs and 267 diseases, forming 5018 known drug-disease pairs. Our method, which we call Grammatical Evolution ARM (GEARM), applies the GE optimization technique on the set of rules learned using ARM and which represent hidden relationships among gene targets. The results produced by this combination show a high accuracy of up to 95 % for the extracted rules. Likewise, the suggested approach was able to discover interesting pairs of drugs and diseases with an accuracy of 92 %. Some of these pairs have previously been reported in the literature while others can serve as new hypotheses to be explored.

Keywords: Association Rule Mining, Grammatical Evolution, Optimization, Drug Repositioning, Genes.

1 Background

Finding new uses for approved drugs, referred to as drug repositioning, can reduce the cycle time of drug discovery since their side effects in a clinical environment have already been studied. The currently available computational power has been exploited to improve the effectiveness and efficiency of drug discovery.

© Springer International Publishing Switzerland 2015
C. di Serio et al. (Eds.): CIBB 2014, LNCS 8623, pp. 209–218, 2015.
DOI: 10.1007/978-3-319-24462-4_18

In this respect, Association Rule Mining is a well-known data mining technique which is widely used for the discovery of interesting relations in large data sets.

Most of the previous methods exploit the similarities between drugs on one side and between diseases on the other, independently, and assume at the mean time that similar drugs are likely to combat similar diseases and *vice versa*. Under this assumption, new pairs of drugs and diseases can be linked. Similarity measures are therefore the key for this method. Earlier attempts employed used either drug-based measurements or disease-based measures measurements to predict new therapeutic purposes (Hansen et al 2009), (Cheng et al 2013) Recently, repositioning techniques have incorporated the similarity of both chemical and pathological perspectives in a single framework (Fukuoka et al 2013), (Gottlieb et al 2011) A recent comprehensive work combined all of those drug and disease similarities into one score to infer novel drug indications (Gottlieb et al 2011).

Other techniques are based on finding the set of genes and pathways underlying drug and disease interactions (Li and Lu 2013, Lamb et al 2006).

Our work combines the advantages of these two strategies in a unified framework. In order to formulate new hypotheses of new drug-disease pairs, we propose a data mining method to find the relationships between drug targets and disease targets as a set of association rules by means of grammatical evolution. Having done this, we use the set of rules as a basis for the extraction of new potential drug-disease pairs. Taking benefit from the definition of fitness functions in association rule mining, we exploit the fact that similar drugs that have the same targets are likely to combat the same disease since they seem to have similar hidden patterns. Moreover, we can make assumptions about genes underlying each new drug-disease pair. This should advance our understanding and lead towards the testing of drugs in the lab.

Our aim is to extract hidden relationships between drug targets and disease targets in the form of X → Y, called an *association rule* , (This is not to be understood as a logical implication, but rather as the existence of an association between the two), and utilize the set of rules as underlying mechanisms to find new potential drug-disease pairs.

The existing algorithms for mining association rules are mainly based on the approach suggested by Agrawal et al. (1994) called the Apriori algorithm. A limitation of this algorithm is that it works in two phases and has a very high computational cost. This makes the evolutionary algorithm techniques suitable for the optimization of this process. Genetic algorithms (GA) and genetic programming (GP) are the evolutionary algorithms the most frequently used to extract association rules (Sharma and Tivari 2012)(Luna et al 2012).

Despite GP has been successfully used to generate ARs in different data sets, there are still limitations to evolving association rules using this type of machine learning algorithms. First, the GP algorithm implementation that was used involves building expression trees. The genetic operations are performed on trees, which has high complexity (hence execution time). In particular, when using more complex data, more complicated GP trees will be generated.

In response to these concerns, we have used grammatical evolution (GE) to extract the set of rules. GE is a variation on genetic programming that addresses some of the drawbacks of GP. By using a grammar, substantial changes can be made to the way rules are constructed through simple manipulations of a specified grammar. Furthermore, GE manipulates a string of digits to perform genetic operations instead of a tree. This obviously substantially improves the execution time of the strategy and increase its flexibility. These features are important improvements that GEARM introduces over other techniques.

2 The Data

We have collected drug-gene data, disease-gene data, and known drug-disease data from a previous study (Zhao and Li 2012). These data sets were originally taken from DrugBank, OMIM, and CTD databases respectively.

Online Mendelian Inheritance in Man (OMIM) is a compendium of genes and their related genetic phenotypes including human diseases (Amberger et al 2009).

The Comparative Toxicogenomics Database (CTD) is a curated database from the literature. It provides information about relationships between chemicals and gene targets and their relationships to diseases(Davis et al 2013).

In terms of data preprocessing, we have prepared two binary matrices: (1) a drug-target matrix, and (2) a disease-target matrix. The two matrices are merged and illustrated in Table.1, where each row represents a drug and its corresponding disease respectively and columns are all the genes of interest.

Each entry in the table indicates the existence of an association where 1 means the gene is a target of the corresponding drug (disease), and 0 otherwise. Note that the sets of genes (Gx) in the columns of the two matrices are not necessarily the same. However, the two matrices have the same number of rows since a row represents a known drug-disease (DR,DI) association.

Table 1. (1)The binary matrix of n drug-target genes for each known pair of Drug-Disease (DR,DI)x. (2)The binary matrix of m disease-target genes for each known pair of Drug-Disease (DR,DI)x.

K Pairs	G_1	G_2	...	G_n	G_1	G_2	...	G_m
	n Drug Target Genes				m Disease Target Genes			
$(DR, DI)_1$	0	1	...	1	1	1	...	0
$(DR, DI)_2$	1	1	...	0	0	0	...	1
$(DR, DI)_3$	0	0	...	1	0	1	...	1
...	...	...	...	...	...	...	...	...
$(DR, DI)_k$	0	1	...	0	1	0	...	1

3   The GEARM Methodology for Drug Repositioning

To generate an interpretable set of rules, a suitable grammar must be defined which specifies the antecedent and the consequent of each rule both as combinations of genes.

In this study, the set of known drug-disease pairs (DR,DI) represents the set of transactions (the transaction database) D= { $(DR, DI)_1$; $(DR, DI)_2$; ...;$(DR, DI)_k$ }, while genes (G) represent the items $I = \{G_1, G_2, ..., G_x\}$.

Each antecedent and consequent is a subset of I depending on whether the genes are drug or disease targets.

We will use Context-Free Grammars to represent the association rules that can be generated between drug and disease targets. Formally, a Context-Free Grammar is defined as a quadruple (S, N, T, P), where:

- S is the start symbol of the grammar.
- N is the set of non-terminal symbols.
- T is the set of terminal symbols which are the elements that appear at the end of the process when all the non-terminals are substituted by applying corresponding production rules. Only non-terminals can appear on the left-hand side of the production rules, whereas the right-hand side may consist of any combination of terminals and/or non-terminals.
- P is the set of production rules.

Grammatical Evolution Association Rule Mining (GEARM) was first defined in a previous study on the problem of gene-gene interaction (Boutorh and Guessoum 2014). The general form of the grammar for the association rules between drug and disease targets is defined as follows:

$G = \{S, N, T, P\}$
$S = \{Rule\}$
$N = \{Rule, Antecedent, Consequent, Target\_Ant, Target\_Cons\}$
$T = \{Ga_1, Ga_2, ..., Ga_n, Gc_1, Gc_2, .., Gc_m\}$
$P = \{< Rule >::=< Antecedent >< Consequent >$
$< Antecedent >::=< Target\_Ant > \mid < Target\_Ant >< Antecedent >$
$< Consequent >::=< Target\_Cons > \mid < Target\_Cons >< Consequent >$
$< Target\_Ant >::= Ga_1|Ga_2|...|Ga_n$
$< Target\_Cons >::= Gc_1|Gc_2|...|Gc_m\}$

The Ga_i and Gc_j used in the grammar are genes. We have used Ga_i to denote a gene used in the antecedent and Gc_j to denote a gene used in the consequent, respectively. There are two possible types of rules:

1. *Disease → Drug* : where the genes of the antecedent are disease targets and the genes of the consequent are drug targets, and the rule means that if the Ga_i is related to a specific disease, then the corresponding drug is applied on the Gc_j.

2. *Drug → Disease*: where the genes of the antecedent are drug targets and the genes of the consequent are disease targets, and the rule means that if the drug is applied on Ga_i, then the corresponding disease affected the Gc_j.

We have randomly divided the data set into four parts, and used 3/4 from it for the training and the remaining 1/4 for the tests.

Each rule R with support and confidence higher than the minimum support and minimum confidence respectively, is evaluated first on the training dataset by calculating its fitness (using Equation (1)).

Since the process of finding association rules returns many rules, the definition of a good measure of fitness is necessary to greatly alleviate this burden. The support and confidence are fundamental criteria for measuring rule quality. We define our fitness by combining the two measures, in order to find the rules whose support and confidence are both larger than those of the other rules.

$$Fitness(R) = (a * Support(R)) + (b * Confidence(R)) \qquad (1)$$

The support of the rule is the proportion of genes in the data set which are both in the antecedent and the consequent of this rule, i.e are targets of both the drug and disease, whereas the confidence is a measure of the rule strength (Equation (2)).

$$Confidence(X \to Y) = \frac{Support(X \cup Y)}{Support(X)} \qquad (2)$$

The weights a and b are two parameters decided empirically by selecting the values that gave the best results after executing the program many times with different values of a and b. We have ended up with $a = b = 0.5$ The best set of rules is then checked against the test data set to measure its accuracy which is defined using the True Positive (TP), True Negative (TN), False positive (FP) and False Negative as indicated in Equation (3)

$$Accuracy = \frac{TP + TN}{TP + TN + FP + FN} \qquad (3)$$

3.1 GEARM Process

All the details related to the grammatical evolution technique are described in (O'Neill and Rayen 2003). The GEARM process that we introduce here is captured in Fig.1.

The GEARM process starts by defining specific parameters.

The parameters of the genetic algorithm were chosen based on different combinations of population and generation size. Below is the set of values that have led to the best performance:

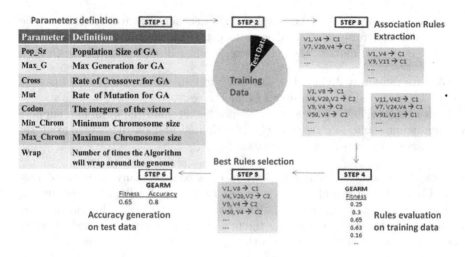

Fig. 1. GEARM process: Step 1: Definition of the parameters of GEARM. Step 2: Dividing the datasets. Step 3: generating a set of association rules from the training data. Step 4: Calculate the fitness of the rules. Step 5: Select the best rules for crossover and mutation. The steps 3-5 are running for Max_G before the best set of rules is obtained. Step 6 : Generate the accuracy for the best set of rules according to the test set data.

- population size (N) = 200 individuals; generation size (NB) = 500;
- crossover rate = 0.9; mutation rate = 0.01.

The GE parameters have also been defined related to the performance of the method. The chosen value for the minimum chromosome size ensures that at least one gene will appear in the antecedent and one in the consequent, so that no part of the rule will be empty. On the other hand, a maximum chromosome size limits the number of genes in the antecedent and the consequent to 3 genes maximum. The codon size is related to the number of genes used in this study.

- wrap count = 2;
- codon = 1850;
- minimum chromosome size = 4;
- maximum chromosome size = 12.

The algorithm begins by generating an initial population of N vectors of integers (N individuals), where the size of each vector is randomly generated between (minimum_chromosome and maximum_chromosome).

All the integer values of the vectors are also generated randomly and are less than the specified codon.

Each individual (i.e. vector) is transformed into an association rule by using the integer values to select a production rule to replace the non-terminal symbol according to the formula $A = C \bmod Na$, where C is an integer of the vector, Na is the number of alternatives of the current non-terminal and A is the selected alternative to replace the non-terminal symbol.

Figure 2 explains in detail an example of the mapping process. The integers of the vector can be wrapped again T times if the end of the genome (the vector) is reached while the mapping process is still incomplete. The mapping process will end when all the non-terminal symbols are replaced by terminals i.e genes, forming a rule which is the final output.

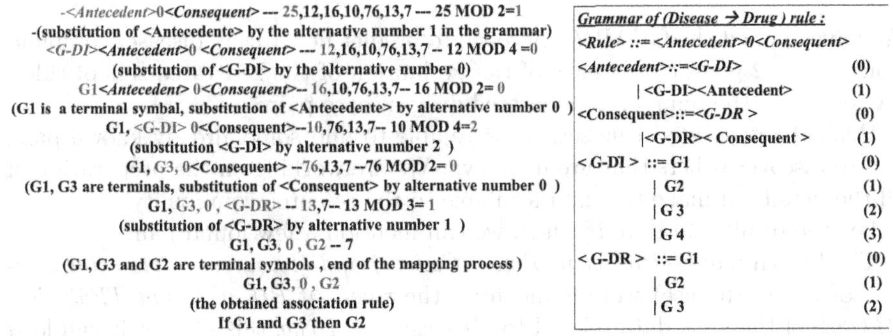

Fig. 2. The mapping process from an integer vector (genotype) to an association rule (phenotype): the grammar in the right is used to define the alternative that substitutes the current non-terminal by applying the *mod* operation on each integer in the vector until the output is made up of only terminal elements. G-DR: Genes target Drug. G-DI: Genes target Disease. Gx: a specific gene.

The generated set of rules is then evaluated on the training data. The fitness is recorded using the two data matrices explained in Section 2.

The best rules are selected for crossover and mutation which are performed on the vector of integers (genomic level). The new generation is then used in the cycle for NB generations after which GEARM stops.

At the end of the process, the set of best rules is selected as the optimal association rules set. This best GEARM set is tested on the test data set and the accuracy is computed (see Fig.1).

3.2   Extracting Drug-Disease Pairs

After generating all the sets of rules, we aim to find the (drug, disease) pairs related to each rule. To that end we propose to take into consideration the minimum number of drugs/diseases related to each gene in the antecedent and the consequent of the rule.

Let us illustrate the idea with an example.
Given the rule *if G_1 and G_2 and G_3 Then G_4 and G_5* which is related to the type *disease $\longrightarrow$ drug.*In order to find the (drug, disease) pairs:

- We first look for diseases that target the three genes of the antecedent part at once. Let us suppose they are DI_1 and DI_2.

– Then we look for the drugs that target the two genes of the consequent part also at once. Let us suppose it is DR_1.

The pairs that can be deduced should combine all the found drugs and diseases. Thus the generated pairs will be *(DR₁, DI₁)* and *(DR₁,DI₂)*.

4   Results

We have applied the GEARM approach to predict new (drug, disease) combinations. Table 2 gives a summary of the evaluation of the generated sets of rules. Table 3 gives the number of generated drug-disease pairs.

From the generated rule set, we were able to find some already known pairs and to discover others that are unknown. We are working on the preparation of all the details to make the data available to the research community.

For the results we have reached, we can mention a few found pairs.

The known pairs *(Icosapent, Breast Cancer) and (Icosapent, Colorectal Cancer)* of (drug, disease) were found from the rule *(if PIK3CA and TP53 then PTGS2)* of the gene data related to *Disease ⟶ Drug*. this same rule can lead to discover new pairs by combining all the disease and drug target genes. It is known that:

– Genes *(PIK3CA)* and *(TP53)* are targeted *Breast Cancer* and *Colorectal Cancer* diseases.
– Gene *(PTGS2)* is targeted by the *Icosapent* drug and *Dihomo-linolenic acid* drug .

In addition to the known pairs that have been found, we have discovered two other pairs *(Dihomo-linolenic acid, Breast Cancer)* and *(Dihomo-linolenic acid, Colorectal Cancer)* as new pairs of drug-disease.

The rule *(if ADRB1 and VEGFA then IL6)* related to *Drug ⟶ Disease*, can find three different pairs, two of which are known, *(Carvedilol, Inflammatory Bowel), (Carvedilol, Rheumatoid Arthritis)*, and one is new *(Carvedilol, Diabetes Mellitus)*. The pairs have been discovered by combining all the drugs and diseases that target all the genes present in the generated rule.

– Genes *(VEGFA)* and *(ADRB1)* target the drug *Carvedilol*
– Gene *(IL6)* target *Inflammatory Bowel, Rheumatoid Arthritis* and *Diabetes Mellitus* diseases.

Table 2. The evaluation results of the generated rules: Number of Rules (Nb_Rules), Average Fitness (Avr_Fit), Average Accuracy (Avr_Accr).

	Nb_Rules	Avr_Fit	Avr_Accr
Dis ⟶ Drg	200	0.389	0.921
Drg ⟶ Dis	200	0.270	0.959

Table 3. The generated Drug-Disease pairs: Number of known pairs (Nb_KwnP), Number of unknown pairs (NB_UnkwnP).

	Nb_KwnP	NB_UnkwnP	Accuracy
Dis $\longrightarrow$ Drg	156	634	92 %
Drg $\longrightarrow$ Dis	261	1151	92 %

5   Conclusion

We have presented in this paper the application of GEARM (the grammatical evolution association rule mining approach) to the problem of drug repositioning.

The sets of rules that have been generated were based on the disease target genes and the drug target genes. The results we have reached show the power of GEARM to find some existing pairs of drug-disease and to extract discovering new unknown pairs which can be a new indication for a drug. The work is still in progress and the approach is under more study to ameliorate its performance and to exploit the drug-disease pairs it can generate.

References

[Amberger et al 2009]Amberger, J., Bocchini, C.A., Scott, A.F., Hamosh, A.: McKu-sick's Online Mendelian Inheritance in Man (OMIM). Nucleic Acids Res. 37, 793–796 (2009)

[1994]Agrawal, R., Srikant, R.: Fast algorithms for mining association rules in large databases. In: 20th International Conference on Very Large Data Bases, Santiago de Chile, pp. 487–499 (1994)

[Boutorh and Guessoum 2014]Boutorh, A., Guessoum, A.: Gramatical evolution asso-ciation rule mining to detect gene-gene interaction. In: International Conference on Bioinformatics Models, Methods and Algorithms, pp. 253–258 (2014)

[Cheng et al 2013]Cheng, F., Li, W., Wu, Z., Wang, X., Zhang, C., Li, J., Liu, G., Tang, Y.: Prediction of polypharmacological profiles of drugs by the integration of chemical, side effect, and therapeutic space. Journal of Chemical Information and Modeling 53(4), 753–762 (2013)

[Davis et al 2013]Davis, A.P., Murphy, C.G., Johnson, R., et al.: The comparative toxicogenomics database: update 2013. Nucleic Acids Res. 41, 1104–1114 (2013)

[Fukuoka et al 2013]Fukuoka, Y., Takei, D., Ogawa, H.: A two-step drug reposition-ing method based on a protein-protein interaction network of genes shared by two diseases and the similarity of drugs. Bioinformation (2), 89–93 (2013)

[Gottlieb et al 2011]Gottlieb, A., Stein, G.Y., Ruppin, E., Sharan, R.: PREDICT: a method for inferring novel drug indications with application to personalized medicine. Molecular Systems Biology 7, 496 (2011)

[Hansen et al 2009]Hansen, N.T., Brunak, S., Altman, R.B.: Generating genome-scale candidate gene lists for pharmacogenomics. Clinical Pharmacology and Therapeu-tics (2), 183–189 (2009)

[Lamb et al 2006]Lamb, J., Crawford, E.D., Peck, D., Modell, J.W., Blat, I.C., Wrobel, M.J., Lerner, J., Brunet, J.-P., Subramanian, A., Ross, K.N., et al.: The connectivity map: using gene-expression signatures to connect small molecules, genes, and disease. Science Signaling 313(5795), 1929 (2006)

[Li and Lu 2013]Li, J., Lu, Z.: Pathway-based drug repositioning using causal inference. BMC Bioinformatics 14(Suppl. 16), S3 (2013)

[Luna et al 2012]Luna, J., Romero, J., Ventura, S.: Design and behaviour study of a grammar guided genetic programming algorithm for mining association rules. Knowledge and Information Systems 32, 53–76 (2012)

[O'Neill and Rayen 2003]O'Neill, M., Rayen, C.: Grammatical evolution: evolutionary automatic programming in an arbitrary language. Genetic Programming, vol. 3. Kluwer Academic Publishers, Boston (2003)

[Sharma and Tivari 2012]Sharma, A., Tivari, N.: A survey of association rule mining using genetic algorithm. International Journal of Computer Applications and Information Technology 1(2), 1–8 (2012), ISSN: 2278-7720

[Zhao and Li 2012]Zhao, S., Li, S.: A co-module approach for elucidating drug-disease associations and revealing their molecular basis. Bioinformatics 28(7), 955–961 (2012)

The Importance of the Regression Model in the Structure-Based Prediction of Protein-Ligand Binding

Hongjian Li[1], Kwong-Sak Leung[1], Man-Hon Wong[1], and Pedro J. Ballester[2,*]

[1] Department of Computer Science and Engineering,
Chinese University of Hong Kong, Shatin, New Territories, Hong Kong
[2] Cancer Research Center of Marseille, INSERM U1068, F-13009 Marseille, France,
Institut Paoli-Calmettes, F-13009 Marseille, France,
Aix-Marseille Université, F-13284 Marseille, France, and CNRS UMR7258,
F-13009 Marseille, France
pedro.ballester@inserm.fr

Abstract. Docking is a key computational method for structure-based design of starting points in the drug discovery process. Recently, the use of non-parametric machine learning to circumvent modelling assumptions has been shown to result in a large improvement in the accuracy of docking. As a result, these machine-learning scoring functions are able to widely outperform classical scoring functions. The latter are characterized by their reliance on a predetermined theory-inspired functional form for the relationship between the variables that characterise the complex and its predicted binding affinity.

In this paper, we demonstrate that the superior performance of machine-learning scoring functions comes from the avoidance of the functional form that all classical scoring functions assume. These scoring functions can now be directly applied to the docking poses generated by AutoDock Vina, which is expected to increase its accuracy. On the other hand, as it is well known that the assumption of additivity does not hold in some cases, it is expected that the described protocol will also improve other classical scoring functions, as it has been the case with Vina. Lastly, results suggest that incorporating ligand- and protein-only properties into a model is a promising avenue for future research.

Keywords: molecular docking, scoring functions, random forest, chemical informatics, structural bioinformatics.

1 Introduction

Protein-ligand docking is a key computational method for structure-based drug design. Docking predicts the preferred conformation and binding affinity of a ligand molecule, typically a small organic molecule, as bound to a protein pocket. Such prediction is not only useful to anticipate whether a ligand binds to a protein target, but also to understand how it binds. The latter is subsequently important to improve the

---

* Corresponding author.

© Springer International Publishing Switzerland 2015
C. di Serio et al. (Eds.): CIBB 2014, LNCS 8623, pp. 219–230, 2015.
DOI: 10.1007/978-3-319-24462-4_19

potency and selectivity of binding. The primary goal is to identify a molecule that binds tightly to the target, so that a small concentration of the molecule is sufficient to modulate the function of the target.

Operationally, docking has two stages: predicting the position, orientation and conformation of a molecule when docked to the target's binding site (pose generation), and predicting how strongly the docked pose of such putative ligand binds to the target (scoring). The single most important limitation of docking is the low accuracy of the scoring functions that predict the strength of binding or binding affinity. Predicting binding affinity is necessary to discriminate between molecules that bind (i.e. those with high binding affinity) and those that do not bind to the target (i.e. those with an affinity so low that does not permit stable binding). This prediction is carried out by a scoring function, mathematically a regression model that relates the atomic-level description of the protein-ligand complex to its binding affinity (such description is often derived from an X-ray crystal structure of the complex, e.g. Figure 1).

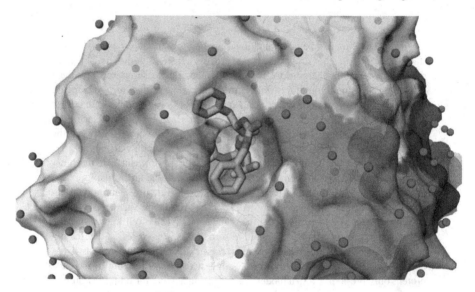

Fig. 1. Example of the X-ray crystal structure of a protein-ligand complex (PDB ID: 7cpa). The ligand molecule is pictured in stick representation colored by atom types. The molecular surface of the protein is colored by spectrum from blue to red with some transparency to show the underlying protein atoms. Water molecules are rendered as red dots. This figure was generated by iview [10], an interactive WebGL visualizer freely available at http://istar.cse.cuhk.edu.hk/iview/ (iview requires no Java plugins, yet supports macromolecular surface construction and virtual reality effects).

Classical scoring functions (empirical, force-field and knowledge-based – see [8] for a detailed description) assume a predetermined theory-inspired functional form for the relationship between the variables that characterise the complex and its predicted binding affinity. The inherent problem of this approach is in the difficulty of explicitly modeling the various contributions of intermolecular interactions to binding affinity. Recently, the use of non-parametric machine learning to circumvent these modelling assumptions has been shown [4][5] to play a key role in the accuracy of scoring functions. For instance, RF-Score [4], the first scoring function using Random Forest (RF) [7] as the regression model, was found to outperform a range of widely-used classical scoring functions by a large margin. RF-Score has recently been used [2] to discover a large number of innovative binders of antibacterial targets. This machine-learning scoring function has now been incorporated [9] into a large-scale docking tool for prospective virtual screening, which is freely available at http://istar.cse.cuhk.edu.hk/idock/. There is also a recent version of RF-Score [1] incorporating interatomic distance-dependent features to improve the characterization of the complex.

The benefits of taking a machine learning approach to this problem have now been demonstrated in many studies. However, there have been a few criticisms as well. For example, the use of oversimplified features in the original version of RF-Score has been pointed out as suboptimal [14], although it is worth noting that this fact did not prevent the method from outperforming classical scoring functions [4] or achieving high hit rates in prospective virtual screening [2]. Furthermore, the combination of machine learning and these features was claimed to learn target properties, which would hamper generalisation to test set complexes with targets dissimilar from those in the training set, a conjecture that was subsequently rebutted [5]. A corollary of this criticism is that there is something about RF-Score features that could prevent the model from predicting well those ligands that are not similar to those in the training set. In the parlance of machine learning, the object of this criticism has therefore been in the selected scheme for data representation.

In this paper, we demonstrate that the superior performance of machine-learning scoring functions comes exclusively from the avoidance of the assumed functional form of classical scoring functions. By fixing the remaining design variables (i.e. using the same features, training set and test set), any performance difference must necessarily come from the choice of regression model. Moreover, in this case the used training data and features are identical, so will be the domain of applicability of the resulting scoring functions. Thus, performance differences between these models will tend to be similar when applied to the same target and ligand types. This research will be carried out in the context of AutoDock Vina [13] as the classical scoring function, whereas RF will be the adopted non-parametric machine learning technique (other techniques are of course possible, but comparing to other techniques is out of the scope of the paper and has been done in the past, e.g. [3]).

2   Methods

2.1   Model 1 – AutoDock Vina

The AutoDock series [13][12][11] is arguably the most widely used docking software by the research community. AutoDock Vina significantly improved [13] the average accuracy of the binding mode predictions offered by AutoDock 4 [11], while running two orders of magnitude faster with multithreading. Vina was an exciting development, not only because of its remarkable pose generation performance in terms of both effectiveness and efficiency, but also because it is an open source tool.

Vina assumes a theory-inspired functional form. While this form also takes the form of a sum of intermolecular energy terms, unlike other classical scoring functions, it is not linear with respect to the set of parameters or weights. Indeed, Vina's score for the kth conformer (e_k) is given by the estimated free energy of binding to the target protein and calculated as:

$$e_k = \frac{e_k - e_{1,intra}}{1 + w_6 N_{rot}} = \frac{e_{k,inter} + e_{k,intra} - e_{1,intra}}{1 + w_6 N_{rot}}$$

Now because this study focuses on co-crystallized ligands, there is only one conformer per molecule (k=1) and thus the intramolecular contribution cancels out giving:

$$e_1 = \frac{e_{1,inter}}{1 + w_6 N_{rot}} \qquad \text{where}$$

$$e_{1,inter} = w_1 Gauss_{1,1} + w_2 Gauss_{2,1} + w_3 Repulsion_1 + w_4 HydroPhobic_1 + w_5 HBonding_1$$
$$\vec{w} = (-0.035579, -0.005156, 0.840245, -0.035069, -0.587439, 0.05846)$$

e_1 is the predicted free energy of binding reported by Vina when re-scoring the structure of a protein-ligand complex. To compare to binding affinities (pK_d or pK_i), the predicted free energy of binding in kcal/mol units is converted into pK_d with pK_d=-0.73349480509e_1 (see for instance [9] for an explanation of how this conversion factor is derived). The values for the six weights were found by minimising the difference between predicted and measured binding affinity using a nonlinear optimisation algorithm (this process was not detailed in the original publication [13]). The first three terms (Gauss1, Gauss2, Repulsion) account for steric interactions, the fourth term (Hydrophobic) is the contribution of hydrophobic effects, and the fifth term (HBonding) accounts for hydrogen bonding (this includes metal ions, which in Vina are treated as hydrogen bond donors). Nrot is the number of active rotatable bonds between heavy atoms in the ligand and it is included in the denominator of the functional form to penalize ligand flexibility. Expressions and further details for the five $e_{k,inter}$ terms can be found in [13][9].

2.2   Model 2 – MLR::Vina

This is a Multiple Linear Regression (MLR) model using the six unweighted Vina terms as features. In order to make the problem amenable to MLR, we made a grid search on the w_6 weight and thereafter ran MLR on the remaining weights as explained in the next section. The scoring function is the calibrated MLR model.

2.3   Model 3 – RF::Vina

While Vina's ability to predict binding affinity is among the best provided by classical scoring functions, it is still limited by the assumption of a functional form. To investigate the impact of this modelling assumption, we used RF to implicitly learn the functional form from the data. A RF is an ensemble of many different decision trees randomly generated from the same training data. RF trains its constituent trees using the CART algorithm [6]. As the learning ability of an ensemble of trees improves with the diversity of the trees [7], RF promotes diverse trees by introducing the following modifications in tree training. First, instead of using the same data, RF grows each tree without pruning from a bootstrap sample of the training data (i.e. a new set of N complexes is randomly selected with replacement from the N training complexes, so that each tree grows to learn a closely related but slightly different version of the training data). Second, instead of using all features, RF selects the best split at each node of the tree from a typically small number (m_{try}) of randomly chosen features. This subset changes at each node, but the same value of m_{try} is used for every node of each of the P trees in the ensemble. RF performance does not vary significantly with P beyond a certain threshold and thus P=500 was set as a sufficiently large number of trees. In contrast, m_{try} has some influence on performance and thus often constitutes the only tuning parameter of the RF algorithm. In regression problems, the RF prediction is given by arithmetic mean of all the individual tree predictions in the forest. Here we built a RF model with the six Vina features using the default number of trees (500) and values of the m_{try} control parameter from 1 to all 6 features. The selected model was that with the m_{try} value providing the lowest RMSE on Out-of-Bag (OOB) data.

2.4   Model 4 – RF::VinaElem

This is essentially model 3 using a total of 42 features: the 36 RF-Score features [4] in addition to the 6 Vina features. Therefore, for a given random seed, a RF for each m_{try} value from 1 to 42 is built and that with the lowest RMSE on OOB data is selected as the scoring function. RF-Score features are elemental occurrence counts of a set of protein-ligand atom pairs in a complex. To calculate these features, atom types are selected so as to generate features that are as dense as possible, while considering all the heavy atoms commonly observed in PDB complexes (C, N, O, F, P, S, Cl, Br, I). As the number of protein-ligand contacts is constant for a particular complex, the more atom types are considered the sparser the resulting features will be. Therefore, a minimal set of atom types is selected by considering atomic number only. Furthermore, a smaller set of interaction features has the additional advantage of leading to computationally faster scoring functions. In this way, the features are defined as the occurrence count of intermolecular contacts between elemental atom types i and j:

$$x_{j,i} \equiv \sum_{k=1}^{K_j} \sum_{l=1}^{L_i} \Theta(d_{cutoff} - d_{kl})$$

where d_{kl} is the Euclidean distance between the kth protein atom of type j and the lth ligand atom of type i calculated from a structure; K_j is the total number of protein atoms of type j and L_i is the total number of ligand atoms of type i in the considered complex; Θ is the Heaviside step function that counts contacts within a d_{cutoff} neighbourhood. For example, $x_{7,8}$ is the number of occurrences of protein nitrogen atoms hypothetically interacting with ligand oxygen atoms within a chosen neighbourhood. This representation led to a total of 81 features, of which 45 are zero due to the lack of proteinogenic amino acids with F, P, Cl, Br and I atoms. Therefore, each complex was characterized by a vector with 36 integer-valued features.

3   Experimental Setup

The PDBbind benchmark [8] is an excellent and common choice for validating generic scoring functions. It is based on the 2007 version of the PDBbind database, which contains a particularly diverse collection of protein-ligand complexes, assembled through a systematic mining of the entire Protein Data Bank, which led to a refined set of 1300 protein-ligand complexes with their corresponding binding affinities. The PDBbind benchmark essentially consists of testing the predictions of scoring functions on the 2007 core set, which comprises 195 diverse complexes with measured binding affinities spanning more than 12 orders of magnitude, while training in the remaining 1105 complexes in the refined set. In this way, a set of protein-ligand complexes with measured binding affinity can be processed to give two non-overlapping data sets, where each complex is represented by its feature vector $\mathbf{x}^{(n)}$ and its binding affinity $y^{(n)}$:

$$D_{train} = \left\{ \left(y^{(n)}, \vec{x}^{(n)} \right) \right\}_{n=1}^{1105}; D_{test} = \left\{ \left(y^{(n)}, \vec{x}^{(n)} \right) \right\}_{n=1106}^{1300}; y \equiv -\log_{10} K_{d/i}$$

This benchmark has the advantage of permitting a direct comparison against the performance of 16 classical scoring functions on the same test set [8]. Furthermore, using a pre-existing benchmark, where other scoring functions had previously been tested, ensures the optimal application of such functions by their authors and avoids the danger of constructing a benchmark complementary to the presented scoring function.

4   Results and Discussion

Performance is commonly measured [8] by the Standard Deviation (SD), Root Mean Square Error (RMSE), Pearson correlation (R_p) and Spearman rank-correlation (R_s) between predicted and measured binding affinity. SD is included to permit comparison to previously-tested scoring functions in this benchmark. RMSE reflects the ability of the scoring function to report an accurate binding affinity number, whereas R_s shows how well it can rank bound ligands according to binding strength. R_p simply shows how linear the correlation is and thus it is a less relevant indicator of the quality of the prediction. In the remaining of the section, a series of tests are presented and discussed.

$$p^{(n)} = f(\vec{x}^{(n)})$$

$$\hat{p}^{(n)} = a + bp^{(n)}$$

$$RMSE = \sqrt{\frac{1}{N} \sum_{n=1}^{N} (y^{(n)} - p^{(n)})^2}$$

$$SD = \sqrt{\frac{1}{N-2} \sum_{n=1}^{N} (y^{(n)} - \hat{p}^{(n)})^2}$$

$$R_p = \frac{N \sum_{n=1}^{N} p^{(n)} y^{(n)} - \sum_{n=1}^{N} p^{(n)} \sum_{n=1}^{N} y^{(n)}}{\sqrt{(N \sum_{n=1}^{N} (p^{(n)})^2 - (\sum_{n=1}^{N} p^{(n)})^2)(N \sum_{n=1}^{N} (y^{(n)})^2 - (\sum_{n=1}^{N} y^{(n)})^2)}}$$

$$R_s = \frac{N \sum_{n=1}^{N} p_r^{(n)} y_r^{(n)} - \sum_{n=1}^{N} p_r^{(n)} \sum_{n=1}^{N} y_r^{(n)}}{\sqrt{(N \sum_{n=1}^{N} (p_r^{(n)})^2 - (\sum_{n=1}^{N} p_r^{(n)})^2)(N \sum_{n=1}^{N} (y_r^{(n)})^2 - (\sum_{n=1}^{N} y_r^{(n)})^2)}}$$

where $p^{(n)}$ is the predicted binding affinity given the feature vector $\vec{x}^{(n)}$, a and b are the intercept and coefficient of the linear correlation between $\{p^{(n)}\}_{n=1}^{N}$ and $\{y^{(n)}\}_{n=1}^{N}$ on the test set, $\hat{p}^{(n)}$ is the fitted value of $p^{(n)}$, and $\{p_r^{(n)}\}_{n=1}^{N}$ and $\{y_r^{(n)}\}_{n=1}^{N}$ are the rankings of $\{p^{(n)}\}_{n=1}^{N}$ and $\{y^{(n)}\}_{n=1}^{N}$, respectively.

4.1   MLR Is Better at Calibrating the Additive Functional Form of Vina's SF

101 versions of model 2 were generated using all the w_6 values from 0 (no influence of N_{rot}) to 1 (maximum influence of N_{rot}) with a step size of 0.01. Since the best models were always between 0.005 and 0.020, we carried out a second grid search in this range with step size 0.001. The best of these models ($w_6 = 0.012$) provided a test set performance with significantly lower error and higher correlation than Vina (see models 1 and 2 in Figure 2). This means that MLR is more suitable to calibrate Vina's scoring function than the originally used nonlinear optimisation algorithm.

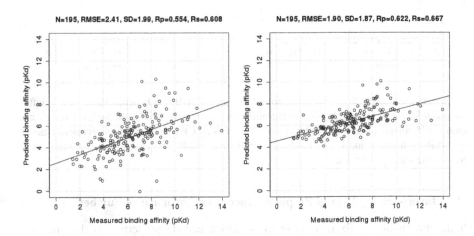

Fig. 2. Performance on the 195 test set complexes in the PDBbind benchmark: model 1 (left) and model 2 (right). Model 1 is AutoDock Vina, whereas model 2 is trained on 1105 of the 1300 complexes used in Vina (i.e. removing the 195 test set complexes that were in Vina's training set) and using the same six Vina features. Both models assume an additive functional form, but differ on the way the weights are estimated from data: an undisclosed nonlinear optimization algorithm (left) and MLR (right).

4.2   Vina's Assumed Functional Form Is Detrimental for Its Performance

Both the linear (model 2) and nonlinear (model 1) optimisation approaches to training Vina assume a quasi-additive functional form. By looking at Figures 2 and 3, it is clear that model 3 performs much better than models 1 and 2. Note that model 3 uses exactly the same features and data sets as the other two models. The only difference between these models is that model 3 implicitly constructs the functional form from the data using RF for regression, whereas the other two Vina models assume a priori form for how the features are combined to form the scoring function. Therefore, these results demonstrate that this performance improvement is entirely due to the avoidance of this commonly-used modelling assumption.

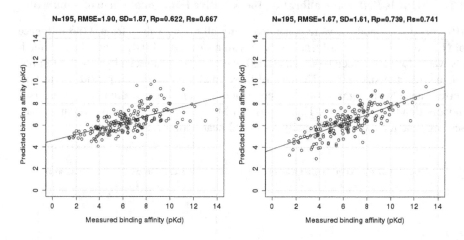

Fig. 3. Performance on the 195 test set complexes in the PDBbind benchmark: model 2 (left) and model 3 (right). Models 2 and 3 were trained on the same data and used the same representation (1105 complexes and six Vina features, respectively), but used different regression models (MLR in model 2 and RF in model 3). Unlike RF, MLR implicitly assumes a functional form for the scoring function.

4.3   Incorporating Ligand Properties Increases Performance Further

Unlike the remaining fixed Vina features, which encode properties of the protein-ligand complex, N_{rot} is exclusively a property of the ligand (the number of rotatable bonds, effectively an estimation of the flexibility of the ligand). When model 3 is run with five features (all but N_{rot}), test set error increases from a best RMSE of 1.67 to 1.74 (Rs correlation drops from 0.741 to 0.706; see Figure 4). This result shows that it is advantageous to add N_{rot} as a model feature. More broadly, this suggests that incorporating ligand properties into the model, such as those that are routinely used in ligand-based QSAR models, may enhance performance further. Likewise, features encoding protein properties could also extend the capabilities of generic scoring functions.

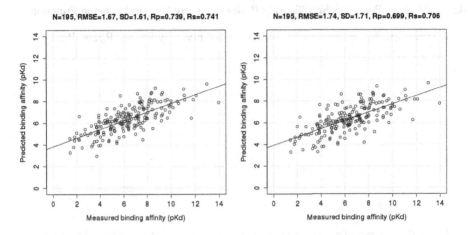

Fig. 4. Performance on the 195 test set complexes in the PDBbind benchmark: model 3 (left) and model 5 (right). Models 3 and 5 were trained on the same data and used the same regression model (RF). However, model 5 uses one feature less (N_{rot}).

4.4   Incorporating Protein-Ligand Features also Increases Performance

Lastly, model 4 performed even better than model 3 (Figure 5). This shows that adding the 36 features from the original RF-Score is beneficial for performance, all other factors being equal. The best test set performance of model 4 was RMSE=1.51, Rp=0.803 and Rs= 0.798.

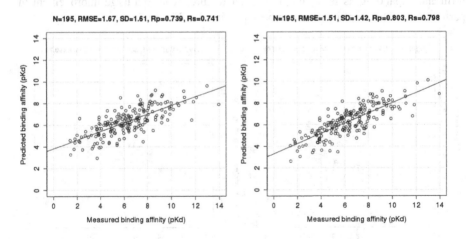

Fig. 5. Performance on the 195 test set complexes in the PDBbind benchmark: model 3 (left) and model 4 (right). Models 3 and 4 were trained on the same data and used the same regression model (RF). However, model 4 uses 36 intermolecular features in addition to the six Vina features used by model 3.

4.5 Model 4 Performs Significantly Better than any Classical Scoring Function

Figure 6 compares the performance of models 1-4 given by RMSE, R_p and R_s.

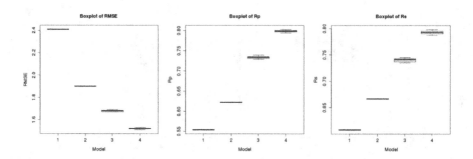

Fig. 6. Performance of each model according to RMSE (left), R_p (middle) and R_s (right).

Models 2-4 were trained on the same 1105 complexes and tested on the 195 complexes in the 2007 core set (Vina is model 1 and was originally trained on all 1300 complexes, so we only tested it on the core set). Models 3 and 4 are stochastic because they are based on RF (a randomized algorithm). Hence, to assess the variability in their response, the same set of 10 random seeds is used to generate 10 versions of the model. The performance of each version is summarized by the boxplots in Figure 6. Because model 1 is off-the-shelf software and model 2's MLR is deterministic, these only provide one set of predictions which are shown as horizontal lines in Figure 6. These results demonstrate that circumventing Vina's assumed functional form and expanding its set of intermolecular features lead to a large improvement in performance. This improvement can be appreciated in Figure 7.

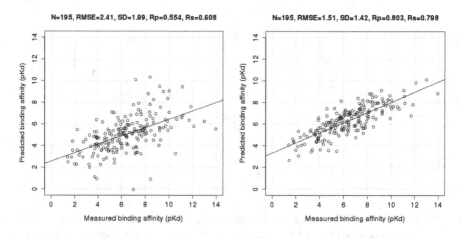

Fig. 7. Performance on the 195 test set complexes in the PDBbind benchmark: AutoDock Vina (model 1; left) and RF::VinaElem (model 4; right). RF::VinaElem constitutes a remarkable improvement on the key requirement of predicting binding affinity.

Next, the performances of RF::VinaElem and AutoDock Vina are compared against that of a broad range of scoring functions on the PDBbind benchmark [8]. Using a pre-existing benchmark, where other scoring functions had previously been tested, ensures the optimal application of such functions by their authors and avoids the danger of constructing a benchmark complementary to the presented scoring function. Table 1 reports the performance of all scoring functions on the independent test set, with RF::VinaElem obtaining the best performance. In contrast, classical scoring functions tested on the same test set obtained significantly higher SD/RMSE errors and lower correlations with measured binding affinities. It is noteworthy that AutoDock Vina is among the most accurate classical scoring functions, which is not surprising given its status as the most widely-used docking software as evidenced by number of citations.

Table 1. Performance of scoring functions on the PDBbind benchmark. These are ordered by highest rank-correlation (Rs) with ties resolved by smaller error (SD). The presented scoring function obtained the best results on this benchmark.

Scoring Function	R_s	R_p	SD
RF::VinaElem	0.798	0.803	1.51
X-Score::HMScore	0.705	0.644	1.83
DrugScore^{CSD}	0.627	0.569	1.96
AutoDock Vina	0.608	0.554	1.99
DS::PLP1	0.588	0.545	2.00
SYBYL::ChemScore	0.585	0.555	1.98
GOLD::ASP	0.577	0.534	2.02
SYBYL::G-Score	0.536	0.492	2.08
DS::LigScore2	0.507	0.464	2.12
DS::LUDI3	0.478	0.487	2.09
GOLD::ChemScore	0.452	0.441	2.15
DS::PMF	0.448	0.445	2.14
SYBYL::D-Score	0.447	0.392	2.19
GlideScore-XP	0.435	0.457	2.14
DS::Jain	0.346	0.316	2.24
GOLD::GoldScore	0.322	0.295	2.29
SYBYL::PMF-Score	0.273	0.268	2.29
SYBYL::F-Score	0.243	0.216	2.35

5   Conclusions and Future Prospects

The use of non-parametric machine learning is a particularly promising and still largely unexplored approach to develop generic scoring functions. This study has shown that assuming a predetermined functional form and restricting to a small set of intermolecular features are the main reasons for the large difference in performance between classical and machine-learning scoring functions. Thus, the latter will eventually translate into more successful docking applications. Importantly, this study strongly suggests that other classical scoring function should experience similar performance gains once an assumed functional form is circumvented using non-parametric machine learning. Finally, the results also suggest that incorporating ligand- and protein-only properties into the scoring function is a promising path to future improvements.

Acknowledgements. This work has been carried out thanks to the support of the A*MIDEX grant (n° ANR-11-IDEX-0001-02) funded by the French Government « Investissements d'Avenir » program, the Direct Grant from the Chinese University of Hong Kong and the GRF Grant (Project Reference 414413) from the Research Grants Council of Hong Kong SAR.

References

1. Ballester, P.J., et al.: Does a More Precise Chemical Description of Protein-Ligand Complexes Lead to More Accurate Prediction of Binding Affinity? J. Chem. Inf. Model. 54(3), 944–955 (2014)
2. Ballester, P.J., et al.: Hierarchical virtual screening for the discovery of new molecular scaffolds in antibacterial hit identification. J. R. Soc. Interface. 9(77), 3196–3207 (2012)
3. Ballester, P.J.: Machine learning scoring functions based on random forest and support vector regression. In: Shibuya, T., Kashima, H., Sese, J., Ahmad, S. (eds.) PRIB 2012. LNCS, vol. 7632, pp. 14–25. Springer, Heidelberg (2012)
4. Ballester, P.J., Mitchell, J.B.O.: A machine learning approach to predicting protein-ligand binding affinity with applications to molecular docking. Bioinformatics 26(9), 1169–1175 (2010)
5. Ballester, P.J., Mitchell, J.B.O.: Comments on "leave-cluster-out cross-validation is appropriate for scoring functions derived from diverse protein data sets": significance for the validation of scoring functions. J. Chem. Inf. Model. 51(8), 1739–1741 (2011)
6. Breiman, L., et al.: Classification and Regression Trees. Chapman and Hall/CRC (1984)
7. Breiman, L.: Random Forests. Mach. Learn. 45(1), 5–32 (2001)
8. Cheng, T., et al.: Comparative Assessment of Scoring Functions on a Diverse Test Set. J. Chem. Inf. Model. 49(4), 1079–1093 (2009)
9. Li, H., et al.: istar: A Web Platform for Large-Scale Protein-Ligand Docking. PLoS One 9(1), e85678 (2014)
10. Li, H., et al.: iview: an interactive WebGL visualizer for protein-ligand complex. BMC Bioinformatics 15(1), 56 (2014)
11. Morris, G.M., et al.: AutoDock4 and AutoDockTools4: Automated docking with selective receptor flexibility. J. Comput. Chem. 30(16), 2785–2791 (2009)
12. Morris, G.M., et al.: Automated docking using a Lamarckian genetic algorithm and an empirical binding free energy function. J. Comput. Chem. 19(14), 1639–1662 (1998)
13. Trott, O., Olson, A.J.: AutoDock Vina: Improving the speed and accuracy of docking with a new scoring function, efficient optimization, and multithreading. J. Comput. Chem. 31(2), 455–461 (2010)
14. Wang, J.-C., Lin, J.-H.: Scoring functions for prediction of protein-ligand interactions. Curr. Pharm. Des. 19(12), 2174–2182 (2013)

The Impact of Docking Pose Generation Error on the Prediction of Binding Affinity

Hongjian Li[1], Kwong-Sak Leung[1], Man-Hon Wong[1], and Pedro J. Ballester[2,*]

[1] Department of Computer Science and Engineering,
Chinese University of Hong Kong, Shatin, New Territories, Hong Kong
[2] Cancer Research Center of Marseille, INSERM U1068, F-13009 Marseille, France,
Institut Paoli-Calmettes, F-13009 Marseille, France, Aix-Marseille Université,
F-13284 Marseille, France; and CNRS UMR7258, F-13009 Marseille, France
pedro.ballester@inserm.fr

Abstract. Docking is a computational technique that predicts the preferred conformation and binding affinity of a ligand molecule as bound to a protein pocket. It is often employed to identify a molecule that binds tightly to the target, so that a small concentration of the molecule is sufficient to modulate its biochemical function. The use of non-parametric machine learning, a data-driven approach that circumvents the need of modeling assumptions, has recently been shown to introduce a large improvement in the accuracy of docking scoring. However, the impact of pose generation error on binding affinity prediction is still to be investigated.

Here we show that the impact of pose generation is generally limited to a small decline in the accuracy of scoring. These machine-learning scoring functions retained the highest performance on PDBbind v2007 core set in this common scenario where one has to predict the binding affinity of docked poses instead of that of co-crystallized poses (e.g. drug lead optimization). Nevertheless, we observed that these functions do not perform so well at predicting the near-native pose of a ligand. This suggests that having different scoring functions for different problems is a better approach than using the same scoring function for all problems.

Keywords: molecular docking, scoring functions, random forest, chemical informatics, structural bioinformatics.

1 Introduction

Protein-ligand docking is a key computational method in structure-based drug design. Docking predicts the preferred conformation and binding affinity of a ligand molecule as bound to a protein pocket. Such prediction is not only useful to anticipate whether a ligand binds to a protein target, but also to understand how it binds or how strongly it binds. Docking is often utilized to identify a molecule that binds tightly to the target, so that a small concentration of the molecule is sufficient to modulate its biochemical function.

---

* Corresponding author.

© Springer International Publishing Switzerland 2015
C. di Serio et al. (Eds.): CIBB 2014, LNCS 8623, pp. 231–241, 2015.
DOI: 10.1007/978-3-319-24462-4_20

Docking has two stages: predicting the position, orientation and conformation of a molecule when docked to the target's binding site (pose generation), and predicting how strongly the docked pose of such putative ligand binds to the target (scoring).

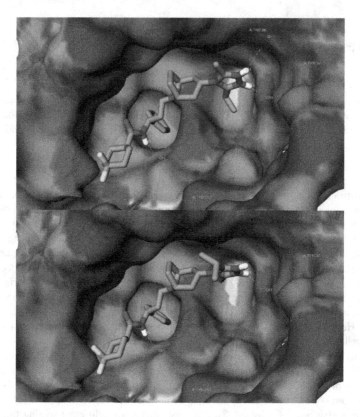

Fig. 1. Example of pose generation error. Top: crystal structure of the MRV ligand bound to the CCR5 Chemokine Receptor (PDB 4MBS). Bottom: re-docked pose of MRV. Hydrogen bonds are pictured as discontinued lines. The Root-Mean Square Deviation (RMSD) between the co-crystallized and the re-docked poses is 2.1 Å, which quantifies pose generation error. The plots were generated with iview [9], an interactive WebGL visualizer freely available at http://istar.cse.cuhk.edu.hk/iview/ (iview requires no Java plugins, yet supports macromolecular surface construction and virtual reality effects).

The single most important limitation of docking is the low accuracy of the scoring functions that predict the strength of binding of a bound ligand. This binding affinity can thereafter be used to select those molecules predicted to bind tightly for wet-lab confirmatory testing. Classical scoring functions [7] assume a functional form that relates the atomic-level description of the protein-ligand complex as specified by an X-ray crystal structure to its binding affinity, usually through Multiple Linear Regression (MLR).

Non-parametric machine learning, which circumvents the need of modelling assumptions implicit in functional forms, has recently been shown [4][5] to introduce a large improvement in the accuracy of scoring functions. RF-Score [4], the first scoring function using Random Forest (RF) [6] as the regression model, was found to outperform a range of widely-used classical scoring functions by a large margin. RF-Score has recently been used [2] to discover a large number of innovative binders of antibacterial targets and has now been incorporated [8] into a large-scale docking tool for prospective virtual screening (http://istar.cse.cuhk.edu.hk/idock/). To avoid confounding factors introduced by pose generation, these studies on scoring accuracy are carried out on data consisting of large sets of X-ray structures of protein-ligand complexes. However, scoring of the docked poses of a molecule is required in those cases where the experimentally-determined pose is not available.

In this paper, we study the impact of pose generation error on classical and machine-learning scoring functions. Furthermore, we investigate which of these scoring functions is more suitable for predicting the near-native pose (i.e. the most similar docked pose to the co-crystallized pose). The numerical experiments will be performed with AutoDock Vina [12], as the classical scoring function, and RF-Score [1, 4] as the machine-learning scoring function. Investigating the generalisation of these results to other non-parametric machine learning techniques previously applied to this problem, such as SVR in [3], is out of the scope of this study, although expected to yield similar outcomes.

2 Methods

2.1 Model 1 – AutoDock Vina

The AutoDock series [12][11][10] is arguably the most widely used docking software by the research community. AutoDock Vina significantly improved [12] the average accuracy of the binding mode predictions offered by AutoDock 4 [10], while running two orders of magnitude faster with multithreading. Vina was an exciting development, not only due to its remarkable pose generation performance, both in terms of effectiveness and efficiency, but also because it is an open source tool that aids docking method development.

Vina's score for the kth pose of a molecule is given by the estimated free energy of binding to the target protein and calculated in Vina as:

$$Vina_k = \frac{e_k - e_{1,intra}}{1 + w_6 N_{rot}} = \frac{e_{k,inter} + e_{k,intra} - e_{1,intra}}{1 + w_6 N_{rot}}$$

where

$$e_{k,inter} = w_1 Gauss_{1,r} + w_2 Gauss_{2,r} + w_3 Repulsion_r + w_4 Hydrophobic_r$$
$$+ w_5 HBonding_r$$
$$e_{k,intra} = w_1 Gauss_{1,a} + w_2 Gauss_{2,a} + w_3 Repulsion_a + w_4 Hydrophobic_a$$
$$+ w_5 HBonding_a$$
$$w_1 = -0.035579, w_2 = -0.005156, w_3 = 0.840245, w_4 = -0.035069,$$
$$w_5 = -0.587439, w_6 = 0.05846$$

e_1 is the predicted free energy of binding reported by Vina when re-scoring the structure of a protein-ligand complex. As usual, to compare to binding affinities (pK_d or pK_i), the predicted free energy of binding in kcal/mol units is converted into pK_d with $pK_d = -0.73349480509e_1$ (see for instance [8] for an explanation of how this conversion factor is derived). The values for these weights were found by minimising the difference between predicted and measured binding affinity using a nonlinear optimisation algorithm (this process was not detailed in the original publication [12]). Further details on each of these Vina terms can be found in [8], including the expressions for intermolecular and intramolecular energetic terms (respectively distinguished with 'r' and 'a' subscripts above).

2.2   Model 2 – MLR::Vina

In studies based on structural data, there is only one 3D geometry or pose of the molecule (k=1), the co-crystallized ligand, and thus the intramolecular contributions in equation 1 cancel out. This means that, once a co-crystallized ligand is redocked, each of the resulting poses is described by 11 Vina terms. Here, a MLR model, effectively a classical scoring function, is built using the 11 unweighted Vina terms as features. In order to make the problem amenable to MLR, we made a grid search on the w6 weight and thereafter run MLR on the remaining weights as explained in the next section.

2.3   Model 3 – RF::Vina

While Vina's ability to predict binding affinity is among the best provided by classical scoring functions, it is still limited by the assumption of additivity in its functional form. Random Forest (RF) [6] can be used to circumvent modeling assumptions. We therefore built a RF model with the 11 Vina features using the default number of trees (500). Instead of using all features, RF selects the best split at each node of the tree from a typically small number (m_{try}) of randomly chosen features. The mtry value with the lowest RMSE on Out-of-Bag (OOB) data is selected.

2.4   Model 4 – RF::VinaElem

This is essentially model 3 using a total of 47 features: the 36 RF-Score features [4] in addition to the 6 Vina features. Therefore, for a given random seed, a RF for each m_{try} value from 1 to 47 is built and that with the lowest RMSE on OOB data is selected as the scoring function. RF-Score features are elemental occurrence counts of a set of protein-ligand atom pairs in a complex. To calculate these features, atom types are selected so as to generate features that are as dense as possible, while considering all the heavy atoms commonly observed in PDB complexes (C, N, O, F, P, S, Cl, Br, I). As the number of protein-ligand contacts is constant for a particular complex, the more atom types are considered the sparser the resulting features will be. Therefore, a minimal set of atom types is selected by considering atomic number only. Furthermore, a smaller set of interaction features has the additional advantage of leading to computationally faster scoring functions. In this way, the features are defined as the occurrence count of intermolecular contacts between elemental atom types i and j:

$$x_{j,i} \equiv \sum_{k=1}^{K_j} \sum_{l=1}^{L_i} \Theta(d_{cutoff} - d_{kl})$$

where d_{kl} is the Euclidean distance between the kth protein atom of type j and the lth ligand atom of type i calculated from a structure; K_j is the total number of protein atoms of type j and L_i is the total number of ligand atoms of type i in the considered complex; Θ is the Heaviside step function that counts contacts within a d_{cutoff} neighbourhood. For example, $x_{7,8}$ is the number of occurrences of protein nitrogen atoms hypothetically interacting with ligand oxygen atoms within a chosen neighbourhood. This representation led to a total of 81 features, of which 45 are zero due to the lack of proteinogenic amino acids with F, P, Cl, Br and I atoms. Therefore, each complex was characterized by a vector with 36 integer-valued features.

3   Experimental Setup

3.1   The PDBbind Benchmark

The PDBbind benchmark [7] is an excellent and common choice for validating generic scoring functions. Based on the 2007 version of the PDBbind database, it contains a particularly diverse collection of protein-ligand complexes from a systematic mining of the entire Protein Data Bank. This procedure led to a refined set of 1300 protein-ligand complexes along with their binding affinities. The PDBbind benchmark essentially consists of testing the predictions of scoring functions on the 2007 core set, which comprises 195 diverse complexes with measured binding affinities spanning more than 12 orders of magnitude, while using the remaining 1105 refined set complexes for training (i.e. both sets have no complexes in common). In this way, a set of protein-ligand complexes with measured binding affinity can be processed to give two non-overlapping data sets, where each complex is represented by its feature vector $\mathbf{x}^{(n)}$ and its binding affinity $y^{(n)}$:

$$D_{train} = \left\{ \left(y^{(n)}, \vec{x}^{(n)} \right) \right\}_{n=1}^{1105} ; D_{test} = \left\{ \left(y^{(n)}, \vec{x}^{(n)} \right) \right\}_{n=1106}^{1300} ; y \equiv -\log_{10} K_{d/i}$$

3.2   Performance Measures

Performance is commonly measured [7] by the Root Mean Square Error (RMSE), Pearson correlation (R_p) and Spearman rank-correlation (R_s) between predicted and measured binding affinity. RMSE reflects the ability of the scoring function to report an accurate binding affinity estimation, whereas R_s shows how well it can rank bound ligands according to binding strength. R_p simply shows how linear the correlation is and thus it is a less relevant indicator of the quality of the prediction. Their expressions are:

$$p^{(n)} = f(\vec{x}^{(n)})$$

$$RMSE = \sqrt{\frac{1}{N}\sum_{n=1}^{N}(y^{(n)} - p^{(n)})^2}$$

$$R_p = \frac{N\sum_{n=1}^{N} p^{(n)} y^{(n)} - \sum_{n=1}^{N} p^{(n)} \sum_{n=1}^{N} y^{(n)}}{\sqrt{(N\sum_{n=1}^{N}(p^{(n)})^2 - (\sum_{n=1}^{N} p^{(n)})^2)(N\sum_{n=1}^{N}(y^{(n)})^2 - (\sum_{n=1}^{N} y^{(n)})^2)}}$$

$$R_s = \frac{N\sum_{n=1}^{N} p_r^{(n)} y_r^{(n)} - \sum_{n=1}^{N} p_r^{(n)} \sum_{n=1}^{N} y_r^{(n)}}{\sqrt{(N\sum_{n=1}^{N}(p_r^{(n)})^2 - (\sum_{n=1}^{N} p_r^{(n)})^2)(N\sum_{n=1}^{N}(y_r^{(n)})^2 - (\sum_{n=1}^{N} y_r^{(n)})^2)}}$$

where f is the scoring function, $p^{(n)}$ is the predicted binding affinity given the feature vector $\vec{x}^{(n)}$, $y^{(n)}$ is the corresponding measured binding affinity, N is the number of test set complexes, and $\{p_r^{(n)}\}_{n=1}^{N}$ and $\{y_r^{(n)}\}_{n=1}^{N}$ are the rankings of $\{p^{(n)}\}_{n=1}^{N}$ and $\{y^{(n)}\}_{n=1}^{N}$, respectively.

The Root-Mean Square Deviation (RMSD) quantifies how different the 3D geometry of the redocked pose is from the corresponding co-crystallized pose of the same ligand molecule (i.e. the pose generation error).

$$RMSD = \sqrt{\frac{1}{N_a}\sum_{n=1}^{N_a}[(x_r^{(n)} - x_t^{(n)})^2 + (y_r^{(n)} - y_t^{(n)})^2 + (z_r^{(n)} - z_t^{(n)})^2]}$$

where N_a is the number of heavy atoms, $(x_r^{(n)}, y_r^{(n)}, z_r^{(n)})$ is the 3D coordinate of the nth heavy atom of the crystal pose, and $(x_t^{(n)}, y_t^{(n)}, z_t^{(n)})$ is the 3D coordinate of the nth heavy atom of the docked pose.

4   Results and Discussion

4.1   Redocking the Ligand of Each Test Set Complex

In this study, each of the 1300 co-crystallized ligands was redocked into the binding site of its target protein using Vina with default settings. Previously, a script was written to automatically define the search space by finding the smallest cubic box that covers the entire ligand and subsequently extending the box by 10Å in all the three dimensions. For each molecule, Vina returned a maximum number of nine docked poses, of which the one with the best Vina score was used. A second script was written to compute their RMSD with respect to the corresponding co-crystallized pose. Because we aimed at investigating the impact of pose generation error on the prediction of binding affinity, a second test set was defined where each of the 195 complexes has its ligand re-docked and its binding affinity predicted by the scoring functions previously trained on the 1105 crystal structures. As a baseline, these scoring functions were also tested on the co-crystallized ligands of the same 195 complexes. It is noteworthy that, in redocked poses, Vina achieved a relatively small pose generation error in the test set (52% of the ligands had a docked pose with RMSD < 2Å).

4.2   Pose Generation Error Slightly Worsens Binding Affinity Prediction

Models 2-4 were trained on the same 1105 complexes and tested on the 195 complexes in the 2007 core set (Vina is model 1 and was originally trained on all 1300 complexes, so we only tested it on the core set). While this training set is composed by crystal structures, there are two versions of the test set: the "crystal" test set with 195 crystal structures and the "docked" test set with 195 re-docked structures (one per complex, generated as explained in the previous subsection). All models were tested on both versions of the test set.

Models 3 and 4 are stochastic because they are based on RF (a randomized algorithm). Hence, to assess the variability in their response, the same set of 10 random seeds were used to generate 10 versions of the model (a different seed per training run). The performance of each model on each test set version, i.e. on co-crystallized poses and redocked poses of the same complexes, is summarized by the boxplots in Figure 2. This performance is measured as the Root Mean Square Error (RMSE), Pearson's correlation coefficient (R_p) and the Spearman's rank-correlation coefficient (R_s) between measured and predicted binding affinity.

Results in Figure 2 show that pose generation error introduces a small degradation in the ability of models 2-4 to rank-order complexes according to predicted binding affinity in all scoring functions (this can be seen in all three plots). In contrast, Vina performed much better on docked poses in terms of RMSE. The latter is a curious result and we are unable to explain it with the information provided in the original paper [12]. On the other hand, it is remarkable that the best scoring function, RF::VinaElem still achieves such a high performance despite pose generation error (see Figure 3). Importantly, since model 3 use the same features as model 1 and a subset of its training set (Vina is trained on all 1300 complexes in PDBbind v2007 refined set, whereas model 3 trains on the 1105 left after removing the 195 complexes in the v2007 test set), RF::Vina performs remarkably better at predicting binding affinity than the widely-used Vina while having the same applicability domain.

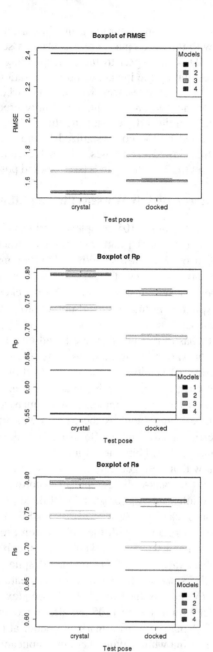

Fig. 2. Performance of each scoring function on the PDBbind v2007 core set test set with co-crystallized ligands (left of each plot) and the same set of test complexes with the re-docked ligand with the lowest Vina score instead (right). Three performance measures are presented: RMSE (top), R_p (middle) and R_s (bottom). Co-crystallized and docked ligands are re-scored with Vina (black), MLR::Vina (red), RF::Vina (green) and RF::VinaElem (blue).

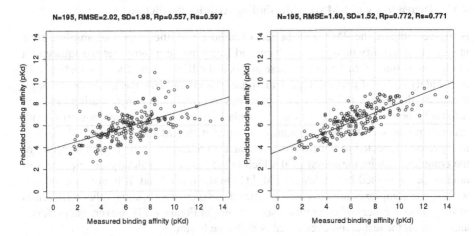

Fig. 3. Performance on the 195 test set complexes in the PDBbind benchmark: AutoDock Vina (model 1; left) and RF::VinaElem (model 4; right). RF::VinaElem constitutes a remarkable improvement on the key requirement of predicting binding affinity when re-scoring redocked poses.

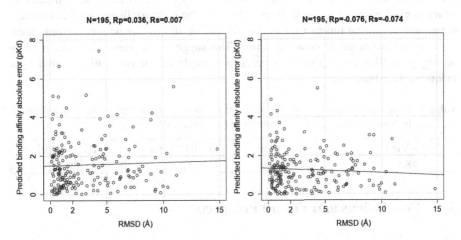

Fig. 4. RMSD from redocking the 195 test set complexes in the PDBbind benchmark: AutoDock Vina (model 1; left) and RF::VinaElem (model 4; right). The y-axis shows the absolute error in predicting pKd for each complex. These results show that both scoring functions are particularly robust to pose generation error. The R_p and R_s stated at the top of these plots quantifies how little the RMSD of the complex generally influences binding affinity prediction, at least when using these scoring functions (these correlations must not be confused with those between predicted and measured pKd in Figure 3).

4.3   Dependency of RMSD with Binding Affinity Prediction

Next, we compare the RMSD of the redocked pose with the individual absolute error in its binding affinity prediction by Vina and RF::VinaElem (note that the square root of the summation of the square of these errors is the RMSE introduced in section 3.2). It is widely believed that the higher the pose generation error the larger the error on predicting that pose will be. Figure 4 plots this information for each scoring function. Strikingly, both scoring functions are particularly robust to pose generation error, with accurate prediction still being obtained in poses with RMSDs of almost 15. This is likely to be connected to uncertainty associated to relating a static crystal structure of the complex with its measured pKd which is the outcome of the dynamic process of binding, as discussed by Ballester et al. [1] On the other hand, it is noteworthy that, while some complexes are very well predicted (pKd error ~ 0), some other have errors of more than 7 orders of magnitude (see left plot in Figure 4). However, the performance over all the test complexes remains high (see Figure 3).

4.4   Native Pose Prediction vs Binding Affinity Prediction

Next, we assess the ability of each scoring function to predict the near-native pose of a molecule as bound to a target (see Table 1). Interestingly, results show that the least accurate predictor of binding strength in this study (Vina) is the best at predicting which docked pose is geometrically the closest to the co-crystallized pose. In contrast, the presented scoring functions, all better at binding affinity prediction, perform much worse than Vina at native pose prediction. This suggests that these two tasks, binding affinity prediction and native pose prediction, might not be optimally covered by a unique scoring function.

Table 1. Percentage of test set complexes where the pose with the highest binding affinity as predicted by a scoring function (1-4 here) also has the lowest RMSD:

1 - Vina	2 - MLR::Vina	3 - RF::Vina	4 - RF::VinaElem
48%	30%	27%	30%

5   Conclusions and Future Prospects

This study has revealed that errors in pose generation generally introduce a small decline in the accuracy of machine-learning scoring functions. Despite this, RF::VinaElem retained the highest performance on PDBbind v2007 core set in the common scenario where one has to predict the binding affinity of docked poses instead of those for co-crystallized poses (usually because a crystal structure of the ligand is not available). Nevertheless, we observed that the presented RF-based scoring functions do not perform particularly well at predicting the near-native pose of a ligand. In the future, we intend to investigate scoring functions tailored to this related problem as well as further investigating why the RMSD of a pose generally has such a small influence on binding affinity prediction.

Acknowledgements. This work has been carried out thanks to the support of the A*MIDEX grant (n° ANR-11-IDEX-0001-02) funded by the French Government « Investissements d'Avenir » program, the Direct Grant from the Chinese University of Hong Kong and the GRF Grant (Project Reference 414413) from the Research Grants Council of Hong Kong SAR.

References

1. Ballester, P.J., et al.: Does a More Precise Chemical Description of Protein-Ligand Complexes Lead to More Accurate Prediction of Binding Affinity? J. Chem. Inf. Model. 54(3), 944–955 (2014)
2. Ballester, P.J., et al.: Hierarchical virtual screening for the discovery of new molecular scaffolds in antibacterial hit identification. J. R. Soc. Interface. 9(77), 3196–3207 (2012)
3. Ballester, P.J.: Machine learning scoring functions based on random forest and support vector regression. In: Shibuya, T., Kashima, H., Sese, J., Ahmad, S. (eds.) PRIB 2012. LNCS, vol. 7632, pp. 14–25. Springer, Heidelberg (2012)
4. Ballester, P.J., Mitchell, J.B.O.: A machine learning approach to predicting protein-ligand binding affinity with applications to molecular docking. Bioinformatics 26(9), 1169–1175 (2010)
5. Ballester, P.J., Mitchell, J.B.O.: Comments on "leave-cluster-out cross-validation is appropriate for scoring functions derived from diverse protein data sets": significance for the validation of scoring functions. J. Chem. Inf. Model. 51(8), 1739–1741 (2011)
6. Breiman, L.: Random Forests. Mach. Learn. 45(1), 5–32 (2001)
7. Cheng, T., et al.: Comparative Assessment of Scoring Functions on a Diverse Test Set. J. Chem. Inf. Model. 49(4), 1079–1093 (2009)
8. Li, H., et al.: istar: A Web Platform for Large-Scale Protein-Ligand Docking. PLoS One 9(1), e85678 (2014)
9. Li, H., et al.: iview: an interactive WebGL visualizer for protein-ligand complex. BMC Bioinformatics 15(1), 56 (2014)
10. Morris, G.M., et al.: AutoDock4 and AutoDockTools4: Automated docking with selective receptor flexibility. J. Comput. Chem. 30(16), 2785–2791 (2009)
11. Morris, G.M., et al.: Automated docking using a Lamarckian genetic algorithm and an empirical binding free energy function. J. Comput. Chem. 19(14), 1639–1662 (1998)
12. Trott, O., Olson, A.J.: AutoDock Vina: Improving the speed and accuracy of docking with a new scoring function, efficient optimization, and multithreading. J. Comput. Chem. 31(2), 455–461 (2010)

Special Session: Large-Scale and HPC Data Analysis in Bioinformatics: Intelligent Methods for Computational, Systems and Synthetic Biology

High-Performance Haplotype Assembly

Marco Aldinucci[3], Andrea Bracciali[1], Tobias Marschall[4,5], Murray Patterson[6],
Nadia Pisanti[2], and Massimo Torquati[2]

[1] Computer Science and Mathematics, Stirling University, UK
abb@cs.stir.ac.uk
[2] ERABLE team, INRIA, Computer Science Department,
University of Pisa, Pisa, Italy
{pisanti,torquati}@di.unipi.it
[3] Computer Science Department, University of Torino, Torino, Italy
aldinuc@di.unito.it
[4] Center for Bioinformatics, Saarland University, Germany
t.marschall@mpi-inf.mpg.de
[5] Computational Biology and Applied Algorithmics,
Max Planck Inst. for Informatics, Germany
[6] Lab. Biométrie et Biologie Evolutive, University Lyon, Lyon, France
murray.patterson@univ-lyon1.fr

Abstract. The problem of *Haplotype Assembly* is an essential step
in human genome analysis. It is typically formalised as the *Minimum
Error Correction* (MEC) problem which is NP-hard. MEC has been
approached using heuristics, integer linear programming, and fixed-
parameter tractability (FPT), including approaches whose runtime is exponential in the length of the DNA fragments obtained by the sequencing
process. Technological improvements are currently increasing fragment
length, which drastically elevates computational costs for such methods.
We present PWHATSHAP, a multi-core parallelisation of WHATSHAP, a
recent FPT optimal approach to MEC. WHATSHAP moves complexity
from fragment length to fragment overlap and is hence of particular interest when considering sequencing technology's current trends. PWHAT-
SHAP further improves the efficiency in solving the MEC problem, as
shown by experiments performed on datasets with high coverage.

1 Introduction

The differences among genomes of distinct individuals of the same species are
called *polymorphisms*. Given two DNA sequences, a *Single Nucleotide Polymorphism* (SNP) is a variation of a single nucleotide occurring at a specific position
in the two sequences. SNPs may occur in genomes of different individuals of the
same species or in different copies of chromosomes of the same individual. The
different forms that a chromosome may exhibit are called *alleles*. The Human
genome consists of two copies of each chromosome, i.e. it is *diploid*. Each copy
comes from one of the two parents.

Genomic data obtained from a sequencing experiment of a human genome is
a mixture of the two copies of the chromosomes in the form of many DNA fragments, called *reads*, which may exhibit one of the forms, i.e. alleles, of parental

© Springer International Publishing Switzerland 2015
C. di Serio et al. (Eds.): CIBB 2014, LNCS 8623, pp. 245–258, 2015.
DOI: 10.1007/978-3-319-24462-4_21

chromosomes. *Haplotyping* is the task of phasing the SNPs, i.e., determining which one of the two alleles they come from.

Haplotyping is an essential task for genome annotation and for several kinds of downstream (comparative) genome analyses, such as finding patterns in human genetic variations for population genomics, or associating genetic variants to diseases, response to drugs, and environmental effects.

When SNP phasing is performed directly on raw sequencing reads, it is referred to as *haplotype assembly* or *read-based phasing*. In this case, reads are first mapped to a reference genome and are then assigned to one of the two haplotypes based on the SNPs they cover. For each SNP position, reads that indicate different alleles must be assigned to different haplotypes. The result is a partition of the reads in two classes according to their originating haplotype. Unfortunately, in real data, such a partition may not exist, due to sequencing errors and also due to reads being misplaced in the mapping phase. For this reason, the task of haplotype assembly becomes a computational optimisation problem where one has to minimise the number of *adjustments* to the data needed to define a bipartition that is then a candidate to represent the correct haplotypes. In the literature, several optimisation problems that formalise haplotype assembly are considered. *Minimum Fragment Removal* (MFR) removes the minimum number of conflicting fragments and hence focuses on mapping errors (that is, misplaced reads in the mapping phase). *Minimum Error Correction* (MEC), asks for the minimum number of characters (nucleotides) to be corrected in the input reads. *Minimum Error Removal* (MER) removes a minimum number of characters from the reads, where removed characters are handled as if the read would not cover these positions at all. MEC and MER have been proved to be equivalent [11], and, since they can be reduced from MAX-CUT [9], are NP-hard. Published algorithms to solve MEC include statistical/heuristic approaches, integer linear programming, or are *exact fixed-parameter tractable algorithms* [10], whose complexity is exponential in the *number of SNPs* per read. Due to ever-increasing read lengths, leading to more SNPs per read, provided by evolving sequencing biotechnologies, methods that are exponential in the read length will perform worse with future-generation longer reads.

In this paper, we present an optimised, parallel implementation of WHAT-SHAP, which was introduced by some of the authors in [20]. WHATSHAP focuses on solving wMEC, a weighted version of MEC. Our choice of WHATSHAP is due to the fact that, remarkably, it is the *first* exact fixed-parameter tractable algorithm for solving wMEC which, instead of being exponential in read length, is instead exponential only in the *sequencing coverage*, i.e. the maximum number of different reads that cover a single SNP position. This makes WHATSHAP particularly appealing with respect to the other currently available proposals, in the light of developments of future generation sequencing techniques, which will provide longer reads.

In wMEC, each SNP value comes with an associated *confidence degree*, which can be set to a combination of the confidence of the base call for that specific position, i.e. which allele the read comes from, and of the confidence of the mapping of the whole read within the chromosome. The confidence degree associated to each SNP is used as the cost of flipping/ignoring that SNP value in order to remove errors. In this way, by minimising the total weight of corrected SNPs, the optimisation problem corrects the most probable sequencing and mapping errors and can be viewed as a maximum likelihood approach. This improves the accuracy of WHATSHAP in comparison to methods that solve the unweighted MEC problem. The weighted variant of MEC was first suggested by [12], and in [21] the authors proposed a heuristic for a special case of wMEC where they also present experiments show that wMEC is more accurate than MEC.

WHATSHAP is still a computationally demanding algorithm. Experimental results show that single-chromosome datasets with a coverage up to 20 can be treated in about 2 hours on a *single* core of an Intel Xeon E5-2620 CPU. The analysis of a whole genome may require the solution of several independent instances of the haplotype assembly problem. In this context, the possibility a high-performance parallel WhatsHap appeared worth exploring.

The main contribution of this paper is to introduce PWHATSHAP, an optimised parallel version of WHATSHAP. We will focus on the parallelisation of the single chromosome haplotype assembly instance on a multi-core machine. Whole-genome approaches can be built on top of "embarrassingly parallel" instances of PWHATSHAP on a range of different architectures.

PWHATSHAP has been engineered by relying upon *skeletons* and *parallel design patterns* provided by the FastFlow framework [2], a methodological approach that allows WHATSHAP to be parallelised with minimal changes to the original sequential code, while minimising the usage of typically slow classical mutual exclusion mechanisms.

Obtained results show a clear performance increase, allowing us to handle larger data sets, with bigger coverage, such as the ones that will be provided by future generation sequencing technologies.

Haplotype assembly and WHATSHAP will be recapped in the next two sections, then PWHATSHAP and obtained performance results will be presented.

2   wMEC Model for Haplotype Assembly

The input dataset for this problem is a set of reads mapped to a reference genome. Arbitrarily re-labelling the alleles to 0 and 1 for each SNP position, the input data is represented as a matrix, having a row for each read and a column for each SNP position. Each element of the matrix reports the value of a given SNP in a read.

More formally, the input dataset is represented as an $n \times m$ matrix F, with n the number of reads and m the number of SNP sites. The elements $f_{i,j}$ of F take values from the set $\{0, 1, -\}$, telling whether, at position j, the read i has the SNP value of the allele 0, or of the allele 1. A value of "$-$" indicates that the respective read does not cover the SNP position. In this case, we say that the read is *not active* at that position. In addition, a *confidence value* (or *weight*) $v_{i,j}$ is associated to each active $f_{i,j}$ as part of the input to the problem. The weight $v_{i,j}$ is the confidence degree of the correctness of the value of $f_{i,j}$, and in the optimisation problem wMEC, it represents the cost of flipping $f_{i,j}$.

A *conflict* between two reads r_p and r_q is a SNP position where the two reads are active and have different values. In the absence of errors, a conflict between two reads implies that the two reads come from different alleles. In this framework, a *correct haplotype assembly* consists of a bipartition of the rows of F (the reads), into two *conflict free* sets R and S. Each conflict free set contains the complete set of reads assigned to the same haplotype. Unfortunately, such a bipartition into conflict free sets usually does not exist in real data sets due to sequencing and mapping errors. The problem thus becomes that of detecting a minimum-weight set of error corrections that allow for a conflict free bipartition. For instance, without correcting errors, no conflict-free bipartition exists for (the rows of) the following 3×2 matrix F of coverage 3 (see column 2), where subscripts are the costs $v_{i,j}$:

$$ F = \begin{pmatrix} 1_9 & 1_9 \\ 0_3 & 1_8 \\ - & 0_8 \end{pmatrix} $$

However, the minimum cost, conflict-free bipartition $R = \{1, 2\}$, $S = \{3\}$ can be obtained by correcting $f_{2,1}$, i.e. flipping it to 1 at a cost of 3.

Several heuristic approaches to solve MEC have been put forward in the last ten years, such as the greedy approaches of [19,15] to assemble the haplotype of a genome, a method to sample a set of likely haplotypes under the MEC model [7], and the much faster follow-up to [7], based on the definition of a graph, analogously to [9], and an iterative greedy heuristic to optimise the MAX-CUT of that graph [6]. The latter outperforms [19,15] while showing similar accuracy to [7]. In [18], reducing MEC to MAX-SAT and using a (heuristic) MAX-SAT solver has been proposed.

All of the abovementioned tools are heuristics – they provide no guarantee on the quality of the solution. To solve the MEC problem to optimality, several exact algorithms have been proposed. To this end, integer linear programming techniques have been developed [11,8]. Fixed-parameter tractable (FPT) algorithms are another way of approaching the MEC problem and have been employed in [13]. However, as noted, these approaches have an exponential complexity in the number of SNPs per read or in the read length, which is going to increase soon and fast with emerging sequencing technologies. Most recently, this problem has been overcome by WHATSHAP [20] which is a FPT algorithm with *coverage* as the cost parameter. It is thus better suited to the current development trends of sequencing technologies. Shortly after WHATSHAP, an equivalent algorithm formulated in terms of belief propagation was independently proposed by Kuleshov [14]. The next section recaps WHATSHAP.

3   WHATSHAP

WHATSHAP is a sequential algorithm, based on dynamic programming, that takes as input the *fragment matrix* F (one row per *read*, one column per SNP position, and values in $\{0, 1, -\}$) and a set of *confidence values* associated to the active positions of the reads, as described in the previous section. It computes, with a dynamic programming method, a minimum-cost conflict-free bipartition of the set of reads.

WHATSHAP builds a *cost matrix* C with as many columns as F (i.e., one column for each SNP). C is constructed incrementally, one column at a time. Let F_j be the set of all reads that are active in the j-th column, let (R, S) be one of the possible bipartitions of F_j, and let $C(j, (R, S))$ be the entry for (R, S) in the j-th column of C. Then, WHATSHAP computes the minimum cost $C(j, (R, S))$ of making (R, S) conflict free, for all possible (R, S).

In general, a read spanning several consecutive positions will induce dependencies across columns, because a single read must be consistently assigned to the same allele throughout all the positions at which it is active (see read 2 in the example of section 2). Therefore, when computing the cost of the bipartitions of F_j for the construction of the j-th column of C, WHATSHAP also needs to consider the (minimum) cost inherited by the construction of *compatible* partitions in F_{j-1}.

Entries $C(1, (R, S))$ in the first column of C, with (R, S) a bipartition of F_1, only depend on the cost of making R and S conflict free (clearly, no inheritance from previous columns). $R \subseteq F_1$ can be made conflict free by flipping all 0s in $f_{1,k}$, with $r_k \in R$, into 1s, at a cost that is equal to the sum of all the weights associated to the 0s that must be flipped, denoted as $W(1)^1_R$. Alternatively, R can be made conflict free by flipping all 1s into 0s, paying $W(1)^0_R$. That is, taking the most advantageous alternative,

$$C(1, (R, S)) = min\{W(1)^1_R, W(1)^0_R\} + min\{W(1)^1_S, W(1)^0_S\}.$$

When considering the j-th column, both the contribution of the column itself (computed in the same way as for the first column), and the cost of a compatible bipartition inherited from previous columns must be taken into account.

Consider, for instance, $C(j, (R, S))$, with $j > 1$ and (R, S) a bipartition of F_j. The *local* contribution of column j is, again, just the cost of the best way to make R and S conflict free over the column j of F (first row in the formula below). To this cost, the cost of keeping (R, S) consistent on all the columns $i < j$ has to be added. This cost is the minimum cost of $C(j - 1, (R', S'))$, for any (R', S') which is "compatible" with (R, S).

A partition (R, S) defined at j and one (R', S') defined at $j-1$ are *compatible*, written $(R, S) \cong (R', S')$, if each element in $F_j \cap F_{j-1}$, i.e. the reads active in both j and $j - 1$, is assigned to the same subset in both (R, S) and (R', S'). It is important to note that, because of the incremental way of proceeding, the cost in the immediately preceding column $j - 1$ summarises all the corrections made in columns 1 to $j - 1$ for keeping (R', S') conflict free. Summing up,

$$C(j, (R, S)) = min\{W(j)^1_R, W(j)^0_R\} + min\{W(j)^1_S, W(j)^0_S\} +$$
$$min\{C(j - 1, (R', S')) \mid (R', S') \cong (R, S)\}$$

The schema of the generic j-th step of the algorithm consists of defining all the possible (R, S) at j and then performing the following three steps:

(a) determine the minimum *local* cost for making the j-th column conflict free by flipping some bits on the column according to their weights and the various correction possibilities;

(b) select the minimum-cost partition amongst those computed at step/column $j-1$ which are compatible with the current partition;

(c) fill in entry $C(j, (R, S))$ with the sum of the outcomes of (a) and (b).

Once the whole matrix C is computed, the result of the wMEC problem is identified by the conflict-free partition (R^*, S^*) of minimum cost in the last column. The actual solution will also comprise all the minimum-cost corrections made throughout the construction of the matrix, which have assigned each read in F to partitions compatible with (R^*, S^*).

The complexity of WHATSHAP algorithm is dominated by the maximum number of bipartitions that must be taken into account at a column. The number of possible bipartitions of the reads at column j is $2^{|F_j|}$, and therefore the complexity is exponential in the amount of active reads that can be found at a position. This critical (i.e. exponential) parameter is therefore the sequencing coverage (see [20] for details).

We conclude this section with a few implementation details of WHATSHAP that are relevant for its paralellisation.

In the construction of the j-th column of C, the possible bipartitons of F_j are considered according to a *Gray code* enumeration, i.e. their binary encodings are ordered in a way that the next entry differs from the previous one by only one bit, e.g. 0001 and 0011, where 0 and 1 indicate the assignment of an active read to either R or S. This implies that two subsequent partitions differ in the position of a *single* read r that moves from set R to set S (or vice versa). This allows for an efficient incremental computation, since, accounting only for the impact of moving r, the computation of the cost of the next partition can be obtained in constant time from the cost of the previous one because updating the values of $W(j)_R^1, W(j)_R^0, W(j)_S^1, W(j)_S^0$ requires constant time.

4   pWHATSHAP: **High-Performance Haplotype Assembly**

The development of a parallel solution for a given problem can be addressed either by developing a parallel algorithm from scratch, or by parallelising an existing sequential algorithm. Our work follows the second approach, where the time complexity of WHATSHAP is a strong motivation for choosing this path. Indeed, WHATSHAP is the first algorithm solving *wMEC* with a complexity which is exponential only *in the sequencing coverage*. As explained, solving the weighted version of the problem caters to its accuracy and exhibiting a complexity independent of the length of the fragments makes it particularly suitable for the current trends in sequencing technology, which will provide fragments of increasing length.

We will here focus on pWHATSHAP, a parallel version of WHATSHAP for a single chromosome. The Multiple instances of haplotype assembly needed for a whole genome are fully independent. Such independent runs can be executed concurrently in an *embarrassingly parallel* fashion, exhibiting best scalability when executed on truly independent platforms (e.g. clusters or cloud resources) where there is no performance degradation due to the concurrent usage of resources, which instead may happen on multi-core architectures. For instance, multiple instances of PWHATSHAP could be supported by cloud infrastructures, rightly considered enabling technologies for bioinformatics and computational biology that provide a large amount of computing power and storage in an elastic and on-demand fashion.

Our PWHATSHAP targets multi-core architectures and relies upon the *Fast-Flow* parallel programming framework [2].

4.1 Technological Background

After decades of increasing clock-frequency and instruction-level parallelism in single core architectures, the current trends for providing high-end performances have steadily focused on increasing the number of cores per chip. Since current multi-core architectures are de-facto small-scale on-chip parallel machines, the most effective way to increase their performance is to use thread-level parallelism. However, legacy sequential code does not necessarily benefit from multi-core architectures, where single-core complexity and clock are typically lower than traditional single-core, and sequential code may perform even worse. Furthermore, parallel programs are inherently more difficult to write than sequential ones due to concurrency issues. Developers, including bioinformatics scientists, are then facing the challenge of achieving a trade-off between high-end performances and time to solution in developing applications and algorithms on current and forthcoming multi-core platforms.

Parallel software engineering addressed this challenge via high-level language extensions and coding patterns aimed at simplifying the porting of sequential codes to parallel architecture, while guaranteeing the efficient exploitation of concurrency [5]. *Parallel design patterns* (PDP) [16] have been recognised to have the potential to induce a radical change in the parallel programming scenario, allowing parallel programs to fully exploit the high parallelism provided by hardware vendors, simplifying programmer's tasks, and making whole application development more efficient. PDPs provide tested and efficient parallel patterns as composable building blocks, which eliminate the need of implementing, tuning and maintaining ad-hoc solutions. The machinery available to application developers is then at a higher level of abstraction with respect to traditional approaches, such as the Message Passing Interface (MPI), where the programmer is fully responsible for the parallel behaviour of an application.

The *FastFlow* parallel framework provides algorithm skeletons and parallel design patterns, enabling a good trade-off between performance, sequential code reuse and time to solution.

Our multi-core parallelisation of WHATSHAP is based on *FastFlow* and exploits the physical shared memory of the underlying architecture, making it unnecessary to move data between threads, a typical source of overhead. However, if this greatly simplifies the parallelisation, it also introduces data sharing and concurrent access related problems. Parallel patterns provided by the *FastFlow* framework solve these problems by defining clear dependencies among different parts of the computations, hence avoiding costly synchronisations.

FastFlow has been demonstrated to be effective in parallelising and redesigning several sequential and concurrent applications, e.g. [1,3,17], and offers an important methodological approach for the parallelisation of WHATSHAP with minimal changes to the original sequential code.

4.2   Parallel WHATSHAP

WHATSHAP follows a dynamic programming approach based on recording so-far computed results in the incremental construction of the solution, i.e. the matrix of all the possible conflict-free partitions of minimum cost. In seeking for a possible decomposition of the algorithm into sub-problems to be solved in parallel by different *executors*, two obvious alternatives are possible: a *vertical* decomposition, where each executor builds a number of columns, i.e. they solve different parts of the genome, and a *horizontal* decomposition, where each executor builds some of the entries of the current column (combinations of the two alternatives could also be considered). The former would constitute a substantial departure from the original structure of WHATSHAP, whose incremental approach induces linear dependencies on columns: each one depends on the results of the previous one, i.e., the minimum-cost compatible partitions of the previous step (see p. 249). Such dependencies make a vertical decomposition difficult, left for future work.

This paper focusses on a horizontal decomposition: each executor evaluates a subset of the possible bipartitions (R, S) of the set F_j of reads that are active in column j.

The first step in the design of PWHATSHAP has been *to profile* the performance of WHATSHAP by measuring the *time cost* of generating the j-th column in the minimum cost matrix C (see p. 249). The time required in the construction of a typical column of a given dimension, i.e. the number of possible bipartitions of the column, depends on the *coverage*, i.e. the number of active fragments of that column (there are $\sim 2^c$ bipartitions for a coverage c). The cost of building columns with a coverage less than 15 is minimal (< 1ms), and does not justify the overhead of a parallel construction. The situation is different for $c > 15$, where the cost of building columns varies from a few milliseconds to a few seconds. In these cases it may be worth adopting an adaptive partitioning, varying the number of executors according to the dimension of the column.

Fig. 1 illustrates the first steps of the parallel construction of the columns of a minimum cost matrix C (Fig. 1(b)) for a fragment matrix F (Fig. 1(a)), with, e.g., read f_1 being 0 in SNP 1 with confidence 5 and read f_2 covering SNPs 1 and 2. In Fig. 1(b), $C(1, (R, S))$ (left matrix) is built by considering all the possible bipartitions (R, S) of the reads active on SNP 1, i.e. f_1, f_2 and f_3, which are represented as binary strings and Gray-code ordered (first three columns). The set of all possible bipartitions is split between two executors (thick horizontal line). Each executor builds C starting from the respective entry points (marked by As) and, in order to maintain as much as possible the original structure of the sequential algorithm, processes bipartitions in Gray code order (see p. 250).

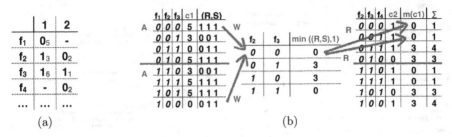

(a)   (b)

Fig. 1. (a): First two columns of a fragment matrix F with associated weights. (b): The (parallel) construction of the cost matrix $C(j, (R, S))$

A bit of care was necessary to properly identify the entry points A in the Gray code sequence.

The costs in $C(1, (R, S))$ (column c1) only depend on making the current partitions conflict free, e.g. for $(\{f_1, f_2, f_3\}, \emptyset)$ (first row) by flipping f_1 to 1 (last three columns) at a cost of 5 (column c1), so that R is conflict free and S empty. In general, for the construction of C_j, the j-th column of C with coverage c_j and k executors, each executor processes approximately $2^{c_j}/k$ possible bipartitions (R, S) of F_j, with k dynamically depending on c_j (and on the hardware configuration). In this phase, each executor computes its own bipartitions in parallel with all other $k - 1$ executors (*map phase*, see Fig. 2).

The construction of C_j depends on the *minimum* costs of the bipartitions in C_{j-1} which are "compatible" with those in C_j, i.e. all those partitions in C_{j-1} which "agree" on the values of common reads, i.e. f_2 and f_3 in Fig. 1(b). This information is recorded on a suitable table (central matrix), where each executor *over-writes* the currently discovered best cost. This may induce write conflicts (Ws in the figure), which have been addressed by constructing local copies of the table for each executor, and then managing their merging by means of a sequential *reduction phase*, executed in pipeline with the *map phase* (Fig. 2). The information recorded in the table is then used to determine the costs in the next column (right matrix) as the sum (Σ) of the minimum cost of compatible bipartitions ($m(c1)$) and minimal corrections on the current bipartition ($c2$). Concurrent *read* accesses (Rs) are of no particular concern.

Interestingly, if more than one minimum exists, the interplay of relative execution speed among parallel executors, may cause non-determinism in the last overwritten minimum, thus providing different solutions of equal minimum cost over different executions. Comparison of different optimal solutions is left as future work.

4.3   Implementation Details

The parallel construction of each column of the minimum cost matrix C has been implemented by using two FastFlow patterns *pipeline* and *task-farm-with-feedback* (Fig.2). The pipeline pattern consists of a 2-stage pipeline whose first stage is a task-farm pattern, with workers (Ws) connected both to the scheduler

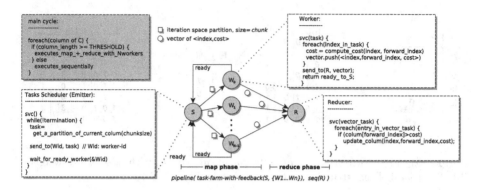

Fig. 2. The FastFlow skeleton used in PWHATSHAP. Each entity is a concurrent thread. The Emitter (S) produces and schedules tasks towards a pool of Workers (Ws). Each Worker sends results to the Reducer (R) and asks for new tasks from S.

thread (S) and the second stage of the pipeline (R). The first stage implements the *map phase* of the proposed parallelisation, where a given number (*chunksize*) of the bipartitions of the fragment set F are computed by each worker in parallel.

The second stage of the pipeline consists of a simple sequential node called *Reducer* (R), which receives *tasks* from all workers (i.e. locally produced results) and then updates the matrix C with the minimum cost found (*reduction phase on all inputs received*). By using these patterns it is possible to exploit: *i)* Emitter–Workers pipeline parallelism: the Emitter computes all possible bi-partitions sending disjoint sub-partitions to Workers using a dynamic scheduling policy; *ii)* parallelism among Workers: computation of local minimum costs in parallel; and *iii)* Workers–Reducer pipeline parallelisms: the Reducer receives multiple results in chunks from each worker.

The proposed parallelisation is quite direct and, importantly, requires minimal changes to the original sequential WHATSHAP code. Furthermore, a high degree of parallelisation is involved due to the many entries of the large *fragment table* F corresponding to many (small) tasks that can be executed in parallel on the available cores.

5   Experimental Results

In this section we report the results of experiments showing how effective the proposed parallelisation is. All the experiments were run using a workstation hosting two E5-2695 Ivy Bridge Intel Xeon processors, each with 12 cores, 64 Gbytes of main memory, Linux Red Hat 4.4.7 with kernel 2.6.32. Each core has two hardware contexts and thus a total of 48 threads are directly supported in hardware. The compiler used was gcc 4.8.2 with optimisation level –O3. For each experiment, CPU frequency was set to the maximum value possible for the considered platform (2.40 GHz, no turbo boost). The parallel version was executed using the shell command `numactl --interleave=all` to exploit all the available memory bandwidth of the 2 NUMA nodes of the hardware platform.

Table 1. Overall speedup and columns considered for each data sets.

max. cov.	n. col.	Time (S) $TSeq$	Time (S) $TPar$	Speedup $\frac{TSeq}{TPar}$
20	23,000	534	171	3.12
22	9,000	568	160	3.55
24	2,600	581	188	3.09
26	600	523	150	3.49

Table 2. Speedup of the single column and % of columns with coverage ≥ 18.

max cov	col. with cov. $\geq 2^{18}$	Time (S) $TSeq$	Time (S) $TPar$	Speedup $(\frac{TSeq}{TPar})$
20	77%	439	88	5.00
22	86%	485	87	5.59
24	83%	503	96	5.25
26	80%	422	70	6.06

Since this paper aims at long reads, synthetic data sets with maximum coverage of 20, 22, 24 and 26 have been generated and used in the presented experiments. Such coverages correspond to quite large data sets. These data sets were produced by generating a single data set with an average coverage of 30 mapped to human genome, and then pruned to smaller coverage data sets (see [20] for details on the construction). Time performance was evaluated by measuring the time elapsed in the computation of subsets (i.e. a given number of columns) of each data set. The dimension of each subset was chosen to guarantee that the entire produced output could be stored in main memory.

First, we ran a set of tests aimed at determining the time spent when computing columns of different coverage. We found that, on the considered platform, it is worth parallelising only columns with a coverage ≥ 18, which we call *higher coverage*. Columns with coverage of 18 have an average computation time of about 7.4ms. Columns with a coverage of less than 18 (*lower coverage*) are processed in less than 1ms on average.

For higher coverage columns, we found that the best execution time was obtained by using all the cores of the platform (24), in particular by using 23 worker threads for the map phase and 1 thread for the reduction phase. Conversely, columns with lower coverage were computed using the same parallel pattern but with just 1 worker thread for the entire *map* phase (see Fig. 2). In this case the parallel skeleton is reduced to a pipeline of 2 sequential stages.

The experimental results obtained from running both the original sequential WHATSHAP and the new parallel PWHATSHAP are summarised in Table 1. For instance, for the data with maximum coverage of 20 (column *max. cov.*), we considered a subset of 23,000 contiguous columns (*n. col.*). The WHATSHAP execution time was 534s (*Tseq*) while the PWHATSHAP time was 171s (*Tpar*), thus obtaining an overall improvement of 3.12 (*Speedup*). For all coverages, the amount of main memory used was fixed to $\sim 63\,\mathrm{GB}$ in all the tested cases. The overall obtained improvement ranges from 3 to 3.5 (see also Fig. 3. Left).

Since it has been observed that the fraction of sequential computation (including both the construction of columns with lower coverage and inherently

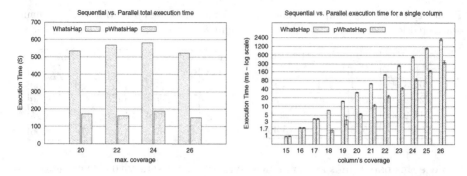

Fig. 3. Left: total execution time varying the maximum columns coverage. **Right**: Average execution time for computing a single column with a given coverage.

sequential parts of the application) amounts to about 20% of the overall computation time (see Table 2), from Amdahl's law [4] it follows that the maximum possible speedup would be at most 5.0 [1].

The results obtained considering only the columns with higher coverage (i.e. the ones we compute in parallel) are summarised in Table 2. In this case, the overall speedup ranges from 5 to 6 times.

The average execution time for computing a single column for several different coverages is reported in Fig. 3. Right (logarithmic scale). The per-column gain obtained, is always in the range 5 - 6.1, both for the smallest coverage (18), which has a very small computation time ($\sim$ 7ms), and for the biggest coverage (26), which requires more than 2s of sequential execution for each column. The fact that we obtained almost the same speedup for very different computation granularities, clearly demonstrates that the limited scalability is not due to the overhead introduced by the parallel run-time code. Instead, we found that the limiting factor is mainly the extensive and non-regular memory access pattern exhibited by WHATSHAP, which does not allow the memory hierarchy of the chosen platform to be fully exploited in concurrent executions. This seems to be connected to the fact that the computation for higher-coverage columns is memory bound. However, further investigation is needed in order to clearly understand how, and if, it is possible to suitably re-organise WHATSHAP's data structures to overcome this issue.

6   Final Considerations

pWHATSHAP aims at further stretching the capabilities of the computational analysis of DNA sequences, targeting high-coverage data sets of long fragments. The performed experiments clearly demonstrated the validity of the proposed

---

[1] Let f be the fraction of the algorithm that is strictly sequential, 1/5 here, then the theoretical maximum speedup that can be obtained with n threads is $S(n) = \frac{1}{f + \frac{1}{n}(1-f)}$, i.e. 5 with $n \to \infty$.

parallelisation of WHATSHAP, with an *overall* speedup of more than 3× for such a fine-grained parallelism problem. Thanks to the design pattern methodology adopted for the parallelisation, such results have been obtained with minimal modifications to the original sequential code. Critical parts of the sequential algorithm amenable to further optimisation have been identified in the process, paving the way for future enhancements of the parallel algorithm. An extensive experimentation on human data sets is planned for future work.

References

1. Aldinucci, M., Bracciali, A., Liò, P., Sorathiya, A., Torquati, M.: StochKit-FF: Efficient systems biology on multicore architectures. In: Guarracino, M.R., et al. (eds.) Euro-Par-Workshop 2010. LNCS, vol. 6586, pp. 167–175. Springer, Heidelberg (2011)
2. Aldinucci, M., Danelutto, M., Kilpatrick, P., Meneghin, M., Torquati, M.: Accelerating code on multi-cores with fastflow. In: Jeannot, E., Namyst, R., Roman, J. (eds.) Euro-Par 2011, Part II. LNCS, vol. 6853, pp. 170–181. Springer, Heidelberg (2011)
3. Aldinucci, M., Torquati, M., Spampinato, C., Drocco, M., Misale, C., Calcagno, C., Coppo, M.: Parallel stochastic systems biology in the cloud. Briefings in Bioinformatics, June 2013
4. Amdahl, G.M.: Validity of the single processor approach to achieving large scale computing capabilities. In: AFIPS 1967 (Spring): Proc. of the April 18-20, pp. 483–485 (1967)
5. Asanovic, K., Bodik, R., Demmel, J., Keaveny, T., Keutzer, K., Kubiatowicz, J., Morgan, N., Patterson, D., Sen, K., Wawrzynek, J., Wessel, D., Yelick, K.: A view of the parallel computing landscape. Communications of the ACM 52(10), 56–67 (2009)
6. Bansal, V., Bafna, V.: HapCUT: an efficient and accurate algorithm for the haplotype assembly problem. Bioinformatics 24(16), i153–159 (2008)
7. Bansal, V., Halpern, A.L., Axelrod, N., Bafna, V.: An MCMC algorithm for haplotype assembly from whole-genome sequence data. Genome Research 18(8), 1336–1346 (2008)
8. Chen, Z.-Z., Deng, F., Wang, L.: Exact algorithms for haplotype assembly from whole-genome sequence data. Bioinformatics 29(16), 1938–1945 (2013)
9. Cilibrasi, R., van Iersel, L., Kelk, S., Tromp, J.: On the complexity of several haplotyping problems. In: Casadio, R., Myers, G. (eds.) WABI 2005. LNCS (LNBI), vol. 3692, pp. 128–139. Springer, Heidelberg (2005)
10. R.G. Downey, M.R. Fellows: Parameterized Complexity, 530 pp. Springer (1999)
11. Fouilhoux, P., Mahjoub, A.R.: Solving VLSI design and DNA sequencing problems using bipartization of graphs. Computational Optimization and Applications 51(2), 749–781 (2012)
12. Greenberg, H.J., Hart, W.E., Lancia, G.: Opportunities for combinatorial optimization in computational biology. INFORMS J. on Computing 16(3), 211–231 (2004)
13. He, D., Choi, A., Pipatsrisawat, K., Darwiche, A., Eskin, E.: Optimal algorithms for haplotype assembly from whole-genome sequence data. Bioinformatics 26(12), i183–i190 (2010)

14. Kuleshov, V.: Probabilistic single-individual haplotyping. Bioinformatics 30(17), i379–i385 (2014)
15. Levy, S., Sutton, G., Ng, P.C., Feuk, L., Halpern, A.L., et al.: The diploid genome sequence of an individual human. PLoS Biol. 5(10), e254 (2007)
16. Mattson, T., Sanders, B., Massingill, B.: Patterns for parallel programming. Addison-Wesley Professional (2004)
17. Misale, C.: Accelerating bowtie2 with a lock-less concurrency approach and memory affinity. In: Proc. of the 22nd International Euromicro Conference PDP 2014: Parallel Distributed and network-based Processing, pp. 578–585 (2014)
18. Mousavi, S.R., Mirabolghasemi, M., Bargesteh, N., Talebi, M.: Effective haplotype assembly via maximum Boolean satisfiablility. Biochemical and Biophysical Research Communications 404(2), 593–598 (2011)
19. Panconesi, A., Sozio, M.: Fast hare: A fast heuristic for single individual SNP haplotype reconstruction. In: Jonassen, I., Kim, J. (eds.) WABI 2004. LNCS (LNBI), vol. 3240, pp. 266–277. Springer, Heidelberg (2004)
20. Patterson, M., Marschall, T., Pisanti, N., van Iersel, L., Stougie, L., Klau, G.W., Schönhuth, A.: Whatshap: Haplotype assembly for future-generation sequencing reads. In: Proc. of 18th ACM Annual International Conference on Research in Computational Molecular Biology (RECOMB), pp. 237–249 (2014)
21. Zhao, Y.-T., Wu, L.-Y., Zhang, J.-H., Wang, R.-S., Zhang, X.-S.: Haplotype assembly from aligned weighted SNP fragments. Computational Biology and Chemistry 29, 281–287 (2005)

Data-Intensive Computing Infrastructure Systems for Unmodified Biological Data Analysis Pipelines

Lars Ailo Bongo[1], Edvard Pedersen[1], and Martin Ernstsen[2]

[1] Department of Computer Science and Center for Bioinformatics,
University of Tromsø, Tromsø, Norway
[2] Now at Kongsberg Satellite Services AS, Tromsø, Norway
larsab@cs.uit.no, edvard.pedersen@uit.no,
martin.ernstsen@ksat.no

Abstract. Biological data analysis is typically implemented using a deep pipeline that combines a wide array of tools and databases. These pipelines must scale to very large datasets, and consequently require parallel and distributed computing. It is therefore important to choose a hardware platform and underlying data management and processing systems well suited for processing large datasets. There are many infrastructure systems for such data-intensive computing. However, in our experience, most biological data analysis pipelines do not leverage these systems.

We give an overview of data-intensive computing infrastructure systems, and describe how we have leveraged these for: (i) scalable fault-tolerant computing for large-scale biological data; (ii) incremental updates to reduce the resource usage required to update large-scale compendium; and (iii) interactive data analysis and exploration. We provide lessons learned and describe problems we have encountered during development and deployment. We also provide a literature survey on the use of data-intensive computing systems for biological data processing. Our results show how unmodified biological data analysis tools can benefit from infrastructure systems for data-intensive computing.

Keywords: data-intensive computing, biological data analysis, flexible pipelines, infrastructure systems.

1   Introduction

Recent advances in instrument, computation, and storage technologies have resulted in large amounts of biological data [1]. To realize the full potential for novel scientific insight in the data, it is necessary to transform the data to knowledge through data analysis and interpretation.

Biological data analysis is typically implemented using a deep pipeline that combines a set of tools and databases [2]. Biological data analysis is diverse and specialized, so the pipelines have a wide range of resource requirements. Examples include the 1000 Genomes project [3] with a dataset of 260 TB analyzed on supercomputers and warehouse-scale datacenters; the many databases [4] and web servers [5] built for

© Springer International Publishing Switzerland 2015
C. di Serio et al. (Eds.): CIBB 2014, LNCS 8623, pp. 259–272, 2015.
DOI: 10.1007/978-3-319-24462-4_22

a specific type of biological process or organism; and simple analyses run using pre-defined Galaxy pipelines [6]. An important challenge when building data analysis tools and pipelines is therefore to choose a hardware platform and underlying data management and processing systems that satisfies the requirements for a specific data analysis problem.

A data analysis and exploration tool architecture typically has the following components:

- A front-end that provides the *user interface* used by the data analysts, including: web applications, pipeline managers [6], web services [7], and low-level interfaces such as file systems, and cloud APIs [8, 9].
- *Data analysis and interpretation services* including: specialized servers, search engines, R libraries such as BioConductor [10], and tool collections such as Galaxy Toolsheds [11].
- *Infrastructure system for data management* including: file systems, databases, and distributed data storage systems.
- *Infrastructure system for parallel and distributed computing* including: queuing systems [12], and the Hadoop software stack [13].
- *Hardware platform*, such as virtual machines, dedicated servers, small clusters, supercomputers, and clouds.

In this paper, we focus on the choice of data management and processing infrastructure systems. In our experience, most biological data analysis uses a file system in combination with a centralized database for data storage and management, and is run either on a single machine or on a small cluster with a queuing system such as Open Grid Engine [12]. This platform has four main advantages. First, the file system-, and database interfaces are stable, and the technology is reliable. Second, many clusters for scientific computing are designed to use a network file system and to run jobs on a cluster using a queuing system. Third, the developers are familiar with these systems and interfaces. Fourth, there are already hundreds of analysis tools implemented for this platform.

An alternative infrastructure must therefore provide better scalability, performance, or efficiency. Or, it must provide services not available on the standard platform. We give an overview data-intensive computing systems, and describe how these can be leveraged for biological data analysis. We describe how we used these systems to transparently extended flexible biological data analysis frameworks with data intensive computing services. We provide lessons learned, and describe the problems we have encountered during development and deployment of the extended pipelines. In addition, we provide a literature survey on the use of data-intensive computing systems in biological data processing.

Our results make it easier to choose data-intensive computing platforms and infrastructure systems for biological data analysis. First, our systems provide data analysis services not available on the standard platform. Second, they do not require modifying biological analysis tools. We believe this combination is essential to increase the use of data-intensive systems in biological data processing.

2   Related Work

There are several specialized infrastructure systems developed for data-intensive computing. Many of these were initially developed and deployed at companies such as Google, Yahoo, Facebook, and Twitter, and then later implemented as open source systems. There are also many new systems under development in academia, the open source community, and the industry. We provide a short description of the features provided by these systems. We limit our description to the most widely used systems and omit many emerging systems, and systems that provide a traditional file system or SQL interface.

Data intensive computing systems are often built on a distributed file systems such as Hadoop Distributed File System (HDFS) [14, 15] that provide reliable storage on commodity component hardware and high aggregate I/O performance. A HDFS cluster co-locates storage and computation resources to avoid the bandwidth bottleneck of transferring large datasets over the network to and from the computation nodes. The main advantage of HDFS for biological data analysis is that the architecture is demonstrated to scale to peta-scale datasets [14, 15], and it is widely used for data-intensive computing in other fields. The main disadvantage is that HDFS does not provide a traditional file system interface, so it is necessary to either rewrite the many data analysis tools that use a POSIX file system interface, to incur an overhead for moving data between HDFS and a local filesystem, or incur the overhead of third-party library such as fuse-dfs [16]. In addition, it is not yet a common platform in scientific computing, so it may be necessary to purchase and build a new cluster with storage distributed on the compute nodes instead of a dedicated storage system.

MapReduce [8, 13] is a widely used programming model and infrastructure system for data-intensive parallel computation. It provides fault-tolerant computation on a HDFS-like file system, and makes it easy to write scalable applications since the system handles data partitioning, scheduling and communication. Biological data analysis applications, especially for next-generation sequencing data, have already been implemented using MapReduce ([17] and [2] provides examples and references). The main advantage of MapReduce is that it scales to peta-scale datasets. In addition, most cloud platforms provide a MapReduce interface. The main disadvantage is that the MapReduce programming model may be too restrictive for some biological applications.

HBase [18, 19] is a column based storage system that provides in-memory caching for low latency random data access, and efficient compression. Biological data analysis applications can use HBase to store data accessed interactively, to implement custom data structures, or to structure data for more efficient compression. Compared to relational databases, HBase does not provide an advanced query engine nor ACID properties. Other systems must implement these on top of HBase if such properties are needed by an application.

An alternative for low-latency query processing is Spark [9, 20]. It offers a richer programming model than MapReduce, including iterative operations. It is well suited to implement machine learning algorithms, and interactive data analysis. Spark uses the Scala programming language, which may be unfamiliar to many developers but bindings exists for Java and Python. Compared to the systems discussed above Spark has just recently become a top-level Apache project, but it is rapidly being adopted by many other open-source and commercial systems.

Several high-level programming models are built on top of MapReduce to make it easier to write data analysis programs, including Pig [21], Hive [22], Cascading [23] and Cassandra [24]. To our knowledge, these are not widely used for biological data analysis (see also the discussion in section 4).

Other data-intensive computing systems for Hadoop, HBase, and Spark includes, Cloudera Impala [25] and Drill [26] that both provide a low-latency SQL query engine (inspired by Dremel [27]), Storm [28] for stream processing, and the Mahout [29] library of machine learning algorithms. To our knowledge, these are also not widely used for biological data analysis (see also the discussion in section 4).

There are many frameworks for specifying and running biological data analysis pipelines. The most widely used systems is Galaxy [6]. It has been integrated with the Hadoop software stack [30].

Table 1. Our use of data-intensive computing systems for biological data processing.

System	Problem	Solution	Issues
Troilkatt	Scalable analysis pipelines	HDFS: scalable storage MapReduce: I/O intensive pipeline processing	Memory management
GeStore	Incremental updates for analysis pipelines	HBase: data structures for generating incremental updates	Hadoop MapReduce job startup time
Mario	Tuning of analysis pipelines	MapReduce: I/O intensive pipeline processing HBase: sparse data structure, low- latency reads and writes	Performance tuning HBase

3   Transparent Data-Intensive Computing

We have extended several data-intensive computing systems to provide services for biological data analysis. In this section, we answer the following questions:

- Why did we choose a particular infrastructure system?
- What problems did the system solve?
- What are the main limitations of the systems for our use?
- What are the lessons learned during development and deployment?

We have used three clusters, with 5, 10, and 64 nodes, for development and deployment of our systems over a period of five years. All were built for data-intensive computing with storage distributed on the compute nodes. We chose the Hadoop software stack, including HDFS for distributed storage and processing. Hadoop is the mostly used data-intensive computing platform, and it has a very active development community. By using Hadoop, we have benefited from improvements to the infrastructure systems, and the addition of new infrastructure systems such as Spark to the ecosystem. In this section, we describe achievements, issues, and lessons learned.

3.1 Troilkatt

Troilkatt is a system for batch processing of large-scale collections of gene expression datasets. We use Troilkatt to process data for the IMP integrated data analysis tool [31]. We built Troilkatt in order to scale our gene expression dataset integration pipeline to process all gene expression datasets for several organisms in NCBI GEO [32]. The pipelines comprise tools for data cleaning, transformation, and signal balancing of these datasets. The integrated data compendium for organisms such as human have ten thousands of datasets. The raw data, pipeline results, and intermediate data in a compendium use tens of terabyte of storage space.

Troilkatt provides infrastructure services for automated genomics compendium management. It consists of five main components (Fig. 1). First, the large genomics compendia maintained by Troilkatt are stored and processed on a cluster. Second, Troilkatt leverages the Hadoop software stack for reliable storage and scalable fault-tolerant data processing. Third, the Troilkatt runtime system provides data management including versioning and a library of tools for downloading and processing data. Fourth, Troilkatt can execute a large collection of external tools and scripts for data processing. Finally, a command line based user interface provides an administrator interface for managing compendium content and steering data processing,

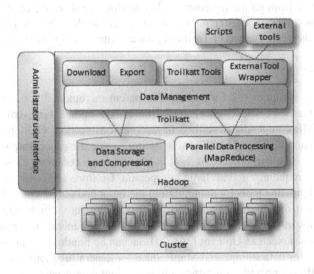

Fig. 1. Troilkatt architecture.

The IMP pipeline processing is well suited for data-intensive systems since most pipeline tools are I/O intensive. The data is stored in HDFS. We use MapReduce for parallel processing and HBase for meta-data storage. We chose MapReduce since the processing must scale to several tens of terabytes of data. We use one Mapper task per dataset for each pipeline tool, since the datasets can be processed independently and hence in parallel. In addition, many tools in the IMP pipeline process one row in a gene expression table at a time, and are therefore well suited for the MapReduce programming model.

Initially we considered using Hadoop streaming (MapReduce with unmodified binaries). However, many biological analysis tools require specifying multiple meta-database files as command line arguments, which is not possible with Hadoop Streaming. To solve this problem, Troilkatt allows specifying multiple input files as command line arguments using environment variables.

We achieved a system that efficiently executes pipelines for processing large-scale integrated compendia. We did not have to implement data communication between tasks, nor data locality aware mapping of tasks to compute nodes. Our main issue was large in-memory data structures in two pipeline tools. Hadoop MapReduce is run in a JVM, so the maximum heap size must be set at system startup time. The memory usage of the largest tasks therefore limits the number of tasks that can be run in parallel on each node. To achieve a good trade-off between memory usage and parallelism, we had to divide the expression datasets by their memory usage and processes similarly sized datasets together in a separate MapReduce job.

Since all datasets can be processed in parallel, and most organisms have hundreds or thousands of datasets, the parallelism in the pipeline exceeded the available compute resources even on the biggest 64-node cluster. We can therefore efficiently utilize a much larger cluster. The raw datasets are larger than the available storage on the clusters. We must therefore periodically delete raw data downloaded from public repositories. A full update of the dataset therefore requires re-downloading tens of terabytes of data from public repositories. In addition, we share the clusters with other non-MapReduce jobs. A cluster management system such as Mesos [33] may therefore improve the performance of Troilkatt jobs and improve resource utilization.

3.2   GeStore

GeStore [34] is a framework for adding transparent incremental updates to data processing pipelines. We use GeStore to incrementally update large-scale compendia such as the IMP compendia described in the previous section. GeStore is integrated with Troilkatt and Galaxy [35]. We built GeStore since the processing time for a full compendium update can be several days even on a large computer cluster, making it impractical to frequently update large compendia. GeStore adds incremental updates to unmodified pipeline tools by manipulating the input and output files of the tools.

GeStore must detect changes in, and merge updates into, compendia that can be tens of terabytes in size. GeStore must also maintain and generate incremental updates of meta-databases, such as UniProt [36].   These can be hundreds of gigabytes in size. We use HBase for data storage and MapReduce for generating input files and merging output files. HBase provides high-throughput random data accesses required for efficient change detection and merging. GeStore stores meta-database entries as HBase rows, and it updates entries by creating a new version of an HBase table cell. The HBase timestamps enable efficient table scans to find entries that have changed in a period and hence are part of an incremental update. In addition, the flexible schema of

HBase tables is utilized to reduce the work required to maintain plugins when file structure or databases change, allowing several years of database versions to be stored in the same HBase table.

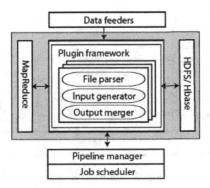

Fig. 2. GeStore architecture.

GeStore provides a plugin system to support many biological tools and file formats (Fig. 2). To add incremental updates for a new tool the plugin maintainer implements a plugin. The plugin specifies how to partition input and meta- data files into entries, which part of an entry are required for the analysis done by a tool, and how to compare the entries to detect updates. In addition, the plugin contains code for writing entries to an incremental input file, and merging incremental output with previous output data. These plugins are relatively small in size; less than 300 lines of code in our most complex plugin.

We achieved up to 82% reduction in analysis time for compendium updates when using GeStore with an unmodified biological data analysis pipeline ([34] has additional experimental results). We found HBase to be well suited for the data management requirements of GeStore. File generation and merging scales to large datasets since we use MapReduce for data processing. In addition, we reduce storage space requirements by storing multiple meta-databases versions in HBase instead of storing all versions as separate files.

The overhead introduced by GeStore is high for incremental updates of small datasets, since the startup time of Hadoop MapReduce jobs is tens of seconds. A system with lower startup time, such as Spark, will significantly reduce this overhead. GeStore overhead can also be large if the implemented plugin is not able to properly detect incremental updates.

3.3   Mario

Mario [37] is a system for interactive iterative data processing. We have designed Mario for interactive parameter tuning of biological data analysis pipeline tools. For such interactive parameter tuning, the pipeline output should quickly be visible for the pipeline developer. Mario combines reservoir sampling, fine-grained caching of derived datasets, and a data-parallel processing model for quickly computing the results of changes to pipeline parameters. It uses the GeStore approach for transparently adding iterative processing to unmodified data analysis tools.

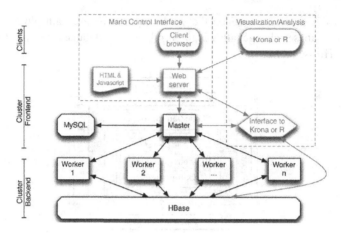

Fig. 3. Mario architecture.

The Mario architecture consists of four main components: storage, logic/computation, the web server and the client/UI (figure 3). The system runs on a cluster of computers, with the master process at the cluster frontend, and the workers at the compute nodes. These are co-located with HBase master at the cluster frontend and the HBase region servers at the compute nodes. The web server and the MySQL server can be located on the cluster frontend, or on separate computers. The user runs the Mario controller, and data visualization and analysis tools on her computer.

Mario must efficiently produce random samples from a stream for reservoir sampling, implement a cache of fine-grained pipeline tool results, and implement parallel pipeline stage processing. We use HBase as storage backend due to its low-latency random read and write capability, its ability to efficiently store sparse data structures, and its scalability. Mario stores all intermediate data records produced during pipeline execution in HBase, and uses the cached data to quickly find the data records that must be updated when pipeline tool parameters are changed. Mario also uses HBase for data provenance and single-pass reservoir sampling. The iterative processing in Mario is similar to Spark Streaming [38]. Mario splits the data randomly into many small parts and distributes these on the cluster nodes. Mario process the parts in parallel. However, it only process a small subset at a time since there are many more parts than processor cores. Mario therefore iteratively updates the output data.

We achieved a system for iterative parallel processing that adds less than 100ms of overhead per pipeline stage, and that does not add significant computation, memory, or storage overhead to compute nodes (additional experimental results are in [37]). We found HBase to be very well suited for efficiently storing and accessing the sparse data-structures used by Mario.

Our main issue was to configure HBase to achieve the required performance. We chose a configuration where HBase region servers allocate 12GB DRAM on the cluster nodes, and we traded reliability for improved write latencies by keeping write-ahead logs in memory (and periodically flushing these to disk). The reduced reliability is acceptable for Mario since it can recompute the intermediate data stored in HBase at low cost, if needed.

4   Discussion

We have assumed that data-intensive computing systems are not widely used for biological data processing. We base this assumption on our own experience, and discussions we have had with bioinformatics users, model developers, and infrastructure maintainers. In this section, we verify this assumption by conducting a literature survey. We do not conduct a comprehensive survey, but we still believe our results provides insights into the usage of data-intensive computing systems for biological data processing by showing the interest for such systems in the bioinformatics literature. The results for the searches described in this section are in Supplementary materials (http://bdps.cs.uit.no/data/cibb14-supplementary.pdf). [2, 17] provides additional references to articles that describe biological data analysis using data-intensive systems.

4.1   Data-Intensive Computing Articles in BMC Bioinformatics

We first examined *Software* articles published in *BMC Bioinformatics*. It is a popular and important journal for bioinformatics analysis tool articles. We also examined *Genome Biology* and the Web Server and Database special issues of *Nucleic Acids Research*, but these journals have few infrastructure articles compared to BMC Bioinformatics. The *Software* articles in that journal also provide detailed implementation details and list of required libraries. We used this information to determine whether the described software uses data-intensive computing systems.

We first got a list of articles from the BMC Bioinformatics website (http://www.biomedcentral.com/bmcbioinformatics/content) by setting *Article* types to *Software*, and *Sections* to *All sections*. We examined the 23 *Software* articles published between August 2014 and November 2014. Of these, two used a HPC platform, one used a data warehouse, and the remaining were implemented for a server or a desktop computer. The results show that most software articles in BMC Bioinformatics describe user interfaces, or data analysis and interpretation services. However, the software in many articles used tools such as BLAST [39] that are ported to both HPC and data-intensive computing platforms. In addition, these articles usually do not describe the backend processing systems used to generate the analyzed data. These may therefore use data-intensive computing systems. To get results that are more specific for our analysis we refined our search to find articles that describe infrastructure systems.

We used the Advanced Search in the Bioinformatics website (http://www.biomedcentral.com/bmcbioinformatics/search). We searched in *All fields* for "infrastructure", and set *Include* to *Software*. This search returned 142 articles in total. We examined 15 articles published between November 2013 and November 2014. Of these, eight described software run on a single server or a desktop computer, six used a cluster for either file storage or as a computational resource, and one system [40] used data-intensive computing techniques. These results indicate that data-intensive computing systems are not widely used for biological data analysis infrastructures.

4.2   Articles About Specific Data-Intensive Computing Systems

In this section, we examine articles describing specific data-intensive computing systems. We want to find which systems that are used for biological data processing, and what these systems are used for. We searched for specific keywords and manually filtered the articles in the returned results to exclude articles that do not describe the system we searched for. We set *Include* to *All article types* to increase the number of articles in the search results. The search results includes articles published before December 2014.

We first searched for "Hadoop" since it is the most widely used data-intensive computing platform. The search returned 24 articles. Of these, the software in 14 articles used systems in the Hadoop stack; three articles describe virtual machine images or provisioning systems that include Hadoop, and the remaining only mention Hadoop in related work. The Hadoop systems were used for search, integrated analysis, data integration, machine learning, and distributed analysis of different data types

We also searched for the Hadoop alternative Azure [41]. There were six articles discussing Azure. The software described in two articles used Azure in combination with MapReduce. Two articles propose to use Azure for processing and storage, one describes a virtual machine provisioning systems, and one is a review article.

We then searched for the names of the data-intensive computing systems described in section 2. For *MapReduce* we found 54 articles. In order to remove articles that just mention MapReduce in the citations, we refined the MapReduce search by limiting the search to *Title+Abstract+Text*. This reduced the number of articles to 27. Of these, 12 describe software that used Hadoop MapReduce and three software that used the MapReduce programming model. Most of these articles are also in the Hadoop search results discussed above. We found four *HBase* articles; of which two [40, 42] describe software that used HBase as a storage backend for sequencing data. The remaining two mention HBase in related work. We found two *Spark* articles. One described how they implemented distributed processing of next-generation sequencing data [43] using Spark. *Cassandra*, *Hive*, and *Mahout* were not used by the software in any articles in the search results, but are discussed in the related work in two articles. We did not find *Impala* in any articles.

We also searched for the *Pig*, *Cascading Drill*, and *Storm* systems, but these return many false positives. We therefore refined the search to: "Pig Apache", "Cascading SQL", "Drill Apache", and "Storm Apache". *Pig* and *Cascading* are both discussed in the related work in two articles. We did not find any articles for *Drill* and *Storm*.

The above results show that MapReduce is the most popular system for data-intensive biological data processing, and that most MapReduce tool implementations use the Hadoop stack. We also found a few articles where HBase and Spark were used. MapReduce, Spark, and HBase are all core systems. We did not find any Bioinformatics articles that described tools that use higher-level systems. This may suggest that the services and abstractions provided by these are not well suited for biological data. The systems that were not mentioned in any articles (Impala, Drill, and Storm) are all recently developed, and may therefore have been unstable when the bioinformatics tools described in the articles were implemented.

4.3   Usage Trends

We also examined whether data-intensive computing systems are becoming more used for biological data analysis. To answer this question we counted the number of articles mentioning MapReduce and Hadoop. The MapReduce design [44] was published in 2004, the open-source Hadoop MapReduce project [13] was started in 2005, and Hadoop became popular around 2008. Since then, an increasingly number of articles mention MapReduce and Hadoop each year, especially in the last two years (Table 2). The results indicate that data-intensive computing systems are becoming more popular for biological data processing, but that it takes a few years from the systems become popular until bioinformatics tools use them.

Table 2. Number of articles per year for keywords MapReduce and Hadoop (many articles are in both results). 2014 does not include articles published in November and December.

Year	Hadoop	MapReduce
2009	1	1
2010	4	7
2011	4	7
2012	4	8
2013	4	15
2014	7	14
Total	24	52

5   Conclusion

We have transparently extended flexible biological data analysis frameworks to utilize data intensive computing infrastructure systems. Our results show that even unmodified biological data analysis tools can benefit from these for: (i) scalable fault-tolerant computing for large-scale data; (ii) incremental updates in order to reduce the resource usage required to update large-scale data compendium; and (iii) interactive data analysis and exploration.

We have identified several limitations of the infrastructure systems we used. However, by using new systems recently added to the Hadoop ecosystem we can remove many of these limitations. We will therefore continue using these systems to extend biological data processing pipelines with new services for making data analysis more efficient and for improving the quality of the analysis results.

We believe that centralized storage system performance is becoming a bottleneck for large-scale biological data processing, and that it will become necessary to use data-intensive computing infrastructure systems and a platform with distributed storage to avoid this bottleneck. The systems presented in this paper motivate and make the transition to such platforms and infrastructure systems easier for two reasons. First, they provide data analysis services that are not available on the standard HPC platform. Second, it is not necessary to modify the biological analysis tools.

We believe the combination of novel services and backward compability is essential to increase the use of data-intensive systems in biological data processing.

Troilkatt, GeStore and Mario are all open source:

- https://github.com/larsab/troilkatt
- https://github.com/EdvardPedersen/GeStore
- http://bdps.cs.uit.no/code/mario/

Acknowledgements. Thanks to Kai Li, Olga Troyanskaya, and Alicja Tadych for their help developing and deploying Troilkatt. Thanks to our colleagues at the Tromsø ELIXIR node for their expertise in developing and deploying biological data processing systems. Thanks to Einar Holsbø, Giacomo Tartari and Bjørn Fjukstad for their comments and insights while writing this paper.

References

1. Kahn, S.D.: On the Future of Genomic Data. Science (80-) 331, 728–729 (2011)
2. Diao, Y., Roy, A., Bloom, T.: Building Highly-Optimized, Low-Latency Pipelines for Genomic Data Analysis. In: 7th Biennial Conference on Innovative Data Systems Research (CIDR 2015), Asilomar, CA, USA (2015)
3. Clarke, L., Zheng-Bradley, X., Smith, R., Kulesha, E., Xiao, C., Toneva, I., Vaughan, B., Preuss, D., Leinonen, R., Shumway, M., Sherry, S., Flicek, P.: The 1000 Genomes Project: data management and community access. Nat. Methods 9, 459–462 (2012)
4. Fernández-Suárez, X.M., Rigden, D.J., Galperin, M.Y.: The 2014 Nucleic Acids Research Database Issue and an updated NAR online Molecular Biology Database Collection. Nucleic Acids Res. 42 (2014)
5. Benson, G.: Editorial: Nucleic Acids Research annual Web Server Issue in 2014. Nucleic Acids Res. 42, W1–W2 (2014)
6. Goecks, J., Nekrutenko, A., Taylor, J.: Galaxy: a comprehensive approach for supporting accessible, reproducible, and transparent computational research in the life sciences. Genome Biol. 11, R86 (2010)
7. Oinn, T., Addis, M., Ferris, J., Marvin, D., Senger, M., Greenwood, M., Carver, T., Glover, K., Pocock, M.R., Wipat, A., Li, P.: Taverna: a tool for the composition and enactment of bioinformatics workflows. Bioinformatics 20, 3045–3054 (2004)
8. Dean, J., Ghemawat, S.: MapReduce: a flexible data processing tool. Commun. ACM 53, 72 (2010)
9. Zaharia, M., Chowdhury, M., Das, T., Dave, A., Ma, J., McCauley, M., Franklin, M.J., Shenker, S., Stoica, I.: Resilient distributed datasets: a fault-tolerant abstraction for in-memory cluster computing. In: Proc. of the 9th USENIX conference on Networked Systems Design and Implementation. USENIX Association (2012)
10. Gentleman, R.C., Carey, V.J., Bates, D.M., Bolstad, B., Dettling, M., Dudoit, S., Ellis, B., Gautier, L., Ge, Y., Gentry, J., Hornik, K., Hothorn, T., Huber, W., Iacus, S., Irizarry, R., Leisch, F., Li, C., Maechler, M., Rossini, A.J., Sawitzki, G., Smith, C., Smyth, G., Tierney, L., Yang, J.Y.H., Zhang, J.: Bioconductor: open software development for computational biology and bioinformatics. Genome Biol. 5 (2004)

11. Blankenberg, D., Von Kuster, G., Bouvier, E., Baker, D., Afgan, E., Stoler, N., Taylor, J., Nekrutenko, A.: Dissemination of scientific software with Galaxy ToolShed. Genome Biol. 15, 403 (2014)
12. Open Grid Scheduler, http://gridscheduler.sourceforge.net/
13. Hadoop homepage, http://hadoop.apache.org/
14. Shvachko, K., Kuang, H., Radia, S., Chansler, R.: The Hadoop Distributed File System. In: 26th Symposium on Mass Storage Systems and Technologies. IEEE (2010)
15. Ghemawat, S., Gobioff, H., Leung, S.-T.: The Google file system. ACM SIGOPS Operating Systems Review, 29 (2003)
16. MountableHDFS, http://wiki.apache.org/hadoop/MountableHDFS
17. Taylor, R.C.: An overview of the Hadoop/MapReduce/HBase framework and its current applications in bioinformatics. BMC Bioinformatics 11 (2010)
18. Apache HBase, http://hbase.apache.org/
19. Chang, F., Dean, J., Ghemawat, S., Hsieh, W.C., Wallach, D.A., Burrows, M., Chandra, T., Fikes, A., Gruber, R.E.: BigTable: A Distributed Storage System for Structured Data. ACM Trans. Comput. Syst. 26, 1–26 (2008)
20. Apache Spark, https://spark.apache.org/
21. Gates, A.F., Natkovich, O., Chopra, S., Kamath, P., Narayanamurthy, S.M., Olston, C., Reed, B., Srinivasan, S., Srivastava, U.: Building a high-level dataflow system on top of Map-Reduce: the Pig experience. In: Proc. of the VLDB Endowment, pp. 1414–1425 (2009)
22. Thusoo, A., Sarma, J.S., Jain, N., Shao, Z., Chakka, P., Anthony, S., Liu, H., Wyckoff, P., Murthy, R.: Hive: a warehousing solution over a map-reduce framework. In: Proc. of VLDB Endowment, pp. 1626–1629 (2009)
23. Cascading, http://www.cascading.org/
24. Lakshman, A., Malik, P.: Cassandra: a decentralized structured storage system. ACM SIGOPS Oper. Syst. Rev. 44, 35 (2010)
25. Impala, http://www.cloudera.com/content/cloudera/en/products-and-services/cdh/impala.html
26. Apache Drill, http://incubator.apache.org/drill/
27. Melnik, S., Gubarev, A., Long, J.J., Romer, G., Shivakumar, S., Tolton, M., Vassilakis, T.: Dremel: interactive analysis of web-scale datasets. In: Proc. VLDB Endow., pp. 330–339 (2010)
28. Storm, https://storm.incubator.apache.org/
29. Mahout homepage, https://mahout.apache.org/
30. Pireddu, L., Leo, S., Soranzo, N., Zanetti, G.: A Hadoop-Galaxy adapter for user-friendly and scalable data-intensive bioinformatics in Galaxy. In: Proc. of 5th ACM Conference on Bioinformatics, Computational Biology, and Health Informatics, pp. 184–191 (2014)
31. Wong, A.K., Park, C.Y., Greene, C.S., Bongo, L.A., Guan, Y., Troyanskaya, O.G.: IMP: a multi-species functional genomics portal for integration, visualization and prediction of protein functions and networks. Nucleic Acids Res. 40, W484–W490 (2012)
32. Barrett, T., Troup, D.B., Wilhite, S.E., Ledoux, P., Evangelista, C., Kim, I.F., Tomashevsky, M., Marshall, K.A., Phillippy, K.H., Sherman, P.M., Muertter, R.N., Holko, M., Ayanbule, O., Yefanov, A., Soboleva, A.: NCBI GEO: archive for functional genomics data sets–10 years on. Nucleic Acids Res. 39, D1005–D1010 (2010)
33. Hindman, B., Konwinski, A., Zaharia, M., Ghodsi, A., Joseph, A.D., Katz, R., Shenker, S., Stoica, I.: Mesos: a platform for fine-grained resource sharing in the data center. In: Proc.of the 8th USENIX Conference on Networked Systems Design and Implementation. USENIX Association (2011)

34. Pedersen, E., Willassen, N.P., Bongo, L.A.: Transparent incremental updates for Genomics Data Analysis Pipelines. In: an Mey, D., Alexander, M., Bientinesi, P., Cannataro, M., Clauss, C., Costan, A., Kecskemeti, G., Morin, C., Ricci, L., Sahuquillo, J., Schulz, M., Scarano, V., Scott, S.L., Weidendorfer, J. (eds.) Euro-Par 2013. LNCS, vol. 8374, pp. 311–320. Springer, Heidelberg (2014)

35. Pedersen, E., Raknes, I.A., Ernstsen, M., Bongo, L.A.: Integrating Data-Intensive Computing Systems with Biological Data Processing Frameworks. In: Euromicro Conference on Parallel, Distributed and Network-Based Processing (2015)

36. Magrane, M., Consortium, U.: UniProt Knowledgebase: a hub of integrated protein data. Database (Oxford). 2011, bar009 (2011)

37. Ernstsen, M., Kjærner-Semb, E., Willassen, N.P., Bongo, L.A.: Mario: Interactive tuning of biological analysis pipelines using iterative processing. In: Lopes, L., et al. (eds.) Euro-Par 2014, Part I. LNCS, vol. 8805, pp. 263–274. Springer, Heidelberg (2014)

38. Zaharia, M., Das, T., Li, H., Hunter, T., Shenker, S., Stoica, I.: Discretized streams. In: Proc. of Twenty-Fourth ACM Symposium on Operating Systems Principles, pp. 423–438. ACM Press (2013)

39. Altschul, S.F., Gish, W., Miller, W., Myers, E.W., Lipman, D.J.: Basic local alignment search tool. J. Mol. Biol. 215, 403–410 (1990)

40. Killcoyne, S., del Sol, A.: FIGG: simulating populations of whole genome sequences for heterogeneous data analyses. BMC Bioinformatics 15, 149 (2014)

41. Azure: Microsoft's Cloud Platform, http://azure.microsoft.com/en-us/

42. O'Connor, B.D., Merriman, B., Nelson, S.F.: SeqWare Query Engine: storing and searching sequence data in the cloud. BMC Bioinformatics 11(Suppl. 1), S2 (2010)

43. Roberts, A., Feng, H., Pachter, L.: Fragment assignment in the cloud with eXpress-D. BMC Bioinformatics 14, 358 (2013)

44. Dean, J., Ghemawat, S.: MapReduce: simplified data processing on large clusters. In: Proc. of Operating Systems Design & Implementation. USENIX (2004)

A Fine-Grained CUDA Implementation of the Multi-objective Evolutionary Approach NSGA-II: Potential Impact for Computational and Systems Biology Applications

Daniele D'Agostino[1], Giulia Pasquale[1], and Ivan Merelli[2]

[1] Institute of Applied Mathematics and Information Technologies - National Research Council of Italy, Via de Marini 6, 16149 Genova, Italy
{dagostino,pasquale}@ge.imati.cnr.it
[2] Institute for Biomedical Technologies - National Research Council of Italy, Via F.lli Cervi 93, 20090 Segrate (Mi), Italy
ivan.merelli@itb.cnr.it

Abstract. Many computational and systems biology challenges, in particular those related to big data analysis, can be formulated as optimization problems and therefore can be addressed using heuristics. Beside the typical optimization problems, formulated with respect to a single target, the possibility of optimizing multiple objectives (MO) is rapidly becoming more appealing. In this context, MO Evolutionary Algorithms (MOEAs) are one of the most widely used classes of methods to solve MO optimization problems. However, these methods can be particularly demanding from the computational point of view and, therefore, effective parallel implementations are needed. This fact, together with the wide diffusion of powerful and low-cost general-purpose Graphics Processing Units, promoted the development of software tools that focus on the parallelization of one or more computational phases among the steps characterizing MOEAs. In this paper we present a fine-grained parallelization of the Fast Non-dominating Sorting Genetic Algorithm (NSGA-II) for the CUDA architecture. In particular, we will discuss how this solution can be exploited to solve multi-objective optimization task in the field of computational and systems biology.

Keywords: Graphics Processing Units, CUDA-accelerated architecture, Multi-objective optimization, Computational and Systems Biology.

1 Introduction

Several computational and systems biology tasks, such as feature classification, gene clustering, tag SNP selection, pathway analysis, sequence alignment, structure prediction, sub-network enrichment and protein-protein interaction analysis can be faced with multi-objective (MO) optimization approaches. In general, almost all optimization problems are multi-objective and the objectives under consideration typically conflict with each other, i.e., fully optimizing a solution

© Springer International Publishing Switzerland 2015
C. di Serio et al. (Eds.): CIBB 2014, LNCS 8623, pp. 273–284, 2015.
DOI: 10.1007/978-3-319-24462-4_23

with respect to a particular objective leads to unacceptable results with respect to all the other objectives. A reasonable method to tackle multi-objective optimization problems is (i) to investigate a set of solutions, each of which satisfies all objectives at an acceptable level without being dominated by any other solution, and then (ii) to choose from this set the "best one" incorporating in the optimization process high level information related to the particular situation considered.

Over the last decades, the application of Genetic (or Evolutionary) Algorithms (GAs or EAs) to find the set of non-dominated solutions (known as Pareto optimal solutions) in multi-objective optimization problems has been largely investigated [1]. Indeed, GAs [2] are suited to solve this class of problems because their population based approach can lead, in a unique run, to an ensemble of solutions - the Pareto optimal set - which can be later investigated for trade-offs.

The Pareto optimal set in fact is composed by solutions that are all non-dominated with respect to each other: this means that while moving from one Pareto solution to another, there is always a certain amount of sacrifice in one or more objectives to achieve a certain amount of gain in the other(s). Pareto optimal solution sets are preferred to single solutions because they can be practical in real-life problems as in computational and systems biology applications, where the final solution is always a trade-off.

Jones et al. [3] reported that 90% of the approaches to multi-objective (MO) optimization aim to approximate the true Pareto front for the underlying problem. A majority of these use a meta-heuristic technique, and 70% of all meta-heuristics approaches are based on evolutionary approaches. Here the attention is focused on the Fast Non-dominated Sorting Genetic Algorithm [4], called NSGA-II because it is an improvement of one of the first proposed MO Evolutionary Algorithms (MOEAs), the NSGA algorithm, presented in [5].

In particular, we focus on the real-coded NSGA-II, i.e. the case in which the individual solutions in the population are represented by a vector of real variables. The extension to the binary-coded algorithm is straightforward and implies only the addiction of two coding-decoding functions to pass from the binary to the real representation and vice-versa.

The paper is structured as follows. Section 2 provides a description of the sequential NSGA-II algorithm, followed by Related Works in Section 3. Section 4 outlines our implementation in CUDA, whose performance are presented in Section 5. Section 6 concludes.

2   The NSGA-II Algorithm

The structure of this algorithm is similar to other MOEAs: in the schema, the general steps of such kind of algorithms are drawn and, for each of them, the particular operations executed by the NSGA-II algorithm are indicated and are described in detail in the following.

- *Initialization*: creation of the first population P_1 extracting a random set of N_S candidate solutions: $P_1 = x_i$ for $i = 1, \ldots, N_S$.

- *Objective evaluation* ($\mathbf{P_1}$): the K objective functions $\mathbf{f}(\mathbf{x_i}) = \mathbf{f_i}$ are computed for each $i = 1, \ldots, N_S$ in the $\mathbf{P_1}$ population.
- *Fitness assignment* ($\mathbf{P_1}$): assign a fitness value to each solution ($\mathbf{x_i}$) in the $\mathbf{P_1}$ population based on its objective function value f_i and other feasibility criteria. This is the first distinctive step characterizing the NSGA-II algorithm, where this operation coincides with the *non-dominated sort*. Moreover, in NSGA-II, in addition to fitness value, a second parameter called *crowding distance* is computed for each individual: this is a measure of how close an individual is to its neighbours and is used as a secondary selection criterion to favourite the choice of more diverse solutions among those with the same fitness value (i.e., in the same front). This is the second distinctive step of NSGA-II. Both of them are described below in more details.
- **for** ($t = 1$ to N_{gen})
  - *Selection*: randomly select N_S solutions from the $\mathbf{P_t}$ population by using a binary tournament selection (N_S random couples of solutions are extracted and then one individual for each couple is chosen) with a comparison operator based on the fitness value and, in NSGA-II, in case of same fitness value, on the computed crowding distance.
  - *Crossover (genetic operator)*: combine the N_S selected solutions through a crossover operator to generate the $\mathbf{Q_t}$ population; NSGA-II uses the Simulated Binary Crossover (SBX) operator [6].
  - *Mutation*: apply a random mutation to each of the solutions in the $\mathbf{Q_t}$ population; NSGA-II uses polynomial mutation [6].
  - *Objective evaluation* ($\mathbf{Q_t}$): the K objective functions $\mathbf{f}(\mathbf{x_i}) = \mathbf{f_i}$ are computed for each $i = 1, \ldots, N_S$ in the $\mathbf{Q_t}$ population.
  - *Merge*: $\mathbf{P_t}$ and $\mathbf{Q_t}$ populations are merged into a unique $\mathbf{R_t}$ population of size $2 * N_S$.
  - *Fitness assignment* $\mathbf{R_t}$: assign each solution $\mathbf{x_i}$ in the $\mathbf{R_t}$ population to a front, based on the non-dominated sorting method described below. Compute the crowding distance for each individual.
  - *Filling*: select N_S solutions from the $2 * N_S$ in the R_t population to generate the P_{t+1} population; in NSGA-II the new generation is filled by keeping whole fronts starting from the first one until the front which would make the population size exceed is reached; at this point the individuals in the front are sorted in decreasing crowding distance order and the missing individuals that fulfil the P_{t+1} population are kept starting from the first one in the front.

Non-dominated Sort

A solution A dominates another solution B if and only if $f_i(A) \leq f_i(B)$ for all the K objective functions and $f_j(A) < f_j(B)$ for at least one of the objective functions (i.e. $i, j \in \{1, \ldots, K\}$). With this definition the Non-dominated sort subdivides the solutions into fronts. The first front is made up by the individuals which are not dominated by any other in the population.

Then, the subsequent fronts are computed in an iterative loop where, at each iteration, the individuals belonging to the last determined front are subtracted from the domination count of the individuals dominated by them; in this way the next front will be composed by those individuals for which the domination count becomes null, i.e., those individuals which are not dominated by any other in the population except from those which have already been assigned to superior fronts.

The algorithm used for this operation is the fast non-dominated sort described in [4], that reduces the computational complexity of the original NSGA algorithm from $O(KN_S^3)$ to $O(KN_S^2)$.

Crowding Distance Computation
Crowding distance computation is a parameter-less diversity preservation method introduced with the NSGA-II algorithm to ensure diversity in the population of solutions. It is an intra-front measure, and the basic idea is to find a measure of the objective space around an individual which is not occupied by any other solution in the same front.

3   Related Works

GPU-based MOEAs have been introduced to multi-objective optimization on regression testing in [7], where NSGA-II is implemented in OpenCL to solve a multi-objective test suite minimization problem. In this study, which considers only binary representation, the objective computation is turned into a matrix multiplication that can be implemented on GPU easily and more efficiently.

The algorithm proposed in [8] focuses instead on test case prioritization, a problem for which binary representation is not suitable. In this implementation framework on CPU+GPU architecture, the CPU manages the NSGA-II evolution process, except objective evaluation and crossover which are parallelized on the GPU.

The NSGA-II implementation proposed in the present paper is based on the works presented in [9] and [10].

In [9] a parallel MOEA based on the CUDA platform is presented. In particular, the CUDA implementation is limited to the objective evaluation and the crossover operation, while our goal was to extend this approach to move most of the algorithm on the GPU, i.e., also the selection, mutation and part of the fitness assignment operation (the dominance checking). Also [10] presents an interesting CUDA parallelization of a binary and real-coded NSGA-II of the selection, crossover and mutation operators, that was considered in our work.

The main aspect to consider in fact, according to the evaluation discussed in [9] and confirmed by our experiments, is that more than 95% of the execution time of the NSGA-II algorithm is spent in performing the two procedures of dominance checking and non-dominated sort, which are very time consuming. In particular sorting solutions based on the dominance covers more than 90% of the total computation time and is the hardest operation to parallelize, because of its iterative nature. Therefore, the key role that a suitable parallelization of this step could play in the overall improvement of the performance of the algorithm results obvious and this is our goal. At the best of our knowledge the present work is the first result presenting a full parallelization of the NSGA-II algorithm for the CUDA platform.

4   CUDA Implementation of NSGA-II

Being this algorithm a milestone in the field of evolutionary multi-objective optimization, several implementations are available in different languages and development environments, as for example in Matlab [11].

The C implementation of the multi-objective NSGA-II provided by the Kanpur Genetic Algorithms Laboratory (KanGAL) at Indian Institute of Technology, Kanpur (KanGAL Group)[1] has been used as the starting point for the parallelization proposed in this work.

We decided to keep the main loop of the program, which iterates over the generations, on the CPU and to offload computations on the GPU launching one (or more) kernels for each step involved in the iteration. This choice is justified by the fundamental limit that it is not possible to create synchronization barriers among different thread blocks, thus each time a synchronization is required between threads in different blocks the computation must be split into different kernels.

Listing 1.1. The pseudocode of the CUDA implementation of NSGA-II

```
set_parameters (...);

malloc_host(Pt_host, popsize, Rt_host, 2*popsize);
malloc_dev(Pt_dev, popsize, Rt_dev, 2*popsize);

upload(Pt_dev, Pt_host, popsize, Rt_dev, Rt_host, 2*popsize);
initialize(Pt_dev, popsize);
evaluate(Pt_dev, popsize, 0, popsize);
non_dominated_sort(Pt_dev, Pt_host, popsize, popsize);

for (t=2 to max_gen) {
    selection_and_crossover(Pt_dev, popsize, Rt_dev, 2*popsize);
    mutation(Rt_dev, 2*popsize, popsize, popsize);
    evaluate(Rt_dev, 2*popsize, popsize, popsize);
    merge(Rt_dev, 2*popsize, 0, Pt_dev, popsize);
    non_dominated_sort(Rt_dev, Rt_host, 2*popsize, popsize);
    fill(Pt_dev, popsize, Rt_dev, Rt_host, 2*popsize);
}

download(Pt_host, Pt_dev, popsize, Rt_host, Rt_dev, 2*popsize);

free_host(Pt_host, Rt_host);
free_dev(Pt_dev, Rt_dev);
```

The pseudocode of our implementation is presented in Listing 1.1. As regards the GPU implementation of the operations (i.e. *initialize, evaluate, mutation, merge, non_dominated_sort, selection_and_crossover* and *fill*), we preferred to exploit the cuRAND, cuBLAS, Thrust and Nvidia Performance Primitives

---

[1] http://www.iitk.ac.in/kangal/codes.shtml

(NPP) libraries when possible and convenient. Otherwise, custom kernels were developed. For the sake of brevity we present only the pseudocode of the *non_dominated_sort* in Listing 1.2.

Listing 1.2. The pseudo-code of the CUDA implementation of the functions for Non-dominated sort and crowding distance computation

```
non_dominated_sort(pop, pop_host, popsize, num_ind) {
    /* S_dev and n_dev computation not shown */
    /* copy objectives on the host for crowding distance comp */
    /* initialize crowding distance to 0 */
    /* 1st front computed
        (i) thresholding n_dev to find zeros
        (ii) copying their indices in felements
            and their count in fsize */

    assign_crowd_dist(pop_host, popsize, pop->fcounter);

    while(fsize[pop->fcounter]>0 &&
          foffset[pop->fcounter+1]<num_ind) {
        calc_next_front <<<grid,block>>>(n_dev, S_dev, popsize,
            felements_dev, fsize_dev, foffset_dev, pop->fcounter);
        assign_crowd_dist(pop_host, popsize, pop->fcounter-1);
        cudaDeviceSynchronize();
    }
    /* copy back the computed crowding distance on the device */
}

calc_next_front(n, S, size, felements, fsize, foffset, front) {
    S_loc, n_loc = n[tid]; // tid=thread's id
    index, new_front=0;

    for( k=0 to fsize[front-1] ) {
        S_loc = S[felements[foffset[front-1]+k]*size+tid];
        n_loc -= S_loc;
        new_front += ( (n_loc==0) * (S_loc==1) );
    }

    if (new_front) {
        index = atomicAdd(fsize+front, 1);
        felements[foffset[front]+index] = tid;
    }
    n[tid] = n_loc;
}
```

In Listing 1.1, the variables *Pt_dev, Pt_host, and Rt_dev, Rt_host* are, respectively, the parent and mixed (parent + child) populations on the device and on the host, while *popsize* is the number of solutions composing the population. The *set_parameters* function sets the problem and simulation parameters. The *malloc_host/free_host* and *malloc_dev/free_dev* functions allocate/free the data structures containing the populations and the auxiliary arrays respectively on the device global memory and on the host. The *upload/download* functions copy such data structures respectively from the host to the device memory and vice-versa.

It is worth noting that the NSGA-II sequential implementation is based on list-like data structures that, in order to efficiently port the algorithm on a CUDA architecture, was converted from arrays-of-structs to structs-of-arrays, avoiding the use of lists.

As said before, the most interesting function is the *non_dominated_sort*, because in literature we did not find previous attempts to parallelize the ranking of non-dominated solutions through the fast non-dominated sorting procedure. This is the core of NSGA-II and, although it is the hardest to parallelize because of its iterative nature, it occupies alone more than 90% of the total computational time needed by the algorithm. Three are the most important parts:

1. Dominance checking with computation of the S matrix: $S_{ij} = 1$ if j dominates i, 0 otherwise. The n array contains the sum of the rows of $S : n_j$ is the number of elements dominating j.
2. first front identification, i.e. the elements for which $n == 0$.
3. other fronts identification. It is an iterative cycle: at each iteration, a kernel on the device determines the next front based on the current one while a function on the host computes the crowding distance among its elements. A page-locked mapped memory is used.

5   Results

There are many problems in the field of computational and systems biology that can be fruitfully addressed using multi-objective optimization. In particular, we identified three examples, tag single nucleotide polymorphisms (SNPs) selection, sub-network enrichment and protein-protein interaction that have concerns about multiple objectives to achieve and that create sets of Pareto-optimal solutions.

For example, while performing SNP genotyping, the extent of linkage disequilibrium is a critical factor to consider for optimizing the coverage of the analysis. A subset of SNPs (called tag SNPs), indeed, is sufficient for capturing alleles of bi-allelic and even multi-allelic variants. Selecting tag SNPs can be formulated as a multi-objective optimization problem that minimizes the total amount of tag SNPs, maximizes tolerance for missing data, enlarges and balances the detection power of each allele class. The use of NSGA-II to solve this problem has been demonstrated to grant enough flexibility to extract different sets of tag SNPs for different platforms and scenarios [12]. Compared to conventional methods, this method explores larger search space and requires shorter convergence time. In particular, a small number of additional tag SNPs can provide sufficient tolerance and balanced power given the low missing and error rates of today's genotyping platforms.

Multi-objective optimization can be successfully applied also to systems biology for the analysis of network, in particular when omics data are mapped on graphs to enable their integration. A well-established approach to data integration, in fact, is to exploit gene/protein networks to map multi-omic information, allowing their statistical analysis relying on pathways. Nonetheless, few approaches have been recently described in literature and a small number of software packages are available. An example of this kind of approach is the work of Mosca et al. [13], which exploits multi-objective optimization procedure in order to drive the identification of sub-networks that are enriched according to several statistical estimators.

The network-based analysis of omic data with MO optimization described in [13] can be applied to different types of experimental designs and biological interactions. For example, this approach can be used to perform network-based comparisons among several data sets: using transcriptomic data from many types of tumours its possible to identify strong and coherent differential expressed pathways. On the other hand, it is possible to identify sub-networks of genes that accumulate the highest number of mutations/variations in cancer cell line. The possibility of exploiting NSGA-II to optimize the selection of vertices with specific omic peculiarities allows the identification of sub-networks that play significant biological processes in tumours.

The last example we present concerning the use of NSGA-II to predict protein contact maps. The algorithm proposed by Marquez et al. [14] provides a set of rules to infer whether there is contact between a pair of residues or not. Such rules are based on a set of specific amino acid properties. These properties determine the particular features of each amino acid represented in the rules. Considering the capability of the algorithm to explore the space using a MOEA approach, this method shows better accuracy and coverage rates than other contact map predictor algorithms.

Simulations on benchmark multi-objective optimization problems have been executed, to provide a fair comparison of results. In particular, we considered the benchmarks described in [4]: a single real variable, called SCH1 and presented in Equation 1, and a multiple real variables problem called KUR and described in Equation 2.

$$f_1(x) = x^2$$
$$f_2(x) = (x - 2)^2 \tag{1}$$

$$f_1(x_1, x_2, x_3) = -10 \sum_{i=1}^{3} e^{-0.2\sqrt{x_i^2 + x_{i+1}^2}}$$
$$f_2(x_1, x_2, x_3) = \sum_{i=1}^{3} (\left| x_i^{0.8} \right| + 5\sin(x_i^3)) \tag{2}$$

We compared preliminary results obtained with our CUDA implementation to that provided by using the NSGA-II C implementation published on the KanGAL website.

As regards the qualitative assessment of results, we show the plots of the Pareto-front to which our CUDA implementation converged in order to compare them to the Pareto-front provided by the KanGAL implementation for both the problems.

In the SCH1 case we can see that our results (Figure 1) are fully coincident not only with those of the equivalent sequential implementation (not shown) but also with those of the KanGAL implementation (Figure 2). The convergence rate is maintained and the two implementations behave exactly in the same way.

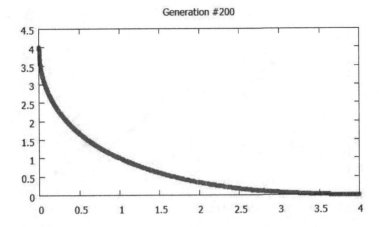

Fig. 1. Final Pareto set of optimal solutions of the SCH1 problem, where our CUDA implementation converged in 200 generations. The horizontal axis reports the first objective function values, the vertical axis reports the second objective function values

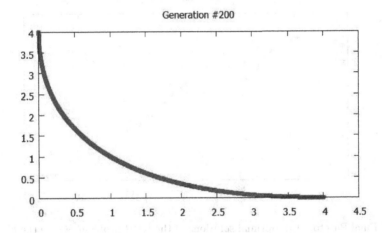

Fig. 2. Final Pareto set of optimal solutions of the SCH1 problem, where the KanGAL implementation converged in 200 generations. The same axes as Figure 1 are maintained

The KUR problem is the second considered: it results clear, looking at Figures 3 and 4, that our implementation does not converge to a well distributed front as the KanGAL implementation. The causes of this behaviour are under investigation. The algorithm seems to converge to the correct Pareto set; the inhomogeneity should probably be attributed to the step of the algorithm at which the individuals in the mixed population are chosen by the *fill* procedure to form the next generation parent population.

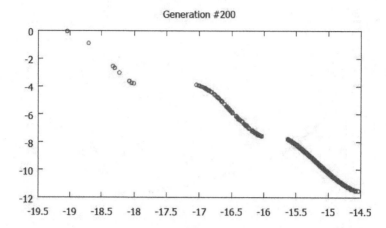

Fig. 3. Final Pareto set of optimal solutions of the KUR problem, where our CUDA implementation converged in 200 generations. The horizontal axis reports the first objective function values, the vertical axis reports the second objective function values

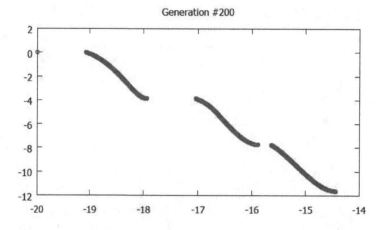

Fig. 4. Final Pareto set of optimal solutions of the KUR problem, where the KanGAL implementation converged in 200 generations. The same axes as Figure 3 are maintained

As regards the performance, they are shown in Table 1 for the SCH1 problem, because those of the KUR problem are very close. As said before, the KanGAL implementation is strictly sequential and not suited for parallelization, in particular because it uses (two) lists as data structures. The first list contains the elements in the population not yet ranked, while the second contains the current front being determined. Each element is picked from the first list and compared to all other elements and, if it is not dominated by anyone, it is pushed in the second list. When all elements have been compared, the second list constitutes the front.

Using a Nvidia GeForce GTX 580 device, our implementation provides speedups of about 7-10x against the KanGAL implementation running on an Intel Xeon E5645 CPU, which at the time of the purchase had almost the same price. It is worth noting that no CUDA streams are used in these runs and the computational resources of the graphics card (e.g. in terms of the number of cores per Streaming Multiprocessor) limit the full exploitation of the parallelism of the implementation.

An interesting result is that our parallelization compared with a sequential implementation of the fast non-dominated sorting procedure derived from the KanGAL one that uses arrays instead of lists presents speedups of the order of 130x.

Table 1. Execution times (in milliseconds) of the proposed CUDA implementation, compared to the NSGA-II implementation using lists or arrays. tot is the time of execution of a complete generation of the genetic algorithm, sort is the time of the non_dominated_sort function (see Listing 1.2) and other is the time of all other functions involved in a generation (see Listing 1.1).

Population	NSGA-II with lists			NSGA-II with array			NSGA-II in CUDA		
	tot	sort	other	tot	sort	other	tot	sort	other
1024	7.7	7.2	0.5	48.1	47.7	0.5	2.1	1.5	0.6
2048	29.6	28.6	1.0	558.0	557.0	1.0	7.3	6.5	0.8
4096	116.5	114.5	2.0	2340.5	2338.5	2.0	17.9	17.0	0.9

6   Conclusion

Many bioinformatics and systems biology challenges can be formulated as optimization problems and therefore can be addressed using heuristics. This is particularly true if we consider the possibility of optimizing multiple objectives.

In this paper we presented our effort on designing a parallel version of the NSGA-II algorithm for the CUDA architectures. This is not a trivial tasks, since the algorithm is made up by different computational steps and each one needs to be optimized in a customized way. Preliminary results shows interesting performance figures, considering that the present work represents a first attempt to parallelize the whole Fast Non-dominating Sorting Genetic Algorithm (NSGA-II) for the CUDA platform. Moreover, it is worth noting that, even if we have not yet considered constrained optimization problems, this implementation provides also the possibility of constraint handling.

Acknowledgments. This work has been supported by the Italian Ministry of Education and Research (MIUR) through the Flagship (PB05) InterOmics, HIRMA (RBAP11YS7K) and the European MIMOMICS projects.

References

1. Konak, A., Coit, A.W., Smith, A.E.: Multi-objective optimization using generic algorithms: A tutorial. Reliability Engineering and System Safety 91 (2006)
2. Whitley, D.: A Genetic Algorithm Tutorial. Statistics and Computing 4 (1994)
3. Jones, D.F., Mirrazavi, S.K., Tamiz, M.: Multi-objective meta-heuristics: an overview of the current state of the art. European Journal of Operational Research 137 (2002)
4. Deb, K., et al.: A Fast and Elitist Multiobjective Genetic Algorithm: NSGA-II. IEEE Transactions on Evolutionary Computation 6 (2002)
5. Srinivas, N., Deb, K.: Multiobjective Optimization Using Nondominated Sorting in Genetic Algorithms. Journal of Evolutionary Computation 2 (1994)
6. Deb, K., Agarwal, R.B.: Simulated Binary Crossover for Continuous Search Space. Complex Systems 9 (1995)
7. Yoo, S., Harman, M., Ur, S.: Highly scalable multi objective test suite minimisation using graphics cards. In: Cohen, M.B., Ó Cinnéide, M. (eds.) SSBSE 2011. LNCS, vol. 6956, pp. 219–236. Springer, Heidelberg (2011)
8. Li, Z., Bian, Y., Zhao, R., Cheng, J.: A fine-grained parallel multi-objective test case prioritization on GPU. In: Ruhe, G., Zhang, Y. (eds.) SSBSE 2013. LNCS, vol. 8084, pp. 111–125. Springer, Heidelberg (2013)
9. Wong, M.L.: Parallel multi-objective evolutionary algorithms on graphics processing units. In: Proceedings of the 11th Annual Conference Companion on Genetic and Evolutionary Computation Conference (GECCO 2009), pp. 2515–2522 (2009)
10. Arora, R., Tulshyan, R., Deb, K.: Parallelization of binary and real-coded genetic algorithms on GPU using CUDA. In: Proceedings of the 2010 IEEE Congress on Evolutionary Computation (CEC 2010), pp. 1–8 (2010)
11. Seshadri, A.: NSGA-II:A multi-objective optimization algorithm. MATLAB Central, http://www.mathworks.com/matlabcentral/fileexchange/10429-nsga-ii-a-multi-objective-optimization-algorithm
12. Ting, C.K., Lin, W.T., Huang, Y.T.: Multi-objective tag SNPs selection using evolutionary algorithms. Bioinformatics 26(11), 1446–1452 (2010)
13. Mosca, E., Milanesi, L.: Network-based analysis of omics with multi-objective optimization. Mol. BioSyst. 9, 2971–2980 (2013)
14. Márquez-Chamorro, A.E., Divina, F., Aguilar-Ruiz, J.S., Bacardit, J., Asencio-Cortés, G., Santiesteban-Toca, C.E.: A NSGA-II algorithm for the residue-residue contact prediction. In: Giacobini, M., Vanneschi, L., Bush, W.S. (eds.) EvoBIO 2012. LNCS, vol. 7246, pp. 234–244. Springer, Heidelberg (2012)

GPGPU Implementation of a Spiking Neuronal Circuit Performing Sparse Recoding

Manjusha Nair[1,2], Bipin Nair[1], and Shyam Diwakar[1]

[1] Amrita School of Biotechnology, Amrita Vishwa Vidyapeetham, Amrita University,
Amritapuri, Kollam, Kerala, India
[2] Amrita School of Engineering, Amrita Vishwa Vidyapeetham, Amrita University,
Amritapuri, Kollam, Kerala, India
manjushanair@am.amrita.edu, shyam@amrita.edu

Abstract. Modeling and simulation techniques have been used extensively to study the complexities of brain circuits. Simulations of bio-realistic networks consisting of large number of neurons require massive computational power when they are designed to provide real-time responses in millisecond scale. A network model of cerebellar granular layer was developed and simulated here on Graphic Processing Units (GPU) which delivered a high compute capacity at low cost. We used a mathematical model namely, Adaptive Exponential leaky integrate-and-fire (AdEx) equations to model the different types of neurons in the cerebellum. The hypothesis relating spatiotemporal information processing in the input layer of the cerebellum and its relations to sparse activation of cell clusters was evaluated. The main goal of this paper was to understand the computational efficiency and scalability issues while implementing a large-scale microcircuit consisting of millions of neurons and synapses. The results suggest efficient scale-up based on pleasantly parallel modes of operations allows simulations of large-scale spiking network models for cerebellum-like network circuits.

Keywords: Graphic Processing Units, cerebellum, computational Neuroscience, neuron, synapse, adaptive Exponential Leaky Integrate and Fire Model.

1   Introduction

Computational modeling allows us to investigate behavior of the neurons and to frame or test hypothesis about their operations. Neural modeling at the level of ion-channel kinetics using Hodgkin-Huxley models had been useful in characterizing single neuron behavior. It is, however, computationally intensive to model large networks of neurons due to the number of simultaneous differential equations that must be evaluated and due to the abundant system parameters that need to be specified for the neuron being modeled. In order to perform information theoretic analysis of the spike responses, massive amount of data from simulations involving thousands of neurons are required. Well-known simulators such as NEURON[1]and GENESIS[2] are used widely for detailed biophysical simulations of single neurons or for the simulations of small

© Springer International Publishing Switzerland 2015
C. di Serio et al. (Eds.): CIBB 2014, LNCS 8623, pp. 285–297, 2015.
DOI: 10.1007/978-3-319-24462-4_24

network of neurons. Due to the computational overhead of the complex neuronal dynamics addressed by them, they fail to perform large-scale simulations in the timescale of the real network of brain. Hence they have been extended to support distributed simulations of biologically realistic network models[3,4]and are run in multi-CPU environments like multi-core processors and Beowulf clusters. Simulation with spiking neurons gained prominent importance in the computational neuroscience community to study the neuronal dynamics of large-scale microcircuits. Spiking Neural Network simulators like NEST[5] and SpikeNET[6]have also followed the same trajectory but used different computational models. Most of these simulators support networks of realistic connectivity employing multi-threading and message-passing interfaces on clusters of computers. With the newer multi-core processors designed as numeric computing engines and with their general purpose programming interfaces, recent computer hardware have shown significant efficiency improvements. Highly parallel programmable processors like GPUs deliver a high compute capacity at low cost. GPUs are enhanced with a greater arithmetic capability, streaming memory bandwidth and with a richer set of APIs. GPUs provide computing power that is easily and cheaply accessible to individuals who cannot afford clusters and supercomputers. Simple spiking neural network models such as Integrate and fire models without bio-realistic features were simulated in the older generation GPUs [7]. Another study focused on simulating a large scale Izhikevich-based realistic spiking network models having 10^5 neurons and 10^7 synaptic connections on GPUs[8]. More recent works proved the potential power of GPGPU techniques in real-time simulation of the different regions of the brain such as basal ganglia circuitry[9]and cerebellum[10]. The studies have demonstrated the use of GPU for neural network simulations.

We constructed a bio-realistic spiking network of neurons of the cerebellar granular layer. Since cerebellum contains more than half of the total population of the neurons in the entire brain, the implications of large-scale simulation become pertinent. Due to the 'embarrassingly' parallel architecture of different layers of neurons in the cerebellum and due to the modular connection geometry between them, we chose this model as a candidate for parallel simulation. Cerebellum is known to be involved in timing and in controlling the ordered and precise execution of motor sequences [11]. Input layer of the cerebellum has been studied for combinatorial operations. Due to its role in motor articulation control, cerebellar modeling is a main area of focus for many real-time robotic applications. Cerebellar granular layer consists of a large number of neurons that receive information from mossy fibers and spatially encode information which then converges onto Purkinje neurons via parallel fibers. The abundance, fast response time and modular architecture of the granular layer neurons of the cerebellum offered an opportunity as well as a challenge to this modeling process. A simple spiking neuron model in NEURON was tuned previously to predict how spikes were processed in the cerebellar granular layer network [12]. Biophysically realistic models of granule cells [11],[13,14] and Golgi cells [15] are available to map and test the known behaviors of granular layer neurons. Our goal was to understand and implement feasible fast models on GPUs. In this paper, individual granule neurons, Golgi neurons and their excitatory-inhibitory synapses were modeled.

A realistic large-scale model of the cerebellum granular layer [16] was reconstructed. The network was simulated on a Tesla K20C GPU with 2496 cores and SM 3.5 support running at 0.71 GHz. This study aimed to model and analyze a cerebellar network of granular layer neurons on GPUs in order to study the computational relevance of such implementations and to understand the role of parallelism in spatio-temporal encoding in the input layer of the cerebellum and.

2   Materials and Methods

2.1   Single Neuron and Synapse Modeling

Membrane and synaptic properties of two types of neurons in the granular layer were modeled using phenomenological models. Single neurons were modeled using adaptive exponential leaky integrate and fire model, a two-dimensional integrate and-fire model that combines an exponential spike mechanism with an adaptation equation, which was able to correctly predict timing of 96% of the spikes (±2 ms) and closely reconstructed the behavior as seen in a detailed conductance-based model [17]. The equations (1) and (2) of the model were able to generate different firing patterns and were used to simulate firing dynamics for single neurons in the network simulations [18].

$$\frac{dV}{dt} = \frac{-gl*(V-El)+gl*delT*\exp\left(\frac{V-Vt}{delT}\right)+Isyn-w}{C} \tag{1}$$

$$\tau w \frac{dw}{dt} = a * (V - El) - w \tag{2}$$

$$If\ V > 0\ mv,\ V = Vr\ \&\ w = w + b$$

The AdEx model is an extended integrate-and-fire model where the passive properties of the neuron and the action potential mechanisms are combined with the adaptation variable w. In the model, V represents the membrane voltage, C is the membrane capacitance, gl is the leak conductance, El is the resting potential, $delT$ is the slope factor, Vt is the threshold potential, Vr is the reset potential, $Isyn$ is the synaptic current, τw is the time constant, a is the level of sub-threshold adaptation and b represents the spike triggered adaptation. The first equation describes the dynamics of the action potential generation while the second equation describes adaptation in the firing rate of the neuron. The equation followed the dynamics of an RC circuit until V reaches Vt. The neuron fires on crossing this threshold voltage and the downswing of the action potential was replaced by a reset of membrane potential V to a lower value, Vr.

Granule cells in the cerebellum receive on an average 1 to 4 excitatory connections via mossy fiber (MF) inputs and 0 to 4 inhibitory inputs through Golgi cell synapses [14]. Variation in number of synaptic inputs affects spike responses in neurons. Bringing these synaptic behaviors [19] to artificial spiking neurons was essential in understanding the various network dynamics. Excitatory synapses were modeled using AMPA receptor dynamics and inhibitory synapses were modeled using GABA receptor dynamics [20] as indicated in equations (3) and (4).

$$g_{AMPA} = \frac{g_{AMPAMax} \times e^{\frac{-t}{18}} \times \left(1 - e^{\frac{-t}{2.2}}\right)}{0.68} \tag{3}$$

$$I_{AMPA} = (V_m - E_{AMPA}) \times g_{AMPA}$$

$$g_{GABA} = \frac{g_{GABAMax} \times e^{\frac{-t}{25}} \times \left(1 - e^{\frac{-t}{1.0}}\right)}{0.84} \tag{4}$$

$$I_{GABA} = (V_m - E_{GABA}) \times g_{GABA}$$

The synaptic currents I_{AMPA} and I_{GABA} were modeled via ohmic conductance g_{AMPA} and g_{GABA} multiplied by the difference between the membrane potential V_m and the reversal potential of the synapses, which is represented by E_{AMPA} for AMPA synapses and E_{GABA} for GABA synapses respectively. The maximal conductance $g_{AMPAMax}$ and $g_{GABAMax}$ were adjusted to match the number of spikes experimentally [14].

2.2   Glomerular Organisation of the Granular Layer Network

Mossy fibers provide excitatory inputs to the granule cells through glutamatergic synapses and Golgi cell axons inhibit granule cell firing through GABAergic synapses. These synapses are located inside a glomerular structure [21]. Specific connection geometry exist between mossy fibers, granule cells and Golgi cells in the granular layer [22]. In this model, a 3D volume of the granular layer with 100 μ m edge length contained granule cells with density 4 x 10^6 / mm^3 (Fig. 1). Each granule neuron model received one to four excitatory connections from mossy fibers synapses and one to four inhibitory connections through Golgi cells [23]. The number of glomeruli were estimated using the convergence-divergence ratio of the mossy fiber-granule cell connections [24]. Each glomerulus received a mean of 53 dendrites from different granule cells and each granule cell had an average of 4 dendrites, each dendrites not extending to glomeruli farther than 40 μm.

Approximately, 2000 granule cells were inhibited by a Golgi cell. The length of the different axes of the cube was varied to incorporate increased or decreased volume of the 3D space while maintaining the geometric properties and convergence-divergence ratio.

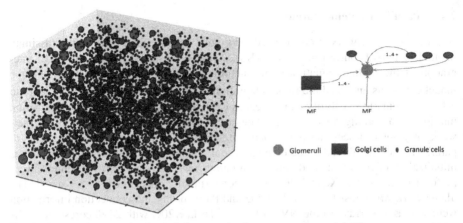

Fig. 1. Glomerular organization of the granular layer network model. Neurons were reconstructed within a spatial cube containing granule cell density 4 x 10^6 / mm ^{3} (LHS). Connectivity between mossy fibers, granule cell dendrites and Golgi cell axon are as indicated (RHS).

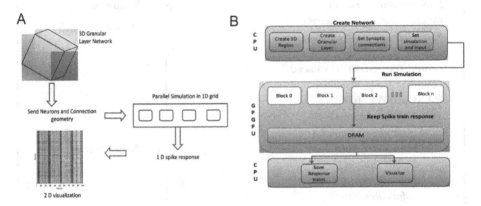

Fig. 2. GPU re-implementation of the granular layer network.(A) The Granular layer was modeled as a 3D cube with the volume occupied by the granule cells, Golgi cells and mossy fibers following precise connection rules. The connection geometry and neuron parameters were sent to the GPU for parallel simulation. The spike responses were collected for 2D visualization. (B) CPU and GPU task subunits for the simulator.

2.3   Simulation Background

The simulations were performed by activating a specific set of glomeruli or by activating entire glomeruli contained in the 3D space. Spike input of frequency 100 Hz were applied to the mossy fiber granule cell relay and the simulations were performed for different time windows ranging from 100 ms to 3 sec. Both *in vitro* like (1 spike per burst) and *in vivo* like (5 spike per burst) inputs were applied. The number of excitatory and inhibitory inputs to the granule cell was calculated at runtime depending on the connection rules and dynamics of the network.

2.4   Parallel Implementation

The cerebellar network reconstructed was both homogeneous and embarrassingly parallel. Standard fourth-order Runge-Kutta method was used for the numerical integration of the voltage equation of the neuron model. All neurons shared the same model equations and used the same integration steps for the computations. In order to achieve automatic scalability and increased efficiency, we adopted a single instruction multiple data paradigm for the simultaneous execution of different parts of the network on graphic processing units. Data-parallel processing mapped data elements to parallel processing threads. The essential serial components of the simulation such as initialization of the inputs and simulation parameters, network construction etc. were performed on an Intel Xeon CPU with 8 cores running on 2.6 GHz clock speed. The 3D network was constructed in the CPU and the neurons and connection information were sent as 1D arrays to the NVIDIA GPU: Tesla K20C with 2496 cores and 5 GB of DRAM. The parallelization can be summarized as follows.

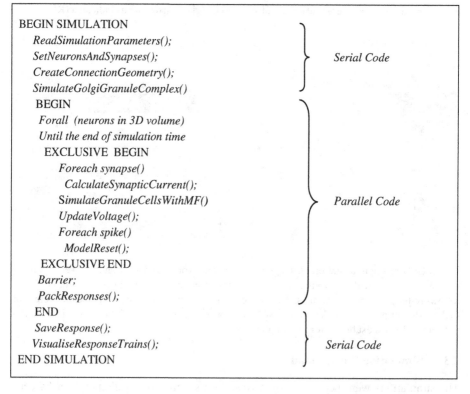

The network of granule cells with 1 to 4 random MF connections with and without inhibition from Golgi cells were then parallel simulated in the GPU (Fig. 2: A & B). The spike responses of the cells were copied back to the CPU and were visualized as 2D raster plots. Each neuron was mapped to one thread of execution in the parallel GPU blocks and both thread level and block level parallelism were explored. The memory requirement for CPU and GPU processes were calculated at runtime and the number of thread blocks were allocated in a scalable manner.

In a traditional CPU, if T time units were required to process each neuron, the time complexity of the entire simulation became directly proportional to M * N * T where M is the number of time steps and N is the number of neurons. In the GPU implementation, for P threads running in parallel where P==N, the total computations for N neurons takes only T time units and hence the total time required became directly proportional to M, the number of time steps, which was a constant. The time complexity of the algorithm was significantly reduced, when executed in parallel.

3   Results and Discussion

3.1   Single Neuron Simulations and Neuronal Firing Dynamics

The adaptive exponential leaky-integrate and fire model (AdEx) generated firing patterns depending on the parameters of the model equations [18]. The scaling and bifurcation parameters of the spiking neuron model were fine-tuned to match the electrophysiological recordings of the granule and Golgi neurons using current clamp protocol. Basic electro-responsiveness properties of both granule neurons and Golgi neurons (Fig. 3A & B) showed the increased firing rate and decreased first spike latency progressively with the injected current. Golgi cells showed spontaneous pace maker activity with a frequency between 1 and 8 Hz at room temperature [25] while granule cells showed no such spontaneous activity at rest but showed regular repetitive firing at current injection [26].

Dynamics of the synapses added significant computational overhead towards the overall time required for completing the simulation. The modeled granule cells contained 1 to 4 mossy fiber excitatory connections and 1 to 4 Golgi cell inhibitory connections. The synaptic dynamics with excitatory and inhibitory inputs were modeled using AMPA and GABA kinetics and the maximal conductance value was adjusted to suit the firing patterns [12] of granule cells during in-vitro (1 spike/burst) and in-vivo (5 spikes/burst) like inputs (Fig. 4: A &B). It was observed that two or more mossy fibers excitation was required to produce a spike output in granule cell. Also, increase in excitation increased the number of spikes while the increase in inhibition reduced the number of spikes. In order to apply tactile inputs which are seen essential for fine motor control, mossy fiber burst input was also applied to the cells (Fig. 4: C). Simulations allowed comparing the effect of these different types of inputs to the network.

For fidelity analysis, single neuron simulations were performed on CPUs and on GPUs and the firing patterns and the frequency of responses were compared. Even though both simulations produced similar numerical reconstructions, single neuron simulations took more time in GPU than in CPU. CPU simulations took 163.21 ms for a single granule cell and 172.35 ms for a single Golgi cell while the GPU simulations took 741.24ms and 1466.51ms respectively with inputs run for a total of 1000ms duration. The result indicated that CPU simulation was approximately 6x faster while GPU simulation was near real-time speeds compared to the biological neuron. The difference in performance time for a single neuron was consequential because CPU is faster on a per-core basis. We presume the delay to arithmetic pipelines and to the need of concurrent threads to sufficiently utilize the parallelism capabilities of the GPU.

3.2   Center-Surround Excitation in the Granular Layer

The distributed processing and plasticity capabilities of the neural network have been known to be dependent on spatial organization [27]. It has been observed that the activation of mossy fiber bundles to granular layer happens on an average in a center-surround manner with decreasing excitation from the center to the periphery for burst stimulation during the sensory inputs [16]. The same activation pattern was also observed when mossy fibers were stimulated with an electrode at specific locations[28]. Center surround structure of the granular layer determines the geometry of activation of the overlying Purkinje neurons of the molecular layer[11]. Centre-surround hypothesis was tested by giving strong excitation in the centre of the granular layer volume and progressively less excitation moving to the periphery (Fig. 5). 5% of glomeruli at the center received 4 MF inputs, 30 % of the surrounding glomeruli received 3 MF inputs, 55% received 2 MF inputs and the remaining 10% in the outer layer received 1 MF input. Granule cell responses to *in vivo* inputs (burst of 5 spikes at 500 Hz) in the centre showed bundle of spikes with shorter delay while the spike bursts in the surround showed reduced spike rate and longer delay[12]. Neurons in the centre responded to spike bursts over a broader frequency region while varying the frequency of the inputs.

3.3   Parallel Network Simulation

A network of granule cells with 1 to 4 random MF connections with and without inhibition from Golgi cells were simulated for 1000 ms time with *in vivo* burst-like input. We considered instantaneous post-synaptic current and hence white Gaussian noise was added to the network. Simulating a volume containing 4096 granule cells and 27 Golgi cells on GPU for 1000 ms took 3492.76 ms to complete the computations and memory transfers while the same in a single CPU took 2534445.25 ms. The GPU simulation was found to be 3.4 times slower than biological neural circuits while the single CPU simulation was 2534 times slower than biological networks. The results indicated the advantage of using GPUs for simulations of large network of neurons. A raster plot of spike responses during the simulation is shown (Fig.6: A). Scalability of the network implementation was tested by increasing the volume of the cube and measuring the computational time required to complete the simulation. Both CPU and GPU time taken for the network was calculated to justify the use of GPUs for large network simulations (Fig. 6: C). 550000 neurons were simulated in GPU and the running time linearly increased with the problem size (Fig. 6: B).

Since each neuron in the granular layer processes the input independent of other neurons in the same layer, our embarrassingly parallel approach of computations took little communication of results between tasks. Hence, no special algorithms were needed to get a working solution. A single large volume of the granular layer was divided into many smaller volumes which are handled by different simultaneously executing blocks of the GPU. Each neuron was mapped to one thread of execution and the simulations of a network of granular layer neurons were performed with

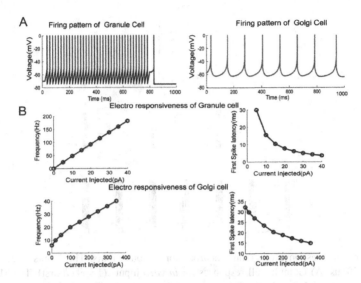

Fig. 3. Single cell electroresponsive properties of granule and Golgi cell. A) Firing patterns of granule and Golgi cell for 10 pA input current. B) The frequency of both granule and Golgi cell firing increased with the increase in current injection while the first spike latency is decreased.

N-parallel threads in a single block. Both block level and thread-level parallelisms were used for the simulations. For single neurons and network of size less than 1024, only thread-level parallelism was used. For networks of larger size, different concurrent blocks were launched. The thread assignment was a multiple of the warp size such that warp scheduling problem was avoided. GPU allowed automatic scalability with the increase in size of the granular layer network without modifications in the program.

Optimal GPU implementation not only depended on parallelization of the underlying algorithm or computations but also on memory optimizations and thread management[29].

Sufficient Parallelism: GPUs gave better performance improvements than CPUs only when sufficient parallelism was employed to hide the latency of the arithmetic pipelines. This was evident from the running time obtained while simulating single neurons on CPUs and GPUs. Since a single neuron was mapped to a single thread in the GPU simulation, the minimum network size was selected as 1024 to sufficiently exploit the parallelism in a single block of the GPU. This utilized the maximum number of threads allocated on a single block in a Tesla K20C GPU card.

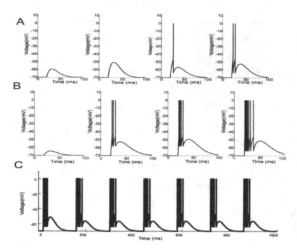

Fig. 4. Simulated response of granule neurons for various inputs. The inputs were provided starting at 20 ms. A) Granule cell responses for *in vitro* inputs (1 spike/burst). The MF inputs were increased from 1 to 4 in the figures from left to right. B) Granule cell responses for *in-vivo* inputs (5 spikes/burst, inter-spike interval 10 ms). The MF inputs were increased from 1 to 4 in the figures from left to right. C) Granule cell responses for tactile inputs (5 spikes/burst, inter-spike interval 10ms, inter-burst interval 100 ms).

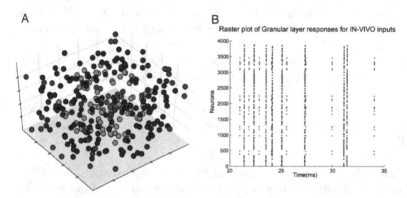

Fig. 5. Network simulation with center-surround excitation. Glomeruli were activated in a center-surround manner. A) The glomeruli in the center received four excitatory and the excitation is reduced progressively going to the periphery. B) The network was simulated with *in vivo* inputs (5 spikes /burst) and the spike responses were shown as the raster plot.

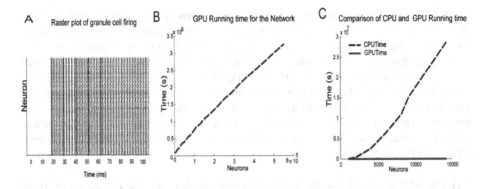

Fig. 6. A): Raster plot of granule cell firing patterns with white Gaussian noise in the network. *In vivo* like input (5 spikes per burst, inter spike interval 10 ms) was provided to the glomerulus through mossy fibers. B) The granular layer network was simulated with 550000 neurons with the granular layer volume increased from $125\mu m^3$ to $125mm^3$. The GPU runtime was found to be increasing linearly with the increase in network size. (C) A similar network was simulated entirely in a single CPU and the running time was compared. GPUs outperformed CPUs and took significantly lesser time to finish the simulation.

Minimizing Warp Divergence. Since all the neurons were simulated for almost same timescales and each neuron received on an average of 4 MF synapses, the wrap divergence and thread wait were not major issues in our current implementation.

Size of Thread Block and Occupancy: The kernel, *SimulateGranuleCellsWithMF()* which was ranked first for optimization based on execution time was chosen for performance improvement since the percentage of total GPU compute time spent executing instances of this kernel was found to be 70%. The register usage was limited by managing register spill-over using local memory (L1). LMEM is slower than registers. A 100% occupancy was not the aim of GPU optimizations since it slowed small network models although it improved runtime on large-scale network models.

4   Conclusion

We were able to reconstruct a cerebellum granular layer microcircuit in order to characterise the activity of a network of neurons sparsely activated by synaptic inputs in the rat cerebellum for the analysis of spatio-temporal geometry affecting signal activation in a central neural circuit. Even though we performed our simulations on a single high end GPU device, this study was a precursor for scaling up of the network model to include different layers of the cerebellum with greater number and more types of neurons to be simulated on clusters of GPUs. GPU based simulations may need to focus on lesser communications and pleasantly parallel or embarrassingly parallel schemes may be apt for large-scale neural simulations rather than fine-grained parallelization. Fixed time-step was more suited for such event-related simulations since variable time-step integration for several neurons caused performance decreases. Occupancy had to be pre-estimated as 100% occupancy was not favorable for small networks since it increased runtime. Sufficient parallelism was essential to compensate latency delays in arithmetic pipelines, which was observed for single neuron and small-sized network simulations.

Extending the current cerebellar model including the molecular and Purkinje layer neurons together with the presently modeled granular layer neurons will make the network a right candidate to explore the other forms of parallelism with dependent computations. As a work in progress, we have started investigating a large-scaled network model on multi-GPU machines. Further studies may be necessary to understand the inherent parallelism and spatial recoding in cerebellar circuits from reconstructed network models.

Acknowledgements. This work derives direction and ideas from the chancellor of Amrita University, Sri Mata Amritanandamayi Devi. This work is supported by NVIDIA CTC grants 2012–13, 2013-14, 2015-16 and partially by DST SR/CSI/49/2010 and SR/CSI/60/2011 and Indo-Italy POC 2012-2013 from DST and BT/PR5142/MED/30/764/2012 from DBT, Government of India.

References

1. Hines, M.L., Carnevale, N.T.: The NEURON simulation environment. Neural Comput. 9(6), 1179–1209 (1997)
2. Bower, J.M.: GEneral NEural SImulation System (2003)
3. Hines, M.L., Carnevale, N.T.: Translating network models to parallel hardware in NEURON, 169(2) (2008)
4. Goddard, N.H., Hood, G.: Large Scale simulation using parallel GENESIS. In: The Book of Genesis, pp. 349–380 (1996)
5. Plesser, H.E., Eppler, J.M., Morrison, A., Diesmann, M., Gewaltig, M.-O.: Efficient parallel simulation of large-scale neuronal networks on clusters of multiprocessor computers. In: Kermarrec, A.-M., Bougé, L., Priol, T. (eds.) Euro-Par 2007. LNCS, vol. 4641, pp. 672–681. Springer, Heidelberg (2007)
6. Delorme, A., Thorpe, S.J.: SpikeNET: An Event-driven Simulation Package for Modeling Large Networks of Spiking Neurons. Netw. Comput. Neural Syst. 14, 613–627 (2003)
7. Bernhard, F.: Spiking Neurons on GPUs (2005)
8. Nageswaran, J.M., Dutt, N., Krichmar, J.L., Nicolau, A., Veidenbaum, A.: Efficient simulation of large-scale Spiking Neural Networks using CUDA graphics processors. In: 2009 Int. Jt. Conf. Neural Networks, pp. 2145–2152, June 2009
9. Igarashi, J., Shouno, O., Fukai, T., Tsujino, H.: Real-time simulation of a spiking neural network model of the basal ganglia circuitry using general purpose computing on graphics processing units. Neural Netw. 24(9), 950–960 (2011)
10. Yamazaki, T., Igarashi, J.: Realtime cerebellum: A large-scale spiking network model of the cerebellum that runs in realtime using a graphics processing unit. Neural Netw., February 2013
11. D'Angelo, E.: Neural circuits of the cerebellum: hypothesis for function. J. Integr. Neurosci. 10(3), 317–352 (2011)
12. Medini, C., Nair, B., D'Angelo, E., Naldi, G., Diwakar, S.: Modeling spike-train processing in the cerebellum granular layer and changes in plasticity reveal single neuron effects in neural ensembles. Comput. Intell. Neurosci. 2012, 359529 (2012)

13. Nieus, T., Sola, E., Mapelli, J., Saftenku, E., Rossi, P., D'Angelo, E.: LTP regulates burst initiation and frequency at mossy fiber-granule cell synapses of rat cerebellum: experimental observations and theoretical predictions. J. Neurophysiol. 95(2), 686–699 (2006)
14. Diwakar, S., Magistretti, J., Goldfarb, M., Naldi, G., D'Angelo, E.: Axonal Na+ channels ensure fast spike activation and back-propagation in cerebellar granule cells. J. Neurophysiol. 101(2), 519–532 (2009)
15. Solinas, S., Forti, L., Cesana, E., Mapelli, J., De Schutter, E., D'Angelo, E.: Computational reconstruction of pacemaking and intrinsic electroresponsiveness in cerebellar golgi cells, vol. 1, December 2007
16. Solinas, S., Nieus, T., D'Angelo, E.: A realistic large-scale model of the cerebellum granular layer predicts circuit spatio-temporal filtering properties. Front. Cell. Neurosci. 4, 12 (2010)
17. Brette, R., Gerstner, W.: Adaptive exponential integrate-and-fire model as an effective description of neuronal activity. J. Neurophysiol. 94(5), 3637–3642 (2005)
18. Naud, R., Marcille, N., Clopath, C., Gerstner, W.: Firing patterns in the adaptive exponential integrate-and-fire model. Biol. Cybern. 99(4–5), 335–347 (2008)
19. Bengtsson, F., Jörntell, H.: Sensory transmission in cerebellar granule cells relies on similarly coded mossy fiber inputs. Proc. Natl. Acad. Sci. U.S.A. 106(7), 2389–2394 (2009)
20. David, J.H., McCormick, A., Wang, Z.: Neurotransmitter Control of Neocortical Neuronal Activity and Excitability. Cereb. Cortex 3(5), 387–398 (1993)
21. Rossi, D.J., Hamann, M.: Spillover-Mediated Transmission at Inhibitory Synapses Promoted by High Affinity α 6 Subunit GABA A Receptors and Glomerular Geometry. Neuron 20, 783–795 (1998)
22. Purve, D.: Neuroscience. Sinauer Associates, Inc., Sunderland (2004)
23. D'Angelo, E., Solinas, S., Mapelli, J., Gandolfi, D., Mapelli, L., Prestori, F.: The cerebellar Golgi cell and spatiotemporal organization of granular layer activity. Front. Neural Circuits 7, 93 (2013)
24. Solinas, S., Nieus, T., D'Angelo, E., Bower, J.M.: A realistic large-scale model of the cerebellum granular layer predicts circuit spatio-temporal filtering properties, 4, 1–17, May 2010
25. Forti, L., Cesana, E., Mapelli, J., D'Angelo, E.: Ionic mechanisms of autorhythmic firing in rat cerebellar Golgi cells. J. Physiol. 574(Pt 3), 711–729 (2006)
26. D'Angelo, E., De Filippi, G., Rossi, P., Taglietti, V., Liu, A., Regehr, W.G., Maejima, T., Wollenweber, P., Teusner, L.U.C., Noebels, J.L., Herlitze, S., Mark, M.D., Brackenbury, W.J., Calhoun, J.D., Chen, C., Miyazaki, H., Nukina, N., Oyama, F., Ranscht, B., Isom, L.L., Filippi, G.D.E.: Ionic Mechanism of Electroresponsiveness in Cerebellar Granule Cells Implicates the Action of a Persistent Sodium Current Ionic Mechanism of Electroresponsiveness in Cerebellar Granule Cells Implicates the Action of a Persistent Sodium Current. J. Neurophysiol., 493–503 (1998)
27. Mapelli, J., D'Angelo, E.: The spatial organization of long-term synaptic plasticity at the input stage of cerebellum. J. Neurosci. 27(6), 1285–1296 (2007)
28. Jonathan Mapelli, E.D., Gandolfi, D.: Combinatorial Responses Controlled by Synaptic Inhibition in the Cerebellum Granular Layer. J. Neurophysiol. 103(1), 250–261 (2010)
29. Hwu, W.W., Kirk, D.B.: Programming Massively Parallel Processors: A Hands-on Approach. Morgan Kaufmann (2009)

NuChart-II: A Graph-Based Approach for Analysis and Interpretation of Hi-C Data

Fabio Tordini[1], Maurizio Drocco[1], Ivan Merelli[3], Luciano Milanesi[3],
Pietro Liò[2], and Marco Aldinucci[1]

[1] Computer Science Department, University of Turin,
Corso Svizzera 185, 10149 Torino, Italy
[2] Computer Laboratory, University of Cambridge,
Trinity Lane, Cambridge CB2 1TN, UK
[3] Institute for Biomedical Technologies - Italian National Research Council,
via F.lli Cervi 93, 20090 Segrate (Mi), Italy

Abstract. Long-range chromosomal associations between genomic regions, and their repositioning in the 3D space of the nucleus, are now considered to be key contributors to the regulation of gene expressions and DNA rearrangements. Recent Chromosome Conformation Capture (3C) measurements performed with high throughput sequencing techniques (Hi-C) and molecular dynamics studies show that there is a large correlation between co-localization and co-regulation of genes, but these important researches are hampered by the lack of biologists-friendly analysis and visualisation software. In this work we present NuChart-II, a software that allows the user to annotate and visualize a list of input genes with information relying on Hi-C data, integrating knowledge data about genomic features that are involved in the chromosome spatial organization. This software works directly with sequenced reads to identify related Hi-C fragments, with the aim of creating gene-centric neighbourhood graphs on which multi-omics features can be mapped. NuChart-II is a highly optimized implementation of a previous prototype developed in R, in which the graph-based representation of Hi-C data was tested. The prototype showed inevitable problems of scalability while working genome-wide on large datasets: particular attention has been paid in order to obtain an efficient parallel implementation of the software. The normalization of Hi-C data has been modified and improved, in order to provide a reliable estimation of proximity likelihood for the genes.

Keywords: Systems Biology, Parallel Computing, Hi-C data, Neighbourhood Graph, Chromosome Conformation Capture.

1 Scientific Background

The representation and interpretation of omics data is complex, also considering the huge amount of information that are daily produced in the laboratories all around the world. Sequencing data about expression profiles, methylation patterns, and chromatin domains are difficult to describe in a systemic view.

© Springer International Publishing Switzerland 2015
C. di Serio et al. (Eds.): CIBB 2014, LNCS 8623, pp. 298–311, 2015.
DOI: 10.1007/978-3-319-24462-4_25

The question is: how is it possible to represent omics data in an effective way? This problem is critical in these years that see an incredible explosion of the available molecular biology information. In particular, the integration and the interpretation of omics data in a systems biology way is complex, because approaches such as ontology mapping and enrichment analysis assume as prerequisite an independent sampling of features, which is clearly not satisfied while looking at long-range chromatin interactions.

In this context, recent advances in high throughput molecular biology techniques and bioinformatics have provided insights into chromatin interactions on a larger scale, which can give a formidable support for the interpretation of multi-omics data. The three-dimensional conformation of chromosomes in the nucleus is important for many cellular processes related to gene expression regulation, including DNA accessibility, epigenetics patterns and chromosome translocations [1]. The Chromosome Conformation Capture (**3C**) technology [2] and the subsequent genomic variants (Chromosome Conformation Capture on-Chip [3] and Chromosome Conformation Capture Carbon Copy [4]) are revealing the correlations between genome structures and biological processes inside the cell and permit to study the nuclear organisation at an unprecedented resolution.

The combination of high-throughput sequencing with the above-mentioned techniques (generally called **Hi-C**), allows the characterisation of long-range chromosomal interactions genome-wide [5]. Hi-C gives information about coupled DNA fragments that are cross-linked together due to spatial proximity, providing data about the chromosomal arrangement in the 3D space of the nucleus. If used in combination with chromatin immunoprecipitation, Hi-C can be employed for focusing the analysis on contacts formed by particular proteins.

In a previous work [6], we developed an R package called *NuChart*, which allows the user to annotate and statistically analyse a list of input genes with information relying on Hi-C data, integrating knowledge about genomic features that are involved in the chromosome spatial organisation. NuChart works directly with sequenced reads to identify the related Hi-C fragments, with the aim of creating gene-centric neighbourhood graphs on which multi-omics features can be mapped. Gene expression data can be automatically retrieved and processed from the Gene Expression Omnibus and ArrayExpress repositories to highlight the expression profile of genes in the identified neighbourhood. The Hi-C fragment visualisation provided by NuChart allows the comparison of cells in different conditions, thus providing the possibility of novel biomarkers identification. Although this software has been proved to be a valid support for Hi-C analysis, the implementation relying on the R environment is a limiting factor for the scalability of the algorithm, which can not cover large genomic regions due to the high computational effort required. Moreover, its overhead in managing large data structures and its weaknesses in exploiting the full computational power of multi-core platforms make NuChart unfit to scale up to larger data sets and highly precise data analysis (which requires many iterations of graph building process).

Normalization. 3C-based techniques employed for the characterization of the nuclear organization of genomes and cell types have widespread among scientific communities, fostering the development of a number of systems biology methods designed to analyse such data. Particular attention is given to the detection and normalisation of systematic biases: the raw outputs of many genomic technologies are affected both by technical biases, arising from sequencing and mapping, and biological factors, resulting from intrinsic physical properties of distinct chromatin states, that make difficult to evaluate the outcomes.

Yaffe and Tanay [7] proposed a probabilistic model based on the observation of the genomic features. This approach can remove the majority of systematic biases, at the expense of very high computational costs, due to the observation of paired-ends reads spanning all possible fragment end pairs. Hu et al [8] proposed a parametric model based on a Poisson regression. This is a simplified, and less computationally intensive normalisation procedure than the one described by Yaffe and Tanay, since it corrects systematic biases in Hi-C contact maps at the desired resolution level, instead of modelling Hi-C data at the fragment-end level. The drawback here is that the sequence information is blurred within the contact map. The first NuChart prototype solved this issue by exploiting Hu et al. solution to assign a score to each read, identifying half of the Hi-C contact instead of normalizing the contact map, thus preserving the sequence information. NuChart-II leverage this solution proposing an *ex-post* normalisation, that is used to estimate a probability of physical proximity between two genes.

Among the related works, the majority of applications rely on the creation of contact maps for the interpretation of Hi-C data, combining Principal Component Analysis and Hierarchical Clustering with this representation. The visualization and exploration of Hi-C data assumes a dramatic importance when analysing Hi-C data. To the best of our knowledge, no other tool proposes a gene-centric, graph-based visualization of the neighbourhood of a gene, as NuChart-II does.

Parallel Computing Tools. Over the years, research on loop parallelism has been carried on using different approaches and techniques that vary from automatic parallelisation to iterations scheduling. In this paper we elected *Fast-Flow* [10] as a viable tool for re-writing NuChart. We then compared the results we have obtained, against *OpenMP* and *TBB*, which represent to a major extent the most widely used and studied frameworks for loop parallelisations.

Intel Threading Building Blocks (TBB) [12] is a library that enables support for scalable parallel programming using standard C++. It provides high-level abstractions to exploit task-based parallelism, independently from the underlying platform details and threading mechanisms. The TBB `parallel_for` and `parallel_foreach` methods may be used to parallelise independent invocation of the function body of a for loop, whose number of iterations is known in advance.

OpenMP [13] uses a directive based approach, where the source code is annotated with pragmas (`#pragma omp`) that instruct the compiler about the parallelism that has to be used in the program. In OpenMP, two directives are used to parallelise a loop: the `parallel` directive declares a parallel region which will be executed concurrently by a pool of threads; the `for` directive is placed within the parallel region to distribute the loop iterations to the threads executing the parallel region.

FastFlow is a parallel programming environment originally designed to support efficient streaming on cache-coherent multi-core platforms [10]. It is realised as a pattern-based C++ framework that provides a set of high-level programming patterns (aka *algorithmic skeletons*). FastFlow exposes a `ParallelFor` pattern [11] to easily deal with loop parallelism.

2   Materials and Methods

Here we present *NuChart-II*, an improved C++ version of the first R prototype software, which enables the user to annotate and visualize Hi-C data in a gene-centric fashion, integrating knowledge data about genomic features that are involved in the chromosome spatial organisation. In the development of this new version of the software, particular attention has been paid at optimising the data structures employed for the management of the information concerning the neighbourhood graph, in order to facilitate the parallel implementation of the algorithm. The computational effort required for the generation of large graphs can now be easily addressed according to the parallelism the application exhibits, properly exploiting the computational power offered by modern multicore architectures. The re-engineering of the software has been conducted on top of FastFlow, using the `ParallelFor` pattern.

The general idea behind this package is to provide a complete suite of tools for the analysis of Hi-C data. A typical Hi-C analysis will start with the preprocessing of FASTQ files with HiCUP, which produces paired-ends reads files in SAM (or BAM) format [1]. These SAM files represent the main input of NuChart-II, along with a list of genes positions along the chromosome and an interval of genomic coordinates that should be analysed in terms of Hi-C contacts for the creation of the neighbourhood graph. Hi-C data are analysed using a gene-centric approach: if a Hi-C connection between two genes is present – i.e. there is a paired-ends read that supports their proximity in the nuclear space – an edge is created between their representative vertices.

The scalability aspect assumes crucial importance in NuChart-II: while the previous R implementation was not suitable to perform thorough explorations due to bottlenecks in memory management and limitations in the exploitation of the available computational resources, the novel algorithm is fully scalable and can be used for exploring Hi-C contact genome-wide. This is functional to our objective of creating a snapshot of the general organisation of the DNA in the nucleus, which is essential, for example, to exploit chromosome conformation data in cytogenetics analysis or to create a metrics of the distance between the different fragment in terms of contacts.

---

[1] see the HiCUP documentation for more details:
http://www.bioinformatics.babraham.ac.uk/projects/hicup/

The novel implementation presented here refines the normalisation, which is now conducted during the edges weighing phase: the weight of an edge is the result of the normalisation process. The weight assumes the role of a "confidence score" that qualifies the reliability of each *gene–gene* contact represented on the neighbourhood graph. Local scores are then rescaled in order to obtain a likelihood of existence for each edge.

2.1 Neighbourhood Graph Construction

We recall that a graph is a formal mathematical representation of a collection of vertices (V), connected by edges (E) that model a relationship among vertices. In this context, vertices represent Genes – i.e. an ordered set of human genes taken from the NCBI RNA reference sequences collection – labelled with genes names. Two genes $g_1, g_2 \in V$ are connected if there exist a paired-ends Hi-C read encompassing both of them. We define this paired-ends Hi-C read as a *connection*, meaning a spatial relationship between two genes. If a connection between two genes g_1, g_2 exists, then exists an edge $e = (g_1, g_2) \in E$. Edges weights – resulting from the normalisation phase described below – provide a likelihood of actual physical proximity for the adjacent genes. The neighbourhood graph can be defined as the induced *subgraph* obtainable starting from a given root vertex v, and including all vertices adjacent to v and all edges connecting such vertices, including the root vertex.

Graph Construction. The graph construction starts from one or more root genes and proceeds until all the nodes of the graph have been visited, or up to the desired "distance" from the root, that determines the levels of the resulting graph: a search at level 1 yields all the genes directly adjacent to the root (which is at level 0); a search at level i returns all directly adjacent genes for each gene discovered up to level $i - 1$, starting from the root.

The procedure exhibits a typical data-parallel behaviour, in which any arbitrary subset of Hi-C reads can be processed independently from each other. This means it can be parallelised in a seamless way by processing those parts elected as main computational cores within a `ParallelFor` loop pattern, whose semantic amounts to execute in parallel the instructions inside the loop, provided they are independent from each other.

Algorithm 1.1 Graph Construction (pseudo-code)

```
1  ParallelNeighboursGraph (root, L_MAX, NTH) {
2      Q = Γ = Graph := ∅
3      C[NTH] = V[NTH] = E[NTH] := ∅
4      lv := 0
5
6      push root in Q
7      while (Q not ∅ and lv < L_MAX) {
8          pop q from Q
9          // find Hi−C Reads for q
10         ParallelFor (r in Reads) {
11             if (r in q[Start, Stop] and r.Chr == q.Chr)
```

```
12                add r to C[thid]
13          }
14          // find neighbour genes for q
15          ParallelFor (c in C[thid]) {
16              intra := 0
17              for_each (g in Genes) {
18                  if (g in c.PairedEnd[Start, Stop]) {
19                      add g to V[thid]
20                      add (q, g) to E[thid]
21                      intra := intra + 1
22                  }
23              }
24              HandleIntergenicCase(Genes,intra)
25          }
26          // level synchronisation
27          Γ := BuildPartialGraph(V[thid], E[thid])
28          for_each (thid in [0, NTH]) {
29              for_each (v in V[thid]) { // next level vertices
30                  if (not v.Visited)
31                      push v in Q
32              }
33          }
34          lv := lv + 1
35          C[thid] = V[thid] = E[thid] := ∅
36      }
37      Graph := BuildGraph(Γ)
38  }
```

Taking inspiration from the work of Hong et al. [14], the algorithm proceeds as a *Breadth First Search*. At the end of each level iteration, the parallel execution is synchronized: at this point thread-local next-level containers are processed and a partial graph is constructed with the genes discovered at the current BFS level. The definitive graph is built in batch at the end of the BFS execution.

The Hi-C reads exploration phase and the genes discovery have been split, in order to avoid mixing the working sets involved in the two phases. This helps minimising the cache thrashing and permits to obtain substantial performance improvements. The pseudo-code of the parallel graph construction is reported in listing 1.1: this high-level approach required some adjustment to the BFS procedure, and the introduction of new thread-local containers needed to handle concurrent write accesses to shared data structures. Specifically, $C[NTH]$, $V[NTH]$ and $E[NTH]$ are used to store per-thread data, where NTH is the number of threads in use and *thid* identifies thread's own container, such that $0 \le thid < NTH$. Q represents our working queue that contains the genes to be processed at the next level exploration. L_MAX determines the maximum distance from the root that has to be reached. Γ is used at every level synchronisation to store partial graphs, that will be merged into a definitive graph at the very end of the graph construction process.

The algorithm starts searching for those Hi-C paired-ends reads whose first pair fragment's start coordinate falls within a gene coordinates. This yields a list of reads containing only chromosome fragments where neighbour genes may be located (rows 10–13): upon this list the search for neighbours takes place, using NTH independent threads over the set of connections (rows 15–24). Each thread looks for genes whose coordinates overlap the second pair fragments' coordinates. When a gene matches the test, the new found gene is added to the thread-local vertices set and an edge is created between the considered vertex and the new one.

Inter-genic cases are optionally handled: when no genes are found among the set of selected reads (i.e *intra* == 0), the search may be expanded to the closest proximal genes (*after* and *before* genes), possibly located within a predefined distance from the second pair fragment's coordinates: when an after gene and a before gene are found (or either one of the two), an edge between the considered vertex and the proximal genes is created, with the proximal genes added to the thread-local vertices set.

The synchronisation starts when all nodes at the current level have been explored: a partial graph Γ, which is initially empty, is updated using nodes and edges discovered during the current exploration, while the working queue Q is set up for the successive level exploration by adding all and only vertices not already visited. At the end of the exploration the full graph is built in batch, by processing partial graphs stored in Γ (row 37).

Edges Weighing. This phase encompasses the normalisation process, which is needed in order to remove systematic biases arising from sequencing and mapping.

Algorithm 1.2 Edges Weighing (pseudo-code)

```
1   ...
2   ParallelFor(edge in Edges) {
3       LenM = GCcM = MapM := ∅ // genomic features matrices
4       X = Y = β := ∅
5       Conv := false
6
7       // populate genomic features matrices
8       ...
9
10      X := RegressorsMatrix(LenM, GCcM)
11      Y := BuildContactMap(edge.Chr1, edge.Chr2)
12
13      while (not Conv) {
14          ApplyLinkFunction(Y)
15          β := ApplyGLM(Y, X, MapM)
16          Conv := CheckConvergence(β)
17      }
18
19      edge.Weight := f(β)
20  }
```

We recall that an edge $e = (g_1, g_2) \in E$, with $g_1 \in V$ and $g_2 \in V$, exists if there is a paired-ends Hi-C read that connects the two genes. For each edge, a contact map (Y) is constructed directly modelling the read count data at a resolution level of 1 megabase. Hi-C data matrix is symmetric, thus we consider only its upper triangular part, where each point of $Y_{i,j}$ denotes the intensity of the interaction between positions i and j. Using local genomic features that describe the chromosome (fragment length, GC-content and mappability) we can set up a *generalized linear model* (GLM) with Poisson regression, with which we estimate the maximum likelihood of the model parameters. This likelihood is then expressed as a the weight of the edge that qualifies the reliability of a *gene–gene* contact.

The model is given by the formula $(Y|X) = g\{X^T\beta\}$. Here β denotes the parameter vector to be estimated and g denotes a known link function. The contact map incorporates the information about the independent variables of our model (i.e. the expected value $\mu = e(Y|X)$); chromosome length and GC-content act as regressors (i.e. the coefficients of the linear combination $g(X^T\beta)$. This model is used to count the occurrences in a fixed amount of space: for this reason the Poisson distribution is used. With the best-fit coefficients returned by the linear regression the weight is computed, so that the edge contains an estimate of the physical proximity, plus the genomic information for both genes, which are preserved and not blurred within the contact map.

The edges weighing phase is an embarrassingly parallel application, where any arbitrary subset of the edges can be processed independently from each other by mean of a parallel loop pattern. This data-parallelism can be properly exploited to boost up performances and drastically reduce execution time, by just calling the function listed in algorithm 1.2 within Fastflow's `ParallelFor`. The GLM with Poisson regression has been implemented adopting the *Iteratively Weighted Least Squares* algorithm (IWLS) proposed by Nelder and Wedderburn [15] using the *GNU Scientific Library* [16]. Listing 1.2 reports a pseudo-code of the function.

The regression is run until a convergence criterion is met. In our case we check that the absolute value of the χ^2 (chi-squared) difference at each iteration is less than a certain threshold τ: $|\chi^2 - \chi^2_{old})| < \tau$. A linear function of the best-fit coefficients stored in β yields the weight of the edge. We then rescale each "local" weight using the feature scaling method and obtain a probability for each connection to exist.

3   Results

The novel makings of NuChart-II have been exploited to verify how Hi-C can be used for citogenetics studies. In particular, we focused on Philadelphia translocation, which is a specific chromosomal abnormality that is associated with chronic myelogenous leukaemia (CML). The presence of this translocation is a highly sensitive test for CML, since 95% of people with CML have this abnormality, although sometimes it occurs also in acute lymphoblastic leukaemia (ALL) and in acute myelogenous leukaemia (AML). The result of this translocation is that a fusion gene is created from the juxtaposition of the ABL1 gene on chromosome 9 (region q34) to part of the BCR ("breakpoint cluster region") gene on chromosome 22 (region q11). This is a reciprocal translocation, creating an elongated chromosome 9 (called der 9), and a truncated chromosome 22 (called the Philadelphia chromosome). The Hi-C technique can be used to study such kind of translocations, and subsequently answer to questions such as "are this kind of chromosomal translocations occurring between nearby chromosomes?", just by exploiting NuChart-II.

With NuChart-II we compared the distance of some couples of genes that are known to create translocation in CML/AML. In particular, our analysis relies on data from the experiments of Lieberman-Aiden [9], which consist in 4 lines of karyotypically normal human lymphoblastoid cell line (GM06990) sequenced with Illumina Genome Analyzer, compared with 2 lines of K562 cells, an erythroleukemia cell line with an aberrant karyotype. Starting from well-established data related to the cytogenetic experiments, we tried to understand if the Hi-C technology can successfully be applied in this context, by verifying if translocations that are normally identified using Fluorescence *in situ* hybridization (FISH) can also be studied using 3C data.

We studied 5 well known couples of genes involved in translocations and we analysed their Hi-C probability contacts in physiological and diseased cells. For validating the presence of an edge in the graph, we used the *p-value* function as a test for quantifying the statistical significance of our experiments. Considering a $p < 0.05$ threshold, we see that ABL1 and BCR (fig. 2) are distant 2 or 3 contacts in sequencing runs concerning GM06990 with HindIII as digestion enzyme (SRA:SRR027956, SRA:SRR027957, SRA:SRR027958, SRA:SRR027959), while they are in close contact in sequencing runs related to K562 with digestion enzyme HindIII (SRA:SRR027962 and SRA:SRR027963). Therefore, there is a perfect agreement between the positive and the negative presence of Hi-C contacts and FISH data. This implies from one side that the DNA conformation in cells is effectively correlated to the disease state and also that Hi-C can be reliable in identifying these cytogenetic patterns. At the same way, AML1 and ETO (fig. 4) are in close proximity in leukaemia cells (SRA:SRR027962 and SRA:SRR027963), while they are far 2 or 3 contacts in normal cells (SRA:SRR027956, SRA:SRR027957, SRA:SRR027958, SRA:SRR027959). Considering the translocation CBFβ-MYH11 (fig. 3), they are distant 2 or 3 contacts in GM06990, while are proximal in K562. We had no appreciable results for NUP214 and DEK translocation and for PML and RARα translocation, which however are more rare in this kind of disease.

These results are of utmost importance for the biomedical community: with the decreasing of sequencing costs, the Hi-C technique can be an effective diagnostic option for cytogenetic analysis, with the possibility of improving the knowledge on chromosomal architecture nuclear organisation. For example, Hi-C can be used to infer non trivial risk markers related to aberrant chromosomal conformation, like the Msc5a loci for breast cancer, which is known to play a critical role in the reorganization of a portion of chromosome 9 by CTCF proteins.

Performance. NuChart-II has been completely re-engineered, with the aim to solve the memory issues that burdened the R prototype: both the graph construction and the edge weighing phases are bounded to the memory size required to hold the data. Concerning the graph construction, it has been tuned to properly use the memory hierarchy and fully exploit cache locality while minimising cache trashing: this now permits to obtain execution times which are incomparable with respect to the R prototype, completing the exploration of the whole genes set in about 113 seconds, that results in a graph of 18450 nodes and 588635 edges. It is worth noting that the original R prototype could not perform such wide exploration, since a neighbourhood graph built up to the 3rd level took about 5 days to complete, resulting in a graph of less than 9000 nodes.

During the weighing task, each worker thread gets a bunch of edges to work on, according to the grain size, and a reference to a static collection of data containing the genomic features to be used within the process. This task performs tight loops doing Floating Point arithmetic calculations on data that fit the L3 cache and can benefit from compiler optimization and vectorization. On the other hand, a number of dynamic memory allocations are necessary during the execution of the normalisation step. The use of a memory allocator not designed for parallel programming causes a serialization of the operations that leads to a reduction of the total execution time. This memory overhead is anyway balanced by the heavy calculation performed by each worker thread: the implementation with FastFlow shows a quasi-linear speedup, and when compared against OpenMP and Intel TBB implementations, the recorded performance is substantially similar. Figure 1 shows some results concerning the sole weighing phase, in terms of speedup and execution time: the three frameworks reach approximately similar levels of speedup and scalability as the number of working threads increases, while Intel TBB begins to suffer for memory allocations when the number of threads is greater than 24, causing its performance to flatten.

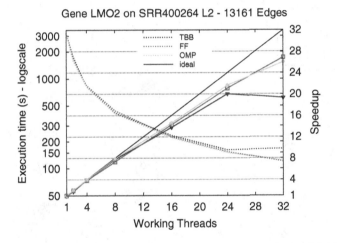

Fig. 1. Execution time and speedup for the weighing phase of gene **LMO2** according to Dixon et al. SRA:SRR400264 experiment: 12361 edges processed.

Outputs. NuChart-II provides both textual output and graphic visualization: textual and tabular outputs are useful to examine the genomic regions explored, and comprise *a)* a list of all the edges resulting from the graph construction, with the weight calculated for each edge; *b)* a list of all discovered genes, with the level (i.e. the distance from the root) where the gene has been found; *c)* a more verbose output of the execution, that reports in detail all the edges of the graph, showing all the genomic information about the two linked genes.

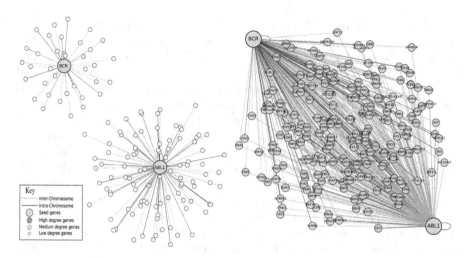

Fig. 2. Neighbourhood graph with genes **ABL1** and **BCR**, according to Lieber-manAiden's SRA:SRR027956 (*left*) and SRA:SRR027962 (*right*) experiments.

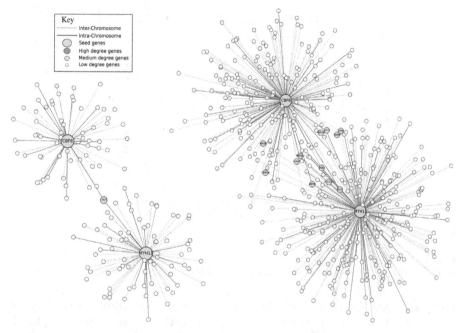

Fig. 3. Neighbourhood graph with genes **CBFB** and **MYH11** according to Lieber-manAiden's SRA:SRR027959 (*left*) and SRA:SRR027963 (*right*) experiments.

NuChart-II supports plotting with *iGraph* and *GraphViz*: these tools perform nicely with small-to-medium sized graphs, but cannot provide useful representation of huge graphs with more than ten thousand edges (as it happens when the deepness of the graph increases or inter-genic contacts are expanded). We are

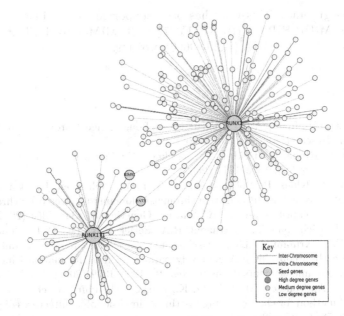

Fig. 4. Neighbourhood graph with genes **AML1** and **ETO** according to Lieber-manAiden's SRA:SRR027962 experiment.

working on viable solutions to address this problem and exploit novel techniques for interactive and dynamic graph visualisation.

4 Conclusion

The added value of this software is to provide the possibility of analysing Hi-C data in a multi-omics context, by enabling the capability of mapping on the graph vertices expression data, according to a particular transcriptomics experiment, and on the edges genomic features that are known to be involved in chromosomal recombination, looping and stability.

The novel implementation of the *NuChart-II* allows the software to scale genome-wide, which is crucial to exploit its full capability for a correct analysis, interpretation and visualisation of the data produced using the Hi-C technique. We think that the possibility of having suitable descriptions of how genes are localised in the nucleus, enriched by genomic features that can characterise the way they are able to interact, can be extremely useful in the years to come for the interpretation of multi-omics data.

We are studying solutions to further improve the new software, in order to address the visualisation problem and make all information easily accessible by direct interaction with the graph.

Acknowledgments. This work has been supported by the Flagship (PB05) InterOmics, MIUR HIRMA (RBAP11YS7K), EU MIMOMICS, EU Paraphrase (GA 288570), EU REPARA (GA 609666) projects.

References

1. Ling, J.Q., Hoffman, A.R.: Epigenetics of Long-Range Chromatin Interactions. Pediatric Research 61, 11R–16R (2007)
2. Dekker, J., Rippe, K., Dekker, M., Kleckner, N.: Capturing chromosome conformation. Science 295, 1306–1311 (2002)
3. Simonis, M., Klous, P., Splinter, E., Moshkin, Y., Willemsen, R., et al.: Nuclear organization of active and inactive chromatin domains uncovered by chromosome conformation capture-on-chip (4C). Nature Genetics 38, 1348–1354 (2006)
4. Dostie, J., Richmond, T.A., Arnaout, R.A., Selzer, R.R., Lee, W.L., Honan, T.A., Rubio, E.D., Krumm, A., Lamb, J., Nusbaum, C., et al.: Chromosome conformation capture carbon copy (5C): A massively parallel solution for mapping interactions between genomic elements. Genome Res. 16, 1299–1309 (2006)
5. Duan, Z., Andronescu, M., Schultz, K., Lee, C., Shendure, J., et al.: A genome-wide 3C-method for characterizing the three-dimensional architectures of genomes. Methods 58(3), 277–288 (2012)
6. Merelli, I., Liò, P., Milanesi, L.: NuChart: an R package to study gene spatial neighbourhoods with multi-omics annotations. PLoS One 8(9), e75146 (2013)
7. Yaffe, E., Tanay, A.: Probabilistic modeling of Hi-C contact maps eliminates systematic biases to characterize global chromosomal architecture. Nature Genetics 43, 1059–1065 (2011)
8. Hu, M., Deng, K., Selvaraj, S., Qin, Z., Ren, B., Liu, J.S.: HiCNorm: removing biases in Hi-C data via Poisson regression. Bioinformatics 28(23), 3131–3133 (2012)
9. Lieberman-Aiden, E., van Berkum, N.L., Williams, L., Imakaev, M., Ragoczy, T., et al.: Comprehensive mapping of long-range interactions reveals folding principles of the human genome. Science 326, 289–293 (2009)
10. Aldinucci, M., Danelutto, M., Kilpatrick, P., Torquati, M.: Fastflow: high-level and efficient streaming on multi-core. In: Pllana, S., Xhafa, F. (eds.) Programming Multi-core and Many-core Computing Systems. Parallel and Distributed Computing, ch. 13. Wiley (2014)
11. Danelutto, M., Torquati, M.: Loop parallelism: a new skeleton perspective on data parallel patterns. In: Aldinucci, M., D'Agostino, D., Kilpatrick, P. (eds.) Proc. of Intl. Euromicro PDP 2014: Parallel Distributed and network-based Processing, IEEE, Torino (2014), http://calvados.di.unipi.it/storage/paper_files/2014_ff_looppar_pdp.pdf
12. "Intel Threading Building Blocks", project site (2013), http://threadingbuildingblocks.org
13. Dagum, L., Menon, R.: OpenMP: An industry-standard api for shared-memory programming. IEEE Comput. Sci. Eng. 5(1), 46–55 (1998)
14. Hong, S., Oguntebi, T., Olukotun, K.: Efficient parallel graph exploration on multi-core cpu and gpu. In: Proceedings of the 2011 International Conference on Parallel Architectures and Compilation Techniques, PACT 2011, pp. 78–88. IEEE Computer Society, Washington, DC (2011), http://dx.doi.org/10.1109/PACT.2011.14

15. Nelder, J.A., Wedderburn, R.W.M.: Generalized linear models. Journal of the Royal Statistical Society, Series A, General 135, 370–384 (1972)
16. Galassi, M., Davies, J., Theiler, J., Gough, B., Jungman, G., Booth, M., Rossi, F.: Gnu Scientific Library: Reference Manual. Network Theory Ltd., February 2003, http://www.worldcat.org/isbn/0954161734

Erratum to: A New Feature Selection Methodology for K-mers Representation of DNA Sequences

Giosuè Lo Bosco[1,2] and Luca Pinello[3,4]

[1] Dipartimento di Matematica e Informatica,
Universitá degli studi di Palermo, Italy
[2] Dipartimento di Scienze per l'Innovazione e le Tecnologie Abilitanti,
Istituto Euro Mediterraneo di Scienza e Tecnologia, Palermo, Italy
[3] Department of Biostatistics, Harvard School of Public Health,
Boston, MA, USA
[4] Department of Biostatistics and Computational Biology,
Dana-Farber Cancer Institute, Boston, MA, USA

Erratum to:
Chapter "A New Feature Selection Methodology
for K-mers Representation of DNA Sequences" in:
C. di Serio et al. (Eds.): Computational Intelligence Methods
for Bioinformatics and Biostatistics, LNCS,
DOI: 10.1007/978-3-319-24462-4_9

The original version of this chapter contained an error. The names of the authors Giosuè Lo Bosco and Luca Pinello were inverted in the original publication. The original chapter was corrected.

The updated original online version for this chapter can be found at
DOI: 10.1007/978-3-319-24462-4_9

Author Index

Printed in the United States
By Bookmasters